W0050957

54 Springer Series in Chemical Physics
Edited by J. Peter Toennies

Springer Series in Chemical Physics

Editors: Vitalii I. Goldanskii   Fritz P. Schäfer   J. Peter Toennies

Managing Editor: H.K.V. Lotsch

Volumes 1–39 are listed on the back inside cover

Deepak Mathur (Ed.)

Physics of Ion Impact Phenomena

With 154 Figures

Springer-Verlag

Berlin Heidelberg New York
London Paris Tokyo
Hong Kong Barcelona
Budapest

Deepak Mathur, Ph.D.
Tata Institute of Fundamental Research
Homi Bhabha Road, Bombay 400 005, India

Series Editors   Professor Dr. Fritz Peter Schäfer

Max-Planck-Institut
für Biophysikalische Chemie
W-3400 Göttingen-Nikolausberg, FRG

Professor Vitalii I. Goldanskii   Professor Dr. J. Peter Toennies
Institute of Chemical Physics   Max-Planck-Institut
Academy of Sciences   für Strömungsforschung
Kosygin Street 4   Böttingerstrasse 6–8
Moscow, 117334, USSR   W-3400 Göttingen, FRG

Managing Editor: Dr. Helmut K. V. Lotsch
Springer-Verlag, Tiergartenstrasse 17,
W-6900 Heidelberg, FRG

ISBN-13:978-3-642-84352-5   e-ISBN-13:978-3-642-84350-1
DOI: 10.1007/978-3-642-84350-1

Library of Congress Cataloging-in-Publication Data
Physics of ion impact phenomena / Deepak Mathur (ed.).   p.   cm. –
(Springer series in chemical physics; v. 54) Includes bibliographical references and index.
ISBN-13:978-3-642-84352-5
1. Electron-ion collisions. 2. Ion bombardment. 3. Chemistry, Physical and theoretical.
I. Mathur, Deepak, 1952 – . II. Series.
QC794.6.C6P487 1991   539.7'57 – dc20   90-27901   CIP

This work is subject to copyright. All rights are reserved, whether the whole or part of the material is concerned, specifically the rights of translation, reprinting, reuse of illustrations, recita-
tion, broadcasting, reproduction on microfilms or in other ways, and storage in data banks.
Duplication of this publication or parts thereof is only permitted under the provisions of the
German Copyright Law of September 9, 1965, in its current version, and a copyright fee must
always be paid. Violations fall under the prosecution act of the German Copyright Law.

© Springer-Verlag Berlin Heidelberg 1991
Softcover reprint of the hardcover 1st edition   1991

The use of registered names, trademarks, etc. in this publication does not imply, even in the
absence of a specific statement, that such names are exempt from the relevant protective laws
and regulations and therefore free for general use.

54/3140 – 5 4 3 2 1 0 – Printed on acid-free paper

Preface

In a ten-year period of unprecedented activity in studies of low energy, gas-phase ion collision phenomena, much progress has been made by researchers from diverse areas of physics and chemistry. Apart from seeking answers to various specific problems of an applied nature, ranging from the physics of thermonuclear fusion reactors to lasers and the chemistry of interstellar space, a principal motivation guiding much of the current activity is the desire to gain insights into the *dynamics* of chemical transformation processes on a microscopic level, collision by collision. The contents of this book provide an overview of contemporary experimental endeavours in the physics of ion-electron and ion-neutral collisions, the emphasis being on fundamental aspects of ionic *interactions* and the quest for qualitatively and quantitatively reliable descriptions of the pathways followed by multiparticle systems when they evolve from one state to another.

The main aim of this volume is to present an uncluttered description of a number of current interests and trends, and generally non-specialized discussions of representative results which provide a flavour of this interdisciplinary subject. The contributors to this volume come from various areas, such as atomic and molecular physics, chemical physics, mass spectrometry, quantum chemistry and physical chemistry. It is hoped that the material presented is in a form that will appeal both to established researchers and to those about to venture into these areas of activity for the first time.

It is a pleasure to acknowledge the wholehearted cooperation of all the contributing authors. I am also grateful to colleagues in my laboratory for their tolerance during the months while this volume was being compiled, and to Helen, Ajay and Paul for showing patience beyond the call of duty.

Bombay
September 1990

D. Mathur

Dedication to Professor J. B. Hasted

It is a pleasure for the contributing authors to dedicate this volume to Professor J. B. Hasted.

John Hasted pioneered experimental research in many areas of ion collision physics in the early 1950s. As a part of Sir Harrie Massey's team at University College London, he helped establish post-war research in many areas. Curve crossing and the adiabatic hypothesis are just two specific aspects of low-energy ion-neutral collision physics which are particularly associated with his name. During four decades of prominent research activity, John Hasted's varied interests have covered studies of low-energy electron scattering from molecules, transport properties of ions in swarms, flowing afterglow and ion trap studies of ionic recombination, the microwave dielectric properties of water and biomolecules, and also areas of activity outside conventional physics.

Apart from his undoubted contributions to atomic and molecular collision physics, John Hasted's colourful personality, his instinct for equality, which always ensures that the newest student and the most radiant luminary receive the same courteous treatment, his special approach to teaching physics and his broad, progressive outlook on nature, art and culture have left indelible marks on the lives of some 80 research students and scores of other associates scattered over five continents.

Most of the authors of this volume have been associated with John at some time or another, and together with numerous former colleagues and friends from all corners of the globe, we acknowledge our indebtedness and gratitude to John Hasted by dedicating this book to him on the occasion of his 70th birthday in February 1991.

To see a World in a Grain of Sand
And a Heaven in a Wild Flower,
Hold Infinity in the Palm of Your Hand
And Eternity in an Hour.

William Blake

Contents

Contributors

A. G. Brenton

Mass Spectrometry Research Unit, University College of Swansea,
Singleton Park, Swansea SA2 8PP, United Kingdom

C. L. Cocke

J. R. Macdonald Laboratory, Department of Physics,
Kansas State University, Manhattan, Kansas 66506, United States of America

M. Hamdan

CNR, Area di Ricerca di Padova, Corso Stati Uniti 4,
35100 Padova, Italy

F. M. Harris

Mass Spectrometry Research Unit, University College of Swansea,
Singleton Park, Swansea SA2 8PP, United Kingdom

L. K. Johnson

Department of Physics, Space Physics and Astronomy,
and Rice Quantum Institute, Rice University, P.O. Box 1892,
Houston, Texas 77251, United States of America

E. Y. Kamber

Department of Physics, Western Michigan University,
Kalamazoo, Michigan 49008, United States of America

N. Kobayashi

Department of Physics, Tokyo Metropolitan University,
1-1 Minami-Ohsawa, Hachioji-shi, Tokyo 192-03, Japan

V. R. Marathe

Tata Institute of Fundamental Research, Homi Bhabha Road,
Bombay 400 005, India

D. Mathur

Tata Institute of Fundamental Research, Homi Bhabha Road,
Bombay 400 005, India

J.B.A. Mitchell

Department of Physics, The University of Western Ontario,
London, Ontario, Canada N6A 3K7

A. Müller

Institut für Kernphysik, Strahlenzentrum der Justus-Liebig-Universität Gießen,
D-6300 Gießen, Federal Republic of Germany

R.F. Stebbings

Department of Physics, Space Physics and Astronomy,
and Rice Quantum Institute, Rice University, P.O. Box 1892,
Houston, Texas 77251, United States of America

1. Introduction

D. Mathur

With 5 Figures

1.1 The Need to Study Ion Impact Phenomena

Studies of the fundamental interactions between positively charged ions and neutral atoms, molecules and electrons constitute a significant, and important, part of contemporary atomic and molecular physics. Much of the importance attributed to this sub-field stems from an awareness that has grown over the last two decades that qualitative and quantitative knowledge of the physics of ionic collisions is a prerequisite for gaining an understanding of a wide variety of natural and man-made phenomena. For instance, the need for ion-electron and ion-atom collision information is now well recognized in seeking to elucidate the mechanisms governing the energy budget of high temperature astrophysical plasmas or within a thermonuclear fusion device, or for attaining an understanding of the processes giving rise to the formation and destruction of atomic and molecular species in cold interstellar space, and in terrestrial and planetary atmospheres. There are numerous other applications also; a comprehensive overview of the applied scope of ion collision physics can be found in the mammoth five-volume treatise [1.1] produced by *Massey* et al. These volumes highlight the importance of the subject in diverse areas, such as geophysics and astrophysics, controlled nuclear fusion, laser chemistry, condensed matter physics and a gamut of other applications, ranging from detectors used in high energy physics, MHD power generation, combustion chemistry and the study of flames and discharges to, unfortunately, military applications of high powered laser systems, high energy particle beams, precision timing and atomic magnetometers and nuclear magnetic resonance gyroscopes.

The central theme in this volume will be fundamental rather than applied, providing a description of a number of contemporary interests and trends in a general fashion, without excessive detail, concentrating on attempts to gain an insight into the fundamental chemical transformation processes in which multiparticle systems evolve from one state to another, or two reactants get transformed into products. Unravelling the mechanisms which govern the dynamics of such transformation processes continues to provide a challenge to physicists and chemists even though, at a superficial level, all that is basically required is an understanding of Coulomb interactions. In this connection, it is pertinent to draw attention to the famous statement made by *Dirac* [1.2] in 1929 that, with the birth of quantum mechanics:

"The underlying physical laws necessary for the mathematical theory of a large part of physics and the whole of chemistry are thus completely known ..."

In reality, of course, the problem of understanding the dynamics of chemical transformation processes is really a 'few-body' problem – a problem in which there are too few particles to be governed by the well-established laws of statistical mechanics, but too many particles for successful, rigorous application of the laws of quantum mechanics. Thus, we have a situation in which although it is arguably possible for us to understand something 'in principle', we have little, or no, predictive capabilities! Of course Dirac appreciated this fact, as can be judged from the continuation of the above oft-quoted statement:

"... and the difficulty is only that the exact application of these laws leads to equations much too complicated to be soluble. It therefore becomes desirable that approximate practical methods of applying quantum mechanics should be developed ..." [1.2].

Much of the contemporary interest in, and fascination for, experimental and theoretical studies of ion impact phenomena which are described in later chapters thus stems from the exciting search for appropriate 'approximations', and for the development of insights into the true nature of chemical transformation processes on a microscopic level, collision by collision.

By way of illustration, let us consider the case of one of the simplest chemical reactions, involving one hydrogen atom and a single hydrogen molecule

$$H_a + (H_b - H_c) \rightarrow (H_a - H_b) + H_c , \qquad (1.1)$$

and attempt to gain information on exactly how this hydrogen transfer reaction proceeds, on a microscopic level. Specifying even the position of the reactant H_2 molecule (comprising atoms H_b and H_c) with respect to the reactant H atom (designated H_a) requires six coordinates, and so a depiction of the changes in energy which accompany this reaction as the H-H_2 distance changes necessitates an absurdly complicated seven-dimensional diagram. Consequently, even for such a system, recourse has to be made to an approximate treatment.

The first approximation we can make is of a theoretical nature: we can choose to consider only those reactions which proceed when the reactant H atom approaches H_2 in a collinear fashion (along the H_b-H_c internuclear axis). In such a case, only two parameters suffice to construct a three-dimensional surface in which energy changes can be plotted as a function of H_a-H_b distance and H_b-H_c distance. It turns out that, in this particular case, such an approximation is extremely good, because the lowest energy pathway happens to be that which results from a collinear collision. Furthermore, high level ab initio quantum-chemical calculations of the energy surfaces of the H-H_2 system can be carried out with a reasonable degree of success using contemporary computational methods.

On the other hand, for most other chemical transformation processes the calculation of energy surfaces continues to be a formidable task. If we consider a chemical reaction in which the reactants and products are considered as a dynamic system comprising n_e electrons and n_n nuclei, the combined motion of all the particles can be described theoretically by solving a nonseparable Schrödinger equation of $3(n_e + n_n - 1)$ variables. The first problem we encounter is that such a description requires a multidimensional configuration space. Moreover, it is also clear, that analytical solutions for such an equation will not be forthcoming; even purely numerical methods, requiring 'reasonable' computational effort, are ruled out in all but a very few cases. It is therefore advantageous to attempt to tackle 'real' problems in a piecewise fashion, by employing techniques which are appropriate only in a given subset of the total configuration space. This necessitates the making of a second approximation, one which is of an experimental nature.

We can choose our reactants in such a fashion that transformation processes of interest will occur with high probability in only a part of an overall complex, multidimensional energy surface. Collision experiments in which one of the reactants is a positively charged ion provide the major example of this approach. Here, those chemical transformation processes can be studied which occur as a result of relatively long-range interactions between reactant species. In such cases asymptotic approximations to complex energy surfaces are often sufficient to explain, and lend substance to, experimental information. An additional advantage of focusing on ion-neutral reactions is that, in many cases, simplifying straight line trajectory approximations, such as the impact parameter method, yield satisfactory results. Ion-neutral collisions leading to single and multiple electron capture, and charge stripping, are examples of reactions which occur readily at large internuclear separations of reactant species, and these are topics which are extensively discussed in later chapters.

Of course, the dynamics of ion-neutral interactions can be expected to be somewhat different from those expected for neutral-neutral reactions of the type represented by (1.1). The major point that has to be considered with respect to the type of studies that are reported in later chapters is that extensive rotational excitation in ion-molecule collisions is avoided by observing collision processes at small centre-of-mass scattering angles. In such cases only long-range electrostatic interactions are being sampled and, for most molecules, particularly non-polar ones, the net torque experienced is very small. For neutral-neutral reactions, the long-range electrostatic forces are considerably less important than the shorter-range valence interactions.

There has, indeed, been remarkable progress in experimental studies of ionic collision phenomena over the last twenty years. There now exists an expanding body of information on how ion-neutral and ion-electron interactions govern the transfer of energy, momentum and angular momentum in the course of transformation of 'reactants' into 'products'. Even hitherto-scarce information is beginning to be available on the dynamics of specific pathways that transformation processes choose to follow.

Presenting a global, uncluttered view of the state of our present knowledge of the physics of ion impact phenomena is the objective of this volume. Despite rapid developments in experimental technique and methodology, enabling experiments of an increasingly sophisticated nature to be carried out, one unifying factor in the diverse subject matter described in the following chapters is the need to have an efficient source of ions of the appropriate mass-to-charge ratio, intensity, energy and spatial characteristics. Since understanding of how positive ions are produced in order to conduct collision studies is a *sine qua non* for the appreciation of later chapters, it is appropriate to review some aspects of contemporary ion production techniques at this point.

1.2 Some Contemporary Ion Production Techniques

The pervasive use of ion beams in many areas of science and technology is reflected in the enormous variety of ion sources that have been developed over the years. Positive ions of atomic and molecular species may be produced by collisions with electrons, photons, energetic neutrals, other ions, or by interactions with static or high frequency electric fields, or by surface ionization phenomena. There is no universal source of ions that can meet the requirements of all applications, and potential users of ion beams have to undertake the exercise of considering the relative importance of factors such as intensity, types of charged species, ion beam quality, efficiency, lifetime, stability and ease of operation in order to select the ion source which is most appropriate for their particular application. A large, and ever increasing, body of information exists in the form of books, review papers and conference proceedings [1.3–11] which describe the science and technology of diverse types of ion sources. Here we shall focus attention on only a small subset of these, concentrating on those contemporary techniques which are of importance in the context of ion beam studies which are described in the chapters to follow.

1.2.1 The Electron Impact (EI) Source

The simplest electron impact (EI) type of ion source is depicted in schematic form in Fig. 1.1. Electrons emitted by a hot tungsten filament and accelerated to an energy in the region of 50–100 eV collide with a vapour of the species A^0 maintained at a typical pressure, p, of ca. 10^{-5}–10^{-3} Torr. The collision can take place either with a 'static' gas target, in which the velocity vectors of A^0 are randomly orientated, or with a beam of A^0 oriented either normal or parallel to the direction of the electron beam. The ions formed can be extracted from the ionization volume with an electrode at a potential V_{ext} of appropriate polarity or by means of a repelling voltage V_{rep} applied on the repeller electrode placed within the electron-neutral interaction zone. This type of source, commonly referred to as a Nier-type source [1.12], is simple to construct, operate and maintain, and has been a standard component in commercial mass spectrometer systems for many decades. It typically produces ions of low charge states, but with fairly low energy widths. As can be seen in Fig. 1.1a, the effective volume, ρ_{eff}, over which ionization occurs is

$$\rho_{\text{eff}} = \pi r^2 l \, .$$

(1.2)

The number, n_A, of neutral particles per unit volume can be deduced from the pressure p and temperature T of A^0 vapour

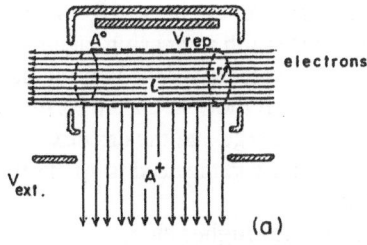

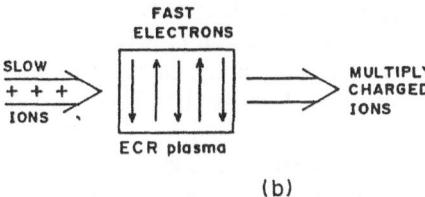

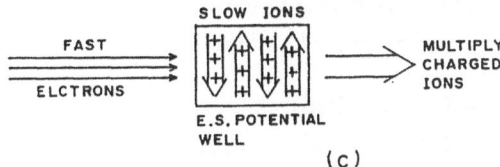

Fig. 1.1 a–c. Schematic representation of some contemporary electron impact ion sources: (a) conventional EI (Nier-type) of ion source; (b) ECRIS (electron cyclotron resonance ion source) and (c) EBIS (electron bombardment ion source)

$$n_A = \frac{p}{kT} \ .$$
(1.3)

Consequently, if all the ions produced in the ionization zone are extracted with unit efficiency, the ion current, I_i, that can be expected from the source will be given by

$$I_i = I_e\, n_A \sigma l \ ,$$
(1.4)

where I_e is the electron current and σ is the electron impact ionization cross section pertaining to the given electron energy that is being used. A useful figure of merit is the efficiency of the ion source, η, which is defined as

$$\eta = \frac{\text{ions/s}}{\text{atoms/s}} \ .$$
(1.5)

Using the ion current formula (1.4) in conjunction with the expression

$$N_A = n_A \bar{v} A$$
(1.6)

for the number of neutral atoms or molecules which cross the ionizing region per second, where $\bar{v}$ is the mean velocity of the neutral particle beam and A is its cross section ($A = \pi r^2$, in terms of Fig. 1.1 a), the efficiency can be expressed in the form

$$\eta = \frac{I_e\, n_A \sigma l}{n_A A \bar{v}} = \frac{I_e\, \sigma l}{A \bar{v}} \ .$$
(1.7)

5

Typical values of η for contemporary IE sources are in the range 10^{-5}–10^{-4}. From (1.7) it is obvious that the efficiency will increase with increasing electron density and effective interaction volume, and if the mean velocity of the neutral particles is decreased. The latter value is limited by the thermal motion. Of the former factors, the electron beam density is limited by the space charge and the effective volume is limited by geometrical constraints of ion source size and the need to spatially confine the extracted beam to reasonable dimensions.

Improvement in the value of η, particularly for higher charge states, has been achieved by modifications such as the application of a magnetic field in a direction which constrains the electrons to oscillate in the interaction zone [1.3], thus increasing the effective interaction path length (and, consequently, the value of ρ_{eff}) or the use of Pierce-type cathodes [1.4]. Efficiencies as high as 10^{-3} have been reported for such sources.

If one requires sufficient fluxes of ions of charge states greater than $3+$ or $4+$, simple EI sources have to be abandoned in favour of more complex sources; two of the most widely used electron impact sources are the EBIS (electron bombardment ion source) and ECRIS (electron cyclotron resonance ion source).

1.2.2 The Electron Bombardment Ion Source (EBIS)

The electron bombardment ion source (EBIS) was initially conceived and built in the mid-1960s by *Donets* and collaborators [1.11] at Dubna, USSR. The basic design of this source stems from the recognition that a highly charged ion cannot be produced in a single collision between an electron and a neutral or singly charged atom; a succession of ionizing collisions are necessary. In order to produce the most probable charge state $\bar{q}$ from a singly charged ion, a definite collision factor, $j\tau$, must be obtained

$$j\tau = \sum_{q=1}^{\bar{q}-1} \sigma_{q \to q+1}^{-1} \, , \tag{1.8}$$

where $\sigma_{q \to q+1}$ is the effective cross section for sequential ionization of the ion with charge state q, j is its current density and τ is the time over which ionizing collisions take place. In order to produce highly charged ions, low energy ions which are initially in low charge states (produced by electron impact) are confined in an electrostatic trap within which they undergo several collisions with fast electrons in such a manner that their average charge state increases sequentially. The basic idea is demonstrated schematically in Fig. 1.1.

The electrostatic trap is produced within the space charge of an intense, cylindrical electron beam, a technique which has been treated in quite some detail by *Hasted* [1.13, 14]. A long solenoidal magnetic field (typically 1–1.5 m long) is required to confine the radial extent of the high density electron beam, and a system of isolated cylindrical electrodes is mounted along the axis of the solenoid to which positive potentials are applied to facilitate axial ion trapping; an electrostatic electrode system is required to extract positive ions after a certain trapping time. A typical sequence of operations with such a source would be as follows:

i) generate a sufficiently extended electron beam of given density and energy;

ii) create an electrostatic field trap along the beam;

iii) inject a definite number of working material ions into the trap over a well-defined period of time;

iv) contain the ions within the trap for a sufficiently long time to produce the required charge state after sequential ion-electron collisions;

and v) extract the multiply charged ions out of the trap, along the electron beam axis, and prepare for the next cycle.

A practical manifestation of such an operating sequence is the EBIS known as KRION-2, which operates in Donet's laboratory in Dubna [1.15, 16]. The drift length used in this source is 1.4 m long, with a residual vacuum better than 10^{-12} Torr. A superconducting magnet is used, and therefore the trap temperature is 4.2 K. Electron beam densities of the order of $600 \, \mathrm{Acm}^{-2}$ are achieved, with a typical electron beam energy of 20 keV. Ion trapping times as long as 10 s have been obtained in KRION-2. Using this device, $10^9 \, \mathrm{C}^{6+}$ ions per pulse and $10^6 \, \mathrm{Xe}^{52+}$ ions per pulse have been produced using a 10 s containment time, 20 keV electron energy and an electron beam density of $150 \, \mathrm{Acm}^{-2}$. Devices with considerably larger electron beam densities have also been reported (in excess of $1000 \, \mathrm{Acm}^{-2}$) in which trapping times of the order of only a few milliseconds are sufficient to produce 10^9 fully-stripped Ne ions per pulse [1.17].

1.2.3 The Electron Cyclotron Resonance Ion Source (ECRIS)

A schematic representation of how multiply charged ions are produced in ECRIS is given in Fig. 1.1, which also shows a certain operational symmetry between ECRIS and EBIS devices. Whereas in EBIS fast electrons are made to interact with slow ions in a trap, the ECRIS idea is to diffuse a plasma of cold, low energy, low charge state ions through a plasma of high energy electrons in order to obtain higher charge states [1.18-20]. Typically, the source comprises a quartz waveguide in which plasma generation as well as ion containment occurs. To facilitate the latter function, magnetic mirror coils are placed at the inlet and ion-extraction ends of the waveguide; a plasma is generated when the appropriate gas is introduced into the inlet, at source pressures of the order of 10^{-3} Torr, by means of a high frequency (several GHz) microwave cavity. As the plasma drifts along the axial magnetic field gradient (to regions where the pressure is of the order of 10^{-6} Torr) the electrons traverse a plane in the waveguide where they are stochastically heated to temperatures of the order of several keV by resonant excitation due to the applied high frequency field, ω_{rf}, and the axial magnetic field, B

$$\omega_{\mathrm{rf}} \simeq \omega_{\mathrm{c}} = \frac{eB}{m} \, , \tag{1.9}$$

where ω_{c} is the electron cyclotron resonance (ECR) frequency. In order to understand the ECR heating of electrons, consider the following [1.20]: take an empty metallic box of unspecified geometry and immerse it in a microwave field. If the dimensions of such a box are larger than the wavelength of the microwaves (typically 3 cm), it behaves like a multimode microwave cavity. On immersion of such a cavity into a magnetic well in which the field varies from, say, 0.2 Tesla to 0.5 Tesla, at some point there will exist a magnetic surface on which the value of the field strength is such that the electrons' gyrofrequency is in resonance with the frequency of the applied microwaves [cf. (1.9)]. Such a resonance is bound to occur because in a multimode cavity there will always be a component of the electric field of the applied microwaves which is perpendicular to the magnetic flux lines. Consequently, as the electrons pass through such a surface, they are accelerated. This phenomena has been termed 'stochastic heating', and, although no reliable general theory is currently available [1.11] to explain this form of electron heating in a quantitative fashion, it has been experimentally shown that keV electron temperatures can be reached with as little as 100 W of applied microwave power [1.20].

As in the case of EBIS, production of highly charged ions requires the product $n_e \tau_i$ to be large, where n_e is the electron density and τ_i is the ion containment time. In ECRIS, the electron density is found to increase with ω_{rf}^2 which, in turn, is limited by the maximum value of applied magnetic field (1.9). The value of τ_i can, however, be increased by building a multistage source [1.18], with each stage possessing its own magnetic mirror configuration and microwave generator. In such devices fast differential pumping is employed to ensure that the gas pressure progressively decreases in going from one stage to the other. The second stage of such a source would typically operate at a pressure of 10^{-7} Torr.

In comparison with EBIS, ECRIS is somewhat less demanding from the viewpoint of experimental technique. To achieve an energy flux of, say, 16 kJcm^{-2}, the average power dissipated in the ion source cylinder would be of the order of 16 kWcm^{-2} in an EBIS compared with only 0.053 kWcm^{-2} in the case of an ECRIS. Moreover, the vacuum requirements in the latter will be much less stringent ($\leq 10^{-6}$ Torr for ECRIS compared with $\leq 10^{-10}$ for an EBIS). A very large number of laboratories across the world have developed, or are currently in the process of developing, ECRIS type of devices; ECRIS is likely to be the workhorse for production of highly charged ions for future ion collision studies.

1.2.4 The 'Ion Hammer' Technique

It is now well established that when high-energy, multiply charged heavy ion projectiles undergo collisions with neutral target atoms, a large number of target electrons can be ejected without imparting much kinetic energy to the target nucleus. Such collisions give rise to the formation of highly charged ions possessing very low kinetic energies. The technique of using a fast heavy ion beam from an accelerator to produce secondary, low energy, multiply charged recoil ions, which was initially developed by *Cocke* [1.21], has come to be known as the 'ion hammer' technique.

To illustrate this type of ion source consider one recent experimental configuration, shown schematically in Fig. 1.2. The geometry of the apparatus is of the crossed-beams type, in which a fast (tens of MeV energy) ion beam from a heavy-ion accelerator intersects, at right angles, a thermal energy neutral atomic or molecular beam. Low energy target recoil ions that are formed as a result of large im-

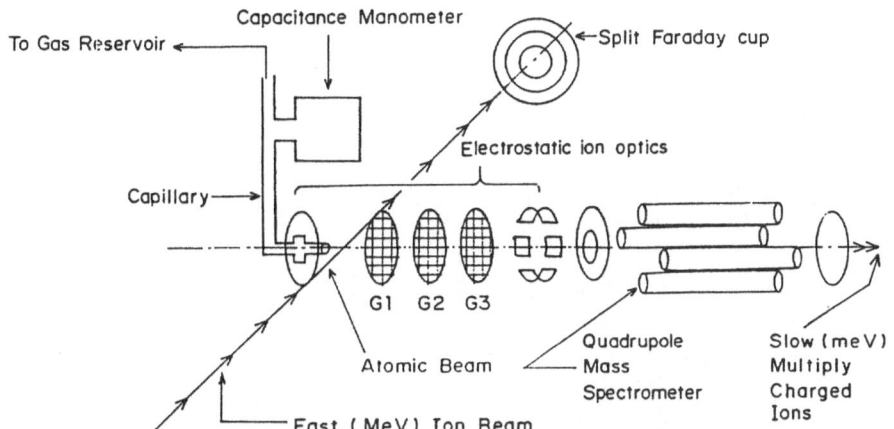

Fig. 1.2. Apparatus used by *Mathur* et al. [1.22] for production of slow (meV energy), multiply charged ions by means of the ion hammer technique

pact parameter ion-neutral collisions are extracted through an ion optical system and a quadrupole mass filter. Typical operating conditions are in the region of 2×10^{-8} Torr with beam load; the neutral beam number densities correspond to a pressure of ca. 10^{-4} Torr, well within 'single collision conditions' as far as collisions between the fast projectiles and the atomic beam are concerned.

In order to assess the effectiveness of this type of ion source, *Mathur* et al. [1.22] have carried out a comparative study of multiple ionization by fast ion impact and by electron impact using the apparatus shown in Fig. 1.2. The mass spectra obtained for multiple ionization of Kr by 96.3 MeV F^{8+} projectiles and by electrons of 700 eV impact energy are shown in Fig. 1.3. The latter value of impact energy was chosen to make the collision velocities approximately the same for both projectiles. Although the observation that fast ion impact leads to a greater degree of ionization may not, in itself, appear surprising, the extent of the difference in the two cases is striking. In the case of Kr^{7+}, for instance, the ion impact technique is nearly six orders of magnitude more efficient. In the case of Ar, Fig. 1.4 shows the same comparison but in a more quantitative fashion. Cross sections for formation of different charge states have been measured with an accuracy which is estimated to be $\pm 50\%$ in the case of ion impact data and is $\pm 20\%$ in the electron impact case.

In assessing the reliability of the comparative data shown in Figs. 1.3 and 1.4, it is pertinent to point out that the relative intensities of the different recoil ion charge states have been corrected for mass discrimination effects introduced by the quadrupole mass filter. This has been done by measuring the cross section value for electron impact single ionization of the rare gases He, Ne, Ar and Kr at an electron energy of 200 eV using the relative flow technique of making such absolute measurements. These measured values are then compared to established, standard absolute cross section values and the dependence of the transmission efficiency of the apparatus on mass-to-charge ratio (covered by the range from He to Kr) is deduced. This

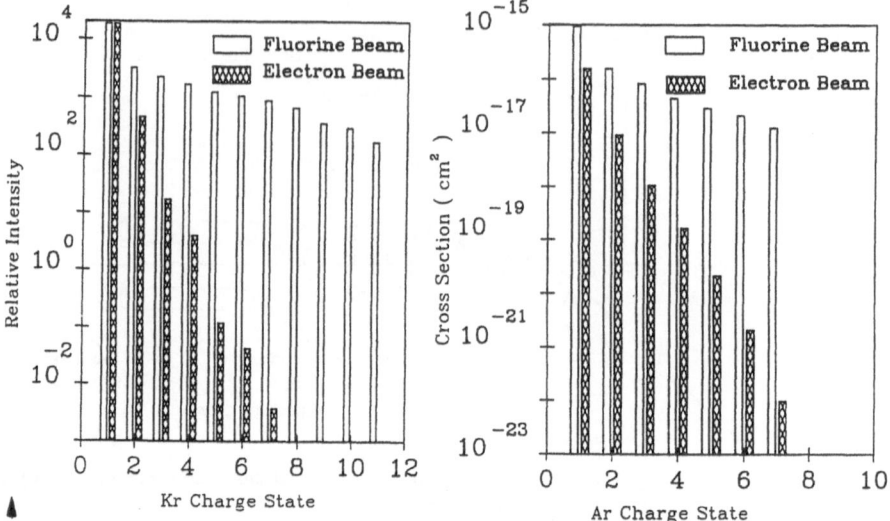

Fig. 1.3. Comparison of multiple ionization of Kr by 96.3 MeV F^{8+} ions and by 700 eV electrons [1.22]

Fig. 1.4. Cross sections for multiple ionization of Ar by 96.3 MeV F^{8+} ions and by 700 eV electrons

functional dependence then enables a correction factor to be applied to the raw mass spectra.

Mathur et al. [1.22] have also investigated the fractions of recoil ions produced in different charge states as a function of the charge state of the fast projectile ion beam. Figure 1.5 shows these fractional dependences in the case of Ar recoil ions obtained with F^{q+} ($q = 5 - 8$) projectiles; these measurements clearly indicate the absence of any pronounced shell effects. It is believed [1.22] that this is a reflection of the importance of *indirect* multielectron ejection processes, such as excitation-autoionization

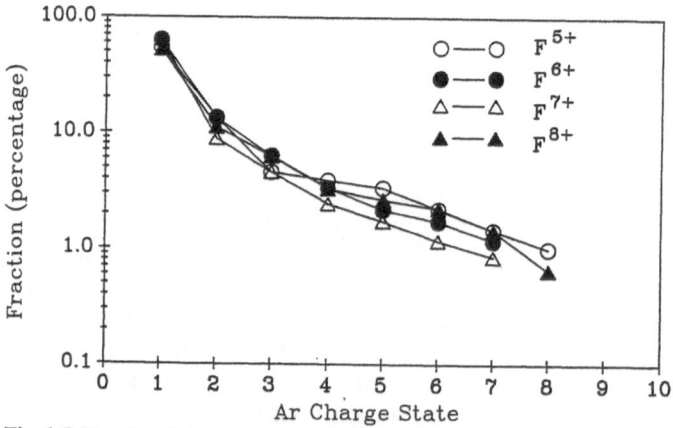

Fig. 1.5. Fractional charge state distribution for Ar recoil ions produced in collisions of F^{q+} for $q = 5, 6, 7$ and 8, at collision energies of 64.2, 74.9, 85.6 and 96.3 MeV, respectively

and inner-shell ionization followed by rapid Auger decay (see Chap. 2). The absence of structure in the vicinity of recoil ion charge state 6 + appears to indicate that such indirect effects, which have not been taken into consideration in any model calculations to date, may play a most significant role in the multiple ionization process in the hammer source. It is clear that in order to develop quantitatively reliable theoretical insight into fast-ion-induced multielectron ejection processes, such mechanisms have to be accounted for in a proper fashion. Achievement of this necessitates more detailed experimental work in which unambiguous distinctions can be made between direct (knock-out) ionization and indirect processes.

Despite the existing lacuna in theoretical understanding of exactly how such sources work, the ion hammer has begun to be successfully utilized in a large number of ion collision experiments (see, for instance, Chap. 3).

References

1.1   H. S. W. Massey, E. W. McDaniel, B. Bederson (eds.): *Applied Atomic Collision Physics*, Vols. 1–5 (Academic, New York 1982)
1.2   P. A. M. Dirac: Proc. Roy. Soc.(London) **A123**, 714 (1929)
1.3   J. B. Hasted: *Physics of Atomic Collisions*, 2nd Ed. (Butterworths, London 1972) pp. 190–205
1.4   L. Vályi: *Atom and Ion Sources* (Wiley, London 1977)
1.5   G. Sidenius: Nucl. Instrum. Methods **151**, 349 (1978)
1.6   T. S. Green: Rep. Prog. Phys. **37**, 1257 (1974)
1.7   G. D. Alton: Nucl. Instrum. Methods **189**, 15 (1981)
1.8   R. G. Wilson, G. R. Brewer: *Ion Beams with Application to Ion Implantation* (Wiley, New York 1973)
1.9   G. Dearnaley, J. H. Freeman, R. S. Nelson, J. Stephen: *Ion Implantation* (North-Holland, Amsterdam 1973)
1.10 L. Liljeby (ed.): Proc. Int. Symp. on the Production and Physics of Highly Charged Ions, Stockholm, Sweden, June 1–5, 1982, Phys. Scr. **T3**, 7–55 (1983)
1.11 S. Bliman (ed.): Proc. Int. Conf. on the Physics of Multiply Charged Ions and International Workshop on E. C. R. Ion Sources, Grenoble, France, September 12–16, 1988, J. de Phys. C1 **50**, 642-892 (1989)
1.12 A. O. Nier: Rev. Sci. Instrum. **11**, 212 (1940)
1.13 J. B. Hasted: "Confinement of Ions for Collision Studies", in *Physics of Ion-Ion and Ion-Electron Collisions*, ed. by F. Brouillard, J. Wm. McGowan, (Plenum, New York 1983) pp. 461–500
1.14 F. A. Baker, J. B. Hasted: Phil. Trans. **A261**, 33 (1966)
1.15 E. D. Donets: IEEE Trans. Nucl. Sci. **NS-23**, 897 (1976)
1.16 E. D. Donets: Phys. Scr. **T3**, 11 (1983)
1.17 J. Arianer, A. Cabrespine, C. Goldstein: IEEE Trans. Nucl. Sci. **NS-26**, 3713 (1979)
1.18 R. Geller: IEEE Trans. Nucl. Phys. **NS-23**, 904 (1975)
1.19 J. Arianer, R. Geller: Ann. Rev. Nucl. Particle Sci. **31**, 19 (1981)
1.20 R. Geller, B. Jacquot: Phys. Scr. **T3**, 19 (1983)
1.21 C. L. Cocke: Phys. Rev. A **20**, 749 (1979)
1.22 D. Mathur, E. Krishnakumar, F. A. Rajgara, U. T. Raheja, C. Badrinathan: Int. J. Mass Spectrom. Ion Proc. **99**, 237 (1990)

2. Ion Formation Processes: Ionization in Ion-Electron Collisions

A. Müller

With 60 Figures

When an ensemble of atoms, of density n_0, is kept at room temperature the collisions that occur are generally not energetic enough to result in ejection of any electrons. This situation is depicted in Fig. 2.1 for sample atoms consisting of an ionic core and only one bound electron. When energy is fed into the gas (Fig. 2.2) the atomic collisions become more violent and some atoms get ionized. The free electrons produced by such collisions can ionize further atoms or recombine with the ions. At a fixed temperature T an equilibrium will be formed with a certain density of electrons $n_e(T)$, ions $n_+(T)$ and atoms $n_0(T)$. Conservation of particles and charge requires $n_e = n_+$ and $n_0(T_0) = n_+(T) + n_0(T)$.

In our simplified system the production rate of ions by electron-atom collisions is proportional to the densities $n_e(T)$ and $n_0(T)$, and the loss rate of ions by electron-ion recombinations is proportional to $n_e(T)$ and $n_+(T)$. Thus, the net change of ion density in time is given by the rate equation

$$\frac{dn_+}{dt} = \alpha(T)n_e n_0 - \beta(T)n_e n_+ . \tag{2.1}$$

The proportionality factors $\alpha(T)$ and $\beta(T)$ are the temperature-dependent rate coefficients for ionization and recombination, respectively. Under equilibrium conditions we have $dn_+/dt = 0$. According to the Saha equation, thermal ionization of atomic hydrogen at a pressure of 1 mbar yields for the ratio $n_+(T)/n_0(T_0)$ a number of the order of 10^{-110} at room temperature. The ratio rises to a magnitude of about 0.1 at $T=10^4$ K. This dramatic increase of n_+/n_0 with temperature causes most of the matter in stars to be fully ionized, which means that an estimated 99 % of all visible matter in the universe is in the so-called fourth state of aggregation, that is, the ionized plasma.

Our knowledge about the behaviour of ionized matter originates from the observation of natural plasmas, like the solar corona and other astrophysical objects, and from the investigation of man-made laboratory plasmas such as gas discharges, explosions, fusion plasmas etc. The properties and the time evolution of plasmas are widely determined by electron-ion collisions, so that knowledge about these collisions is necessary for the understanding of plasma behaviour; and that is why the physics of electron-ion collisions has always received new impetus when a better understanding of plasma phenomena proved to be essential for progress in various fields of applied physics.

The first push to study electron collision processes came about a hundred years ago along with the first experiments making use of simple glow-discharge tubes. Pioneering experiments on electron impact ionization of atoms were carried out by *Lenard* et al. [2.1] and, as early as 1912, *J. J. Thomson* [2.2] published a theoretical estimate of cross sections for electron impact ionization of atomic hydrogen. Considerable new impetus came in the 1950s when the importance of electron-ion collisions for astrophysical plasmas was recognized. The interest in these collisions was spurred even more by the international effort to achieve controlled thermonuclear fusion. Many of the plasma properties, for example the radiation emitted, depend on the state of ionization of ions in these plasmas. For interpretation of spectroscopic observations and development of theoretical models describing the formation and evolution of the plasma, cross sections or rate coefficients for electron impact phenomena are needed. Even the solution of the rate equation (2.1) representing our simplified plasma model (Figs. 2.1, 2) requires knowledge of rate coefficients for ionization and recombination.

Motivated by the data needs and aided by technological advances in ultra-high vacuum and electronics, *Dolder* et al. [2.3] carried out the first successful electron-ion ionization measurements on e + He^+ → He^{2+} + 2e. Not only was this the first ever measurement employing two interacting beams of charged particles, but it also set a standard of quality work for the many subsequent beams experiments which were to come.

Within five years of the He^+ crossed beams measurement, *Hinnov* [2.4] introduced techniques for measurements of rate coefficients of ionization in plasmas. *Baker* and *Hasted* [2.5] introduced trapped-ion techniques for ionization measurements. These two additional basic approaches to ionization rate measurements remain the only ways to study very highly charged ions. However, with the forth-

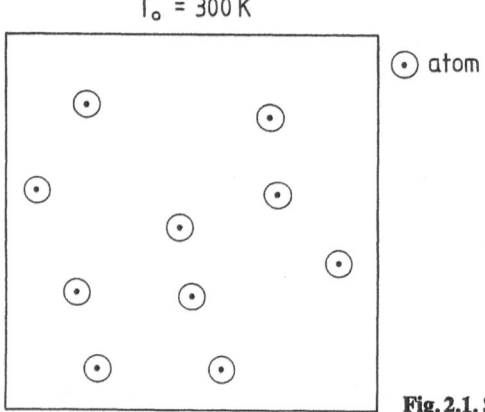

T_0 = 300 K

⊙ atom

Fig. 2.1. Sample of atoms at a temperature T_0 = 300 K

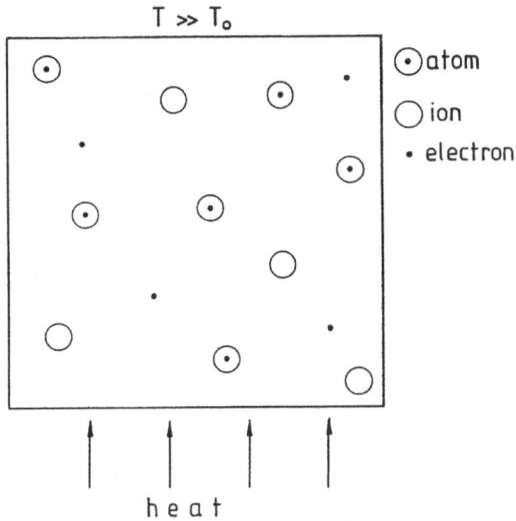

T ≫ T₀

⊙ atom
◯ ion
• electron

Fig. 2.2. Sample of atoms at a temperature $T \gg T_0$

heat

coming storage rings for heavy ions and the integrated electron cooler devices, new progress can be expected for direct experiments on electron impact ionization of very highly charged ions.

A number of reviews have been written and bibliographies compiled. Recent ones are by *Phaneuf* [2.6], *Dunn* [2.7], *Dolder* and *Peart* [2.8], *Salzborn* [2.9] and *Crandall* [2.10]. An especially useful data compilation is that by *Tawara* and *Kato* [2.11]. This chapter introduces the reader to some of the fundamental aspects of electron-impact ionization of ions. It will briefly review experimental and theoretical techniques and highlight some of the important results produced recently on different ionization mechanisms.

2.1 Theoretical Methods

No attempt will be made here to develop the details of collision theory. For detailed discussions the reader is referred to recent review articles and the references therein [2.12–14]. However, some of the general expressions and concepts are given here to allow insight and comparisons.

2.1.1 The Classical Approach of J. J. Thomson

The first and most simple theoretical method to estimate ionization cross sections was employed in 1912 by *J. J. Thomson* [2.2]. Starting from a hydrogen atom he considered the target electron quasi-free, that is, the electron is first treated like a free particle and then a correction for the binding energy is made. Figure 2.3 shows a simplified view of the collision process: a projectile electron with energy $E_e =$

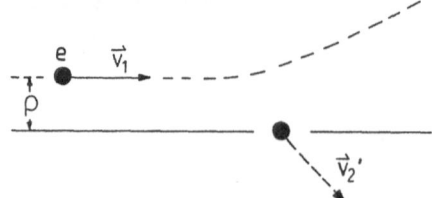

$mv_1^2/2$ approaches with impact parameter ρ a target electron considered at rest. The target nucleus is assumed to be so far away that it does not influence the trajectory of the incoming electron. Classical theory of Coulomb scattering of two charged particles predicts the energy transfer ΔE from the projectile to the target electron, depending on the impact parameter ρ, to be

$$\Delta E = \frac{E_e}{(1 + \rho E_e/a)} \, , \tag{2.2}$$

where $a = e^2/(4\pi\varepsilon_0)$. The target electron could emerge from the collision with energy ΔE if it were really free at the beginning. Since this is not the case, the electron can only leave if ΔE exceeds its binding energy, E_i, in the target atom.

Thomson assumed that the target electron will have zero probability of leaving the atom when $\Delta E < E_i$ and probability $P = 1$ when $\Delta E > E_i$. On the basis of this simple approach the ionization cross section was calculated by integrating over all possible impact parameters weighted by the probability $P(\rho)$

$$\sigma_i = \int_0^\infty P(\rho) 2\pi\rho \, d\rho \, . \tag{2.3}$$

At large ρ the energy transfer ΔE is too small. Hence, the integration has to be carried out only up to a maximum impact parameter $\rho_{\max}$ which is determined from the minimum energy transfer $\Delta E_{\min} = E_i$ sufficient to ionize the target. Thus, $\sigma_i = \pi\rho_{\max}$ and with $\rho_{\max}$ from (2.2) for $\Delta E_{\min} = E_i$ the classical ionization cross section is obtained

$$\sigma_i = \frac{a^2}{\pi} \frac{1}{E_e} \left(\frac{1}{E_i} - \frac{1}{E_e} \right) \, . \tag{2.4}$$

Multiplying (2.4) with E_i^2 ($= Z^2 \cdot 13.6\,\text{eV}$ for hydrogenic atoms) reveals the scaling behaviour of ionization cross sections

$$E_i^2 \sigma_i = f(\xi) \, . \tag{2.5}$$

σ_i depends only on the ionization energy E_i and a general function f of the reduced energy $\xi = E_e/E_i$. Although the results of the classical approach are only qualitative, the predicted scaling rule is useful to estimate cross sections for the ionization of ions along a given isoelectronic sequence.

2.1.2 Quantum Mechanical Approaches

The de Broglie wavelength, $\lambda = h/p_e$, for an electron with velocity v_0 ($= 2.18 \times 10^8$ cm s^{-1}) calculated for the first Bohr orbit of a hydrogen atom is 3×10^{-8} cm and this exceeds the diameter of that orbit by a factor of about 3. Already this comparison shows that quantum calculations have to be used in electron-ion collision phenomena. The accepted non-relativistic theoretical approaches all begin with the Schrödinger equation for the system of (N+1) electrons

$$H(Z, N + 1)\Psi = i\hbar\frac{\partial}{\partial t}\Psi , \tag{2.6}$$

where N and Z are the number of electrons and the nuclear charge of the ion, respectively, and Ψ is the total wavefunction of the (N+1)-electron system.

For ionization of a hydrogenic atom of nuclear charge Z we have

$$H = H_a + H_e + H_{int} , \tag{2.7}$$

where

$$H_a = \frac{p_2^2}{2m} - a\frac{Z}{r_2} \tag{2.8}$$

is the Hamiltonian for the unperturbed one-electron atom,

$$H_e = \frac{p_1^2}{2m} \tag{2.9}$$

is the Hamiltonian of the free incident electron, and

$$H_{int} = \frac{a}{|r_2 - r_1|} - \frac{aZ}{r_1} \tag{2.10}$$

is the interaction Hamiltonian. Here p_j and r_j are the momentum and position vectors of the bound electron ($j = 2$) and the incident electron ($j = 1$), m is the rest mass of an electron. It is assumed that the target nucleus is so heavy that during the collision the center of mass coincides with the nucleus. For an inelastic process H_e is different before and after the collision. In the following sections we briefly discuss some of the most common theoretical approaches which will be used later in comparisons with experimental data.

a) The Plane-Wave Born Approximation (PWBA). The first quantum mechanical approach towards electron impact ionization was made by in 1930 by *Bethe* [2.15]. In the simplest form of the first Born approximation, the unperturbed Hamiltonian, H_0, is the sum of H_a and H_e. The perturbed Hamiltonian, H', therefore includes both electron-electron and electron-nucleus interactions in H_{int}

$$H_0 = H_a + H_e \tag{2.11}$$

$$H' = H_{int} . \tag{2.12}$$

Since H_0 does not include any interaction between the projectile and the target, the appropriate initial wavefunction Ψ_i for H_0 is a product of the atomic wavefunction Φ_0 and that for a free electron ϕ_e, that is, a plane wave. Thus

$$\Psi_i = \phi_e \cdot \Phi_0 = A(k)e^{ik \cdot r_1}\Phi_0(r_2)e^{-iWt/\hbar} . \qquad (2.13)$$

Here, A is a normalization factor and $W = \hbar^2 k^2/(2m) + E_0$ is the total energy, where E_0 is the energy of the atom in its initial (ground) state.

In the case of excitation to state n of the atom, described by a wave function Φ_n with energy E_n, the final total wavefunction is

$$\Psi_f = B(k')e^{ik' \cdot r_1}\Phi_n(r_2)e^{-iW't/\hbar} , \qquad (2.14)$$

with $W' = \hbar^2 k'^2/(2m) + E_n$. The calculation of the differential cross section according to first order perturbation theory requires the computation of the interaction matrix element

$$M_{if} = \langle \Psi_f | H' | \Psi_i \rangle . \qquad (2.15)$$

Bethe carried out these calculations introducing an additional restriction of high electron energies – beyond the one already inherent in the first order perturbation method (which can only be applied when the projectile electron is fast enough to produce only a small perturbation for the atom). The Bethe approach agrees with experiments usually at energies $E_e > 30 \, E_i$. The cross section calculated by Bethe is characterized by a logarithmic dependence on the electron energy

$$\sigma_i = \frac{A}{E_e}[\ln(E_e/B) + C/E_e] , \qquad (2.16)$$

with constants A, B, C which have to be determined for each particular ion separately. While the classical approach leads to a high-energy limit $\sigma_i \approx E_e^{-1} E_i^{-1}$ [see (2.4)] the quantum mechanical treatment reduces the high-energy decrease of σ_i by an additional factor $\ln(E_e/E_i)$. The scaling behaviour of σ_i according to (2.16) is the same as that given by (2.5) and only the function f is different.

b) The Coulomb Born Approximation (CBA). When the target is an ion, particularly with a high charge state, the nuclear interaction term, $-aZ/r_1$, in H_{int} (2.10) is more important than the electronic interaction term $a/|r_2 - r_1|$. In this case, it may be better to include the nuclear term in the unperturbed Hamiltonian, H_0

$$H_0 = H_a + H_e - \frac{aZ}{r_1} , \qquad (2.17)$$

$$H' = \frac{a}{|r_2 - r_1|} . \qquad (2.18)$$

The Hamiltonian H_0 in (2.17) still does not contain any interaction term between bound and incident electrons. Thus, the appropriate initial wavefunction Ψ_i for H_0 is still a product of Φ_0 and $\phi_e(Z)$, where $\phi_e(Z)$ must now satisfy

$$\left[\frac{p_1^2}{2m} - \frac{aZ}{r_1} \right] \phi_e(Z) = \varepsilon_e \phi_e(Z) \,. \tag{2.19}$$

The solution of (2.19) is the hydrogenic Coulomb function with continuum energy $\varepsilon_e = \hbar^2 k^2/(2m)$. The interaction matrix element for excitation now contains only the interelectronic term

$$M_{if} = a \langle \phi_{ef} \Phi_n || r_1 - r_2|^{-1} |\phi_{ei} \Phi_0 \rangle \,. \tag{2.20}$$

The calculations following this scheme for electron impact ionization give much better agreement with experiment than the simpler plane-wave calculations.

c) The Distorted-Wave Born Approximation (DWBA). A further improvement beyond the Coulomb Born approximation can be made by replacing a constant screened charge $\zeta = Z - N$ by a screening function that changes during the collision. In practice, the Schrödinger equation has to be solved for the projectile electron in the field of bound electrons and the nucleus of the target atom. The screened charge distribution in the target atom can be represented by a local or non-local potential. Then, for a local potential $V(r_1)$, instead of (2.17) and (2.18), we have

$$H_0 = H_a + H_e - \left[\frac{aZ}{r_1} - V(r_1) \right] \,, \tag{2.21}$$

$$H' = \frac{a}{|r_1 - r_2|} - V(r_1) \,. \tag{2.22}$$

Now the wavefunction $\phi_e(V)$ for the projectile electron must satisfy the equation

$$\left[\frac{p_1^2}{2m} - \left(\frac{aZ}{r_1} - V \right) \right] \phi_e(V) = \varepsilon_e \phi_e(V) \,. \tag{2.23}$$

The basic numerical technique to evaluate the interaction matrix element M_{if}, and hence the DWBA cross section, is to expand the continuum (distorted) functions $\phi_e(V)$ into partial waves. For low electron energies the number of partial waves which contribute to the cross section is rather limited. This can be understood on the basis of a simple classical consideration. An electron passing an atom at a distance $5a_0$ with a speed $2v_0$ (where a_0 and v_0 are the radius and the electron velocity for the first Bohr orbit in a hydrogen atom) has a classical angular momentum $l = 10\hbar$. With increasing electron energy an increasing number of high-l partial waves has to be included in a DWBA calculation, so the computational effort becomes enormous. Therefore, at high electron energies the plane-wave Born approximation is more favorable.

Since electrons are identical particles it is necessary to consider exchange processes. One can distinguish between two types of exchange in ionization: potential exchange and scattering exchange. Potential exchange denotes exchange between a free electron and a target electron. It is analogous to exchange in a Hartree-Fock bound state calculation and is easily accounted for in a frozen core approximation by including the appropriate non-local potential terms in $V_{\text{DW}} = aZ/r_1 - V$ [see (2.21)]. (In the frozen core approximation the target wave function is fixed during the collision). The second form of exchange, scattering exchange, describes the possible interchange of the two final state continuum electrons. Considerable attention has been devoted to the problem of scattering exchange, especially since it is closely involved with the three-body nature of the final state. Problems arise even at the most elementary level in considerations of scattering exchange. The exchange contribution to the scattering matrix element M_{ex} looks similar to the interaction matrix element M_{if} but with r_1 and r_2 exchanged in the final wavefunction. The total probability for ionization is then given by

$$P_{\text{if}} \propto |M_{\text{if}} + M_{\text{ex}}|^2 = |M_{\text{if}}|^2 + |M_{\text{ex}}|^2 - \delta|M_{\text{if}}||M_{\text{ex}}| . \tag{2.24}$$

Here, $|M_{\text{if}}|^2$ is the direct ionization term, $|M_{\text{ex}}|^2$ the exchange ionization term, and $\delta|M_{\text{if}}||M_{\text{ex}}|$ the interference term. The phase factor δ is related to the relative phases of the direct and exchange matrix elements and depends on the interaction between the two final state continuum electrons. So far there is no detailed theory about the proper determination of δ. In theoretical calculations δ is usually chosen for computational tractability and agreement with experiment rather than for physical necessity. Since most theoretical predictions of direct (knock-out) ionization cross sections are large compared to experimental data, it seems practical to use the Peterkop maximum interference formulation with $\delta = 1$.

d) **The Close-Coupling Approximation.** In this method the incident electron is treated in the same way as the bound electron in a many-electron atom. The total wave function for the collision system is given by an expansion in terms of basis functions Φ_n

$$\Psi(1, 2, \ldots, N, N + 1) = \mathcal{A} \sum_n \Phi_n(1, \ldots, N)\phi_n(N + 1) , \tag{2.25}$$

where $1, 2, \ldots, N, N + 1$ denote the set of variables and quantum numbers needed to identify bound and incident electrons, $\mathcal{A}$ is the antisymmetrization operator, Φ_n is a target wavefunction representing one of the initial and excited states, and ϕ_n is the corresponding wavefunction of the scattering electron. The target wavefunctions are considered known, and ϕ_n is determined by substituting (2.25) into the Schrödinger equation which results in a number of coupled differential equations. The coefficients ϕ_n in their asymptotic form ($r \to \infty$) provide the cross section ($\propto |\phi_n|^2$) for exciting the target to state n. In a real calculation the expansion (2.25) has to be truncated to a modest length in order to limit computation time. The effectiveness of the method depends critically on the choice of target wavefunctions Φ_n included in the expansion. Sometimes pseudo-state wavefunctions with no real corresponding states of the target are used in the expansion to help reduce its length, but there are no general principles to guide the choice of such pseudo-state wave functions. Also, these functions could produce fictitious resonances.

It is difficult to handle ionization in the frame work of the close-coupling approximation. For ionizing collisions, two continuum orbitals must be included in (2.25) and the joint asymptotic behaviour of the two unbound electrons must be determined in order to obtain ionization cross sections. Such asymptotic behaviour is not well understood theoretically. On the other hand, the exchange effect between incident and bound electrons is built into the formulation by the antisymmetrization operator in (2.25).

For a better description of resonance phenomena in the excitation of target electrons one additional modification to (2.25) is usually introduced. In addition to the N-electron target wavefunctions Φ_n, $(N + 1)$-electron bound state functions θ_n $(1, 2, \ldots, N, N + 1)$ are also used to represent the correlation between the projectile and the target electrons; this facilitates the theoretical inclusion of intermediate capture states in the electron-ion collision process. The extended expansion has the form

$$\Psi(1, \ldots, N + 1) = \mathcal{A} \sum_n \Phi_n(1, \ldots, N)\phi_n(n + 1)$$
$$+ \sum_m c_m \theta_m(1, \ldots, N + 1), \qquad (2.26)$$

where c_m are numerical coefficients. The close-coupling approximation in this form, with a truncated set of basis functions, is most appropriate for slow collisions, particularly for the detailed study of threshold behaviour and resonances.

2.1.3 Predictor Formulae

In view of the very complex theoretical approaches, and motivated by the need to provide reasonably accurate estimates of unknown cross sections without lengthy computations, several semi-empirical formulae have been proposed. A compilation of such predictor formulae is given in [2.16]. The most successful one, which has been used extensively in plasma modelling calculations is that of *Lotz* [2.17]

$$\sigma_i(E_e) = \sum_{\nu=1} a_\nu \xi_\nu \frac{\ln(E_e/E_\nu)}{E_e \cdot E_\nu} \left\{ 1 - b_\nu \cdot \exp[-c_\nu(E_e/E_\nu - 1)] \right\}, \qquad (2.27)$$

where E_ν is the binding energy of an electron in the νth of N subshells ($\nu = 1$ means the outermost shell) containing ξ_ν equivalent electrons. The coefficients a_ν, b_ν and c_ν are tabulated for a variety of atoms and ions. They have been adjusted to best reproduce experimental and theoretical data. The uncertainty in the formula is estimated to be less than $\pm 40\,\%$. For ions in charge states $q = (Z - N) \geq 4$, the numbers $a_\nu = 4.5 \times 10^{-14}$ cm^2 eV^2, $b_\nu = c_\nu = 0$ give results which agree within $20\,\%$ with Coulomb-Born-Exchange calculations of *Rudge* and *Schwartz* [2.18] for hydrogen-like, sodium-like, and magnesium-like ions with high Z.

The Lotz formula has to be used with care when inner-shell electrons are ionized and autoionizing intermediate states can be produced. It is then important to account for branching ratios of radiative decay versus autoionization. Among the numerous other predictor formulae for ionization cross sections emphasis is given here to the classical binary encounter calculations of *Gryziński* [2.19] who has also provided the only general approach to multielectron ionization. Gryziński's formula for single ionization has the form

$$\sigma_i(E_e) = \sigma_0 \sum_\nu g_\nu(x) \xi_\nu / E_\nu^2 , \qquad (2.28)$$

with $x = E_e/E_\nu$ and (for electron impact) $\sigma_0 = 6.56 \times 10^{-14}$ cm² (eV)². The shape of the cross section is given by the function

$$g_\nu(x) = \frac{1}{x} \left(\frac{x-1}{x+1} \right)^{3/2} \left[1 + \frac{2}{3} \left(1 - \frac{1}{2x} \right) \ln[2.7 + (x-1)^{1/2}] \right] . \qquad (2.29)$$

The quantities ξ_ν and E_ν are the same as in (2.27).

The classical approach to double ionization of atoms and ions considers two possibilities for the removal of two electrons. The incident electron may hit a target electron which is ejected and then the projectile electron, although now having reduced energy, has a certain probability of knocking a second electron off the same atom. It is also possible that the first ejected electron hits a second target electron which also becomes ionized. The sum of these two contributions in the classical binary encounter approach is given by

$$\sigma_i^{2e} = \frac{\sigma_0^2}{P_1^2 P_2^2} \frac{\xi_1^{5/3}(\xi_1 - 1)}{4\pi R^2} g \left(\frac{E}{P_1 + P_2} \right) . \qquad (2.30)$$

The quantity σ_0 is the same as in (2.28). $P_1(= E_i)$ and P_2 are the binding energies of the first two electrons which can be removed from the target atom or ion, respectively, ξ_1 is the number of equivalent electrons in the outermost shell, R is the gas kinetic radius of this shell and g is a general function given in graphical form by *Gryziński* [2.19]. This function can be well approximated by

$$g(x) = \frac{\ln x}{a_1 x^2 + a_2 x + a_3} + a_4(x-1) + a_5(x-1)^2 , \qquad (2.31)$$

with $a_1 = 1.59$, $a_2 = -3.71$, $a_3 = 16.20$, $a_4 = 2.14 \times 10^{-4}$ and $a_5 = -3.67 \times 10^{-6}$.

All empirical formulae, except the one by *Burgess* and *Chidichimo* [2.20], only include direct ionization processes. Experimental results obtained during the last 10 years show, however, that indirect processes often dominate direct knock-on ionization. Indirect contributions proceeding via intermediate multiply excited states, which can decay by cascades of radiative or autoionization processes, can be so complex that any attempt to represent all ionization cross sections by a universal analytic function must fail.

2.1.4 Threshold Behaviour

The threshold ionization of atoms by electron impact has received considerable attention, both experimentally and theoretically, and is often considered a subfield in itself. Unlike electron-ion excitation cross sections, the ionization cross section is equal to zero at threshold: $\sigma(E_e = E_i) = 0$. The calculation of the threshold behaviour of the ionization cross section $\sigma(E_e)$ as $E_e \rightarrow E_i$, is complicated since in the region of low energies of both outgoing electrons the interelectron repulsion becomes comparatively strong and the three-body character of the collision is enhanced.

According to the classical treatment by *Wannier* [2.21] the threshold cross section is

$$\sigma_i \propto (E_e - E_i)^\alpha \tag{2.32a}$$

with

$$\alpha = \frac{1}{4}\left[\left(\frac{100q + 91}{4q + 3}\right)^{1/2} - 1\right], \tag{2.32b}$$

where q is the charge state of the parent ion. For atoms ($q = 0$) the exponent is $\alpha = 1.127$, a value which has been confirmed theoretically [2.22]. Experimentally, $\alpha = 1.131 \pm 0.019$ was found for He [2.23], providing a convincing verification of the Wannier threshold law.

It is primarily the number of outgoing electrons which determines the exponent in the threshold law. Hence, the threshold law for ionization of a singly charged ion ($q = 1$) can be experimentally studied by double photoionization of the atom ($q = 0$). Such experiments have been successful in verifying Wannier's prediction (2.32) for $q = 1$ (see, for example, *Kossmann* et al. [2.24]). Direct precision measurements of threshold laws for electron impact ionization of ions have not been reported. The main difficulty in electron-ion collision experiments is in the accuracy of cross sections needed for a determination of the exponent in the threshold law. This accuracy is presently limited by the necessary measurement of electron and ion beam overlap (form factors) which appears to have a lowest uncertainty level of about $\pm 1\%$ at best, and which cannot be held constant when the collision energy is changed. Together with low signal/noise ratios at threshold this uncertainty has limited the determination of α to the first digit, that is, small deviations from a linear ionization onset can hardly be quantified.

For multiple electron impact ionization (removal of n electrons from the atom or ion, with $n = 2, 3, \ldots$), *Wannier* [2.25] predicted $\alpha = n$. By more detailed analysis *Klar* and *Schlecht* [2.26] found $\alpha = 2.162$ for electron impact double ionization of a singly charged ion (or triple photoionization of a neutral atom). This prediction is supported by a triple photoionization experiment by *Samson* and *Angel* [2.27].

2.2 Ionization Mechanisms

In the experiments to be discussed later ions are collided with electrons by crossing beams of these particles. The charge of an ion ionized by electron impact is analyzed roughly a microsecond after the collision. This means, an inner-shell vacancy in the ion produced by ionization or excitation has time enough to be filled again by relaxation processes before the ion charge is measured. The time span for radiative decay or autoionization is typically much less than 10^{-8}s, so that even cascades of different relaxation processes happen long before the charge of the ion is analyzed. Hence, the meaning of single or multiple ionization has to be specified and information on the time of observation is useful. In this chapter single or multiple ionization usually specifies the net result of a collision process where the ion charge is detected long after the collision. This also means that it is not a priori possible in such experiments to distinguish between different decay mechanisms which have led to the same final ion charge. In principle, interference of different contributions is possible. However, most experiments which have been carried out so far with ions do not show immediate evidence for the necessity to sum amplitudes rather than probabilities when a cross section is calculated. In this sense different ionization mechanisms contribute to the net experimental result by an additive probability.

2.2.1 Single Ionization

The most straightforward view of direct single ionization is given in Fig. 2.3. The knock-out process leading to single ionization is

$$e + A^{q+} \rightarrow A^{(q+1)+} + 2e , \tag{2.33}$$

where A^{q+} is an ion in charge state q. In addition to direct ionization, a two-step process with inner-shell excitation and subsequent autoionization can contribute to net ionization. This process is called excitation-autoionization

$$e + A^{q+} \rightarrow [A^{q+}]^{**} + e , \tag{2.34a}$$

$$[A^{q+}]^{**} \rightarrow A^{(q+1)+} + e . \tag{2.34b}$$

A schematic picture of this process is shown in Fig. 2.4.

A third ionization mechanism is possible in which the projectile electron is first captured by the ion in a radiationless process, using up the excess energy by the excitation of an inner shell electron. By this resonant capture of the incident electron a state can be populated which is so highly excited that it can emit two electrons, either sequentially or both at a time

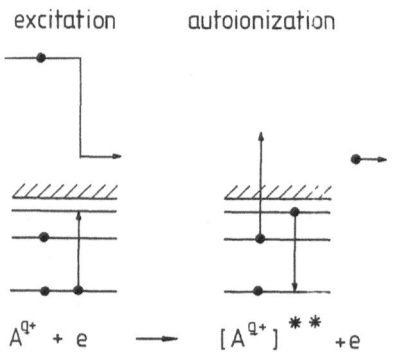

excitation   autoionization

$A^{q+} + e \longrightarrow [A^{q+}]^{**} + e$

Fig. 2.4. Schematic energy level diagrams for the two-step process of excitation-autoionization

$$e + A^{q+} \rightarrow [A^{(q-1)+}]^{***},$$ (2.35 a)

$$[A^{(q-1)+}]^{***} \rightarrow A^{(q+1)+} + 2e.$$ (2.35 b)

The net result is ionization of the parent ion. The first step in the process described by (2.35) is called dielectronic capture. Other terms are also used, depending on the decay channel of the intermediate multiply excited state observed. Widely used terms are resonant excitation, resonant recombination, dielectronic capture or even (incorrectly) dielectronic recombination although the latter implies the subsequent stabilization of the ion $[A^{(q-1)+}]^{***}$ by photon emission after the initial capture process. Each of the first three names has its merits and will probably further be used in the literature. The term dielectronic capture seems to be the most descriptive; however, it turned out recently that even three electrons may be involved in the capture of the incident electron, which would require a term trielectronic capture. In this chapter the term dielectronic (and where required, trielectronic) capture will be used.

Dielectronic capture involves the interaction of one electron in the ion core with the projectile electron. Figure 2.5 presents a simple scheme of the process which leads from an initial electronic state i in the ion A^{q+} to an electronic state j in the ion $A^{(q-1)+}$ which is at least doubly excited. Dielectronic capture is time-reverted autoionization (or Auger electron emission). Hence it becomes clear that dielectronic capture can only occur when the energy of the incident (free) electron matches the resonance condition

$$E_{\text{res}} = E(j) - E(i),$$ (2.36)

where $E(j)$ and $E(i)$ are the total binding energies of all electrons in the excited ion $A^{(q-1)+}(j)$ and the parent ion $A^{(q+)}(i)$, respectively. Due to the principle of detailed balance, the cross section σ_c for dielectronic capture is proportional to the probability of its time-reverse process, that is, the autoionization decay rate A_a of the ion $A^{(q-1)+}(j)$ that leads back to the ion $A^{q+}(i)$ plus a free electron.

$$\sigma_c = \frac{c}{E_e} \cdot \frac{g_j}{2g_i} \cdot \frac{A_a \cdot \Gamma(j)}{(E_{\text{res}} - E_e)^2 + \Gamma(j)^2/4}.$$ (2.37)

dielectronic capture ①

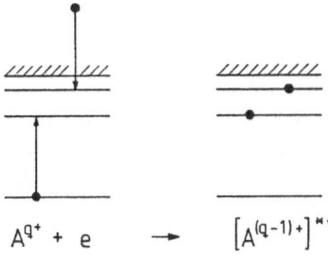

$A^{q+} + e \quad \longrightarrow \quad [A^{(q-1)+}]^{**}$

Fig. 2.5. Schematic energy level diagrams for the dielectronic capture process leading to a multiply excited state

In this formula $c = 7.89 \times 10^{-5}$, σ_c is in cm^2, E_{res} is in eV, E_e is the energy of the projectile electron in eV, g_i and g_j are the statistical weights of the initial electronic state i and the multiply excited state j, respectively, A_a is in s^{-1} and $\Gamma(j)$ is the total width (in eV) of the state j. For the determination of $\Gamma(j)$ all possible radiative and non-radiative decay probabilities have to be summed. The cross section σ_c has the shape of a Lorentzian distribution with the width $\Gamma(j)$. In most of the cases which will be discussed in this chapter, $\Gamma(j)$ is much smaller than the experimental energy spread ΔE_e. Assuming a Gaussian distribution

$$G(E, E_e) = \frac{\Delta E_e}{\pi/2} \exp\left\{ -\frac{1}{2}\left((E - E_e)^2/(\Delta E_e/2)^2\right) \right\} \tag{2.38}$$

of the experimental electron energy E_e, the observed cross section is given by

$$\langle \sigma_c(E_e) \rangle = \int \sigma_c(E) \cdot G(E, E_e)\, dE . \tag{2.39}$$

For $\Delta E_e \gg \Gamma(j)$ the maximum of the observed (convoluted) cross section $\langle \sigma_c \rangle$ is reduced compared to that of σ_c by a factor

$$\left. \frac{\langle \sigma_c \rangle}{\sigma_c} \right|_{E_e = E_{res}} = \frac{\pi}{2} \frac{\Gamma(j)}{\Delta E_e} . \tag{2.40}$$

Hence, the sensitivity of an experiment for the detection of resonances critically depends on the energy spread of the electron beam.

The dependence of a measured cross section $\langle \sigma_c \rangle$ on the experimental energy spread ΔE_e calls for a quantity to characterize the resonance which is independent of the experimental conditions. Such a quantity is the resonance strength

$$S = \int \sigma_c(E)\, dE = \int \langle \sigma_c(E) \rangle\, dE , \tag{2.41}$$

which is given by the area under a resonance curve.

The excited state j formed by dielectronic capture is unstable. It can decay by autoionization as was discussed above

$$A^{(q-1)+}(j) \rightarrow A^{q+}(k) + e , \tag{2.42}$$

or by the emission of a photon

$$A^{(q-1)+}(j) \rightarrow A^{(q-1)+}(p) + h\nu . \tag{2.43}$$

The autoionization can lead to a state k in the ion A^{q+}, the photon emission to a state p in the ion $A^{(q-1)+}$. The states k and p can still be excited states, subject to further decay processes. In particular, when $k = i$ the autoionization is the time-reverse dielectronic capture. The net result of dielectronic capture plus subsequent autoionization with $k = i$ would be observed experimentally as a resonant elastic scattering of an electron from an ion A^{q+}. When k describes an excited state of the ion A^{q+} one observes resonant inelastic electron scattering. The simplest case then is the one where state k can just decay by the emission of a photon. In an experiment looking at line radiation from the state k the combined processes dielectronic capture plus autoionization, would result in resonances in the photon emission (excitation) cross section. The possibility that k is also an autoionizing state will be discussed later.

In the case of photon emission, an excited state p of the ion $A^{(q-1)+}$ is populated. When p is non-autoionizing the new charge state of the ion after capture of the incident electron has become stabilized; however, emission of at least one more photon is necessary to reach the ground state of the ion $A^{(q-1)+}$. The net result of the combined processes of dielectronic capture plus subsequent emission of a photon would then be a dielectronic recombination.

The cross sections for resonant electron scattering and dielectronic recombination can be obtained by multiplying σ_c with branching ratios for the particular decay processes leading to the observed net result. This is true only if there is no interference of the resonant two-step process with the direct channels, that is, non-resonant elastic scattering and radiative electron capture, respectively.

So far, resonances in elastic electron scattering, in stabilized electron-ion recombination and in electron impact excitation of ions have been discussed. All these processes required not more than one electron bound in the parent ion. When the parent ion has at least two electrons an additional class of resonances becomes possible which can be seen in the channel of net single ionization of the ion.

Figure 2.6 shows a possible scheme of decay processes after a dielectronic (or trielectronic) capture event finally resulting in net single ionization of the parent ion. For the sake of simplicity it is assumed that the parent ion has one electron in the lowest level (number 1) and one in the second level (number 2). Dielectronic capture in this example excites the electron in level 1 to level 3 and the initially free electron is captured also into level 3. The resulting intermediate configuration $(2,3,3)$ is displayed in Fig. 2.6 on the left side. In the first autoionization process the two electrons in level 3 (in the present example) interact, one jumps into level 2 and the other one finds itself in the continuum. The resulting configuration $(2,2)$ undergoes a second autoionization process lifting one electron into the continuum while the other makes a transition to level 1.

Dielectronic capture (Fig. 2.5) of the incident electron by the A^{q+} ion and subsequent successive emission of two electrons (double autoionization; Fig. 2.6) leave an ion $A^{(q+1)+}$ and two electrons in the continuum with well defined energies. The

double autoionization ②

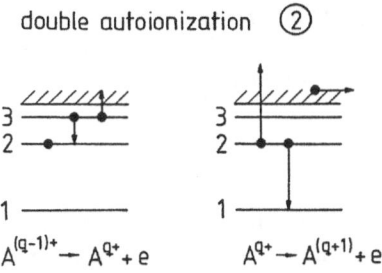

Fig. 2.6. Scheme of double autoionization of a triply excited electron configuration

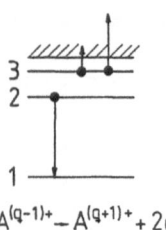

double ionization ③

3 ──●──●──
2 ──┬──
   │
1 ──┴──

$A^{(q-1)+} \rightarrow A^{(q+1)+} + 2e$

Fig. 2.7. Scheme of auto-double-ionization of a triply excited state, that is, emission of two electrons in a single process

combined process $1+2$ thus results in net ionization of the parent ion. It is called resonant-excitation-double-autoionization (REDA).

REDA can only occur when the ion A^{q+} has a suitable autoionizing state which can be populated in the first decay step from the intermediate $A^{(q-1)+}$ ion. The cross section for REDA processes $e + A^{q+}(i) \rightarrow A^{(q-1)+}(j) \rightarrow A^{q+}(k)+e \rightarrow A^{(q+1)+}(s)$ (without consideration of higher order processes involving also radiative transitions in between the single steps) can be written as

$$\sigma_{\text{REDA}} = \sigma_c \frac{1}{\Gamma(j)} \sum_{k,s} \Gamma_a(j \rightarrow k) \frac{1}{\Gamma(k)} \Gamma_a(k \rightarrow s) . \qquad (2.44)$$

Here, $\Gamma(j)$ and $\Gamma(k)$ are total widths of the intermediate autoionizing states j and k; $\Gamma_a(j \rightarrow k)$ and $\Gamma_a(k \rightarrow s)$ are partial autoionization widths for the transitions between the states j, k and s, respectively. The sum extends over all autoionizing intermediate states k and non-autoionizing states s. (In case s is autoionizing, even more complex mechanisms are possible and then (2.44) has to be modified).

Figure 2.7 shows another possibility for a resonant contribution to net single ionization. As in Fig. 2.6, it is assumed that the initial dielectronic capture of the incident electron by the ion $A^{q+}(i)$ produces an ion $A^{(q-1)+}$ with an electron configuration $(2,3,3)$. This can decay by correlated emission of two electrons which also leads to the net loss of one electron from the parent ion A^{q+}. The simultaneous emission of two electrons is called auto-double-ionization or, sometimes, double Auger process. Dielectronic capture (1) $e + A^{q+}(i) \rightarrow A^{(q-1)+}(j)$ and subsequent simultaneous emission of two electrons (auto-double-ionization (3)) $A^{(q-1)+}(j) \rightarrow A^{(q+1)+}(s) + 2e$ leave an ion $A^{(q+1)+}$ and two electrons in the continuum sharing the excess energy with a certain probability distribution. The combined process (1)+(3) is called resonant-excitation-auto-double-ionization (READI). The cross section for READI in lowest order can be written as

$$\sigma_{\text{READI}} = \sigma_c \sum_s \frac{\Gamma_{2a}(j \rightarrow s)}{\Gamma(j)} , \qquad (2.45)$$

where $\Gamma_{2a}(j \rightarrow s)$ is the partial auto-double-ionization width for the transition from state j to state s. The sum extends over all non-autoionizing states s in the ion

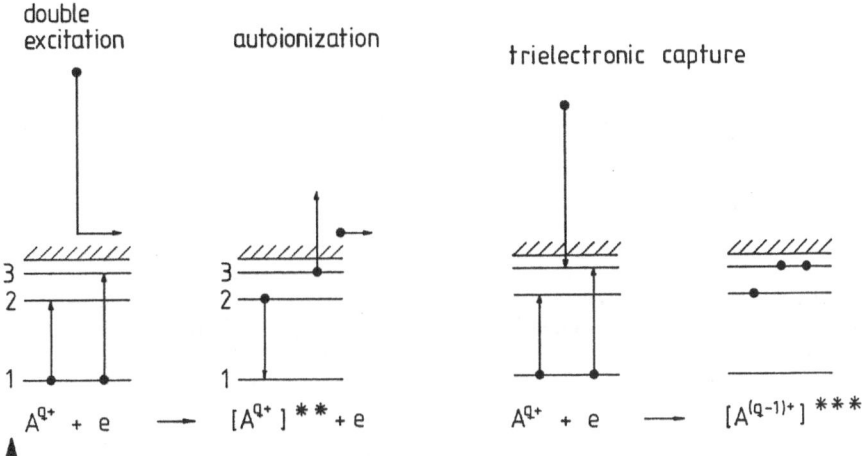

double
excitation   autoionization

trielectronic capture

$A^{q+} + e \longrightarrow [A^{q+}]^{**} + e$ $A^{q+} + e \longrightarrow [A^{(q-1)+}]^{***}$

Fig. 2.8. Scheme of electron impact double excitation with subsequent autoionization

Fig. 2.9. Scheme of trielectronic capture leading to a triply excited state

$A^{(q+1)+}$. Whenever REDA is energetically allowed, READI can also occur, though the probability for READI processes is generally much smaller than for REDA processes. Currently, experiments cannot distinguish between REDA and READI resonances when both are allowed. However, when there is no suitable intermediate autoionizing state k in the ion A^{q+} REDA cannot occur. Then, at sufficiently low incident electron energies, READI may become the only possible resonant contribution to net ionization and resonances found in that energy range can be unambiguously identified with the READI mechanism.

With increasing refinement of experiments higher order contributions to ionization can be detected. Recently it was shown that processes involving double excitations provide exotic possibilities. Figure 2.8 shows a scheme for simultaneous excitation of two electrons in an ion A^{q+}. The resulting doubly excited state may autoionize so that an ion $A^{(q+1)+}$ is found after the collision and net single ionization of the parent ion is observed. It is even possible that the incident electron is captured with simultaneous excitation of two core electrons (Fig. 2.9). This process requires interaction of three electrons (as does the process sketched in Fig. 2.8) and hence it is called trielectronic capture. The resulting ion $A^{(q-1)+}$ is triply excited so that it can decay by double autoionization (Fig. 2.6) or by auto-double-ionization (Fig. 2.7). In both cases a net loss of one electron from the parent ion is observed.

2.2.2 Multiple Ionization

A simple classical picture for double ionization used by Gryziński to estimate cross sections has already been referred to. The concept is sketched in Fig. 2.10. One possibility for direct double ionization in the classical picture is the ionizing interaction of the projectile with two target electrons in successive collisions within the ion. The

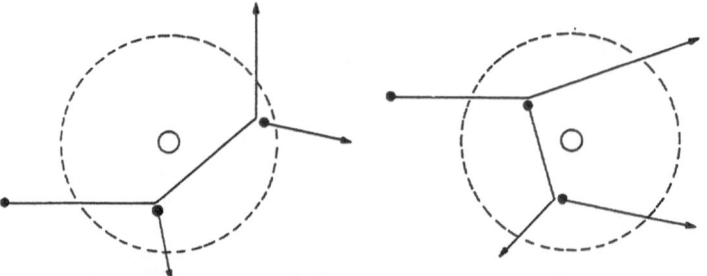

Fig. 2.10. Classical schemes for electron impact double ionization. Left: the projectile electron kicks out two target electrons on its way through the atom; right: the projectile electron knocks on a target electron which then ionizes a second target electron on its way out of the atom

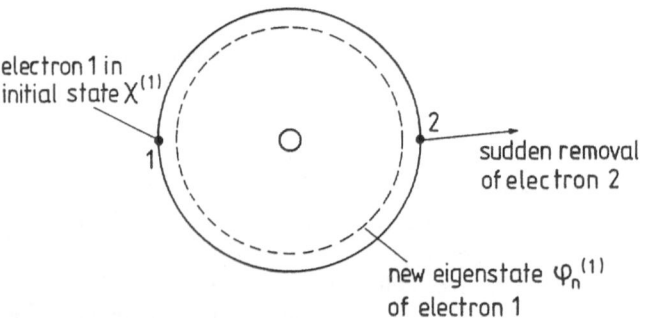

Fig. 2.11. Simplified view of a shake-off process

second classical mechanism is the interaction of the projectile electron with one target electron which is then scattered from a second electron within the same ion and both electrons are set free. The calculation of these two knock-on double ionization mechanisms by Gryziński led to the cross section given by (2.30).

A quantum mechanical formulation related to the second mechanism starts from the picture of a shake-off process, that is, the rearrangement of electron states after a sudden change of screening. A simple picture is shown in Fig. 2.11. Initially, there are two electrons in their lowest orbits. Then, suddenly, electron 2 is removed, for example, by electron impact. "For a moment" electron 1 is still in its initial state $\chi^{(1)}$ which is no longer an eigenstate when the other electron is removed. As a result there is a certain probability P for the remaining electron 1 to go to any one of the new bound eigenstates $\varphi_n^{(1)}$

$$P = \sum_{n=1}^{\infty} |\langle \varphi_n^{(1)} | \chi^{(1)} \rangle|^2 , \qquad (2.46)$$

and a probability P_i to become ionized

$$P_i = 1 - P . \qquad (2.47)$$

Hence, the ratio of the cross sections for net double and net single ionization is given by

$$\gamma = \frac{\sigma_{q,q+2}}{\sigma_{q,q+1}} = \frac{1-P}{P} . \qquad (2.48)$$

Unfortunately, there is no general quantum mechanical formulation of multiple ionization by electron impact. The difficulty is in the description of the asymptotic form of the wavefunction of three outgoing electrons.

In addition to direct multiple ionization

$$e + A^{q+} \rightarrow A^{(q+n)+} + (n+1)e \qquad (2.49)$$

there are also multistep processes leading to the net production of $A^{(q+n)+}$ from A^{q+}. The simplest of these processes is an inner-shell ionization followed by autoionization

$$e + A^{q+} \rightarrow [A^{(q+1)}]^{**} + 2e , \qquad (2.50\,a)$$

$$[A^{(q+1)}]^{**} \rightarrow A^{(q+n)} + (n-1)e . \qquad (2.50\,b)$$

It is also possible to excite an inner-shell electron to a bound level and then let the ion autoionize (at least two electrons have to be emitted in this case)

$$e + A^{q+} \rightarrow [A^{q+}]^{**} + e , \qquad (2.51\,a)$$

$$[A^{q+}]^{**} \rightarrow A^{(q+n)+} + ne . \qquad (2.51\,b)$$

The autoionization processes can occur either sequentially or mixed with auto-double-ionizations. Tracing individual decay paths leading to the observed net charge state of the ionized ion may thus become a difficult task. This is even more true for resonant contributions to multiple ionization. As in net single ionization it is possible that the incident electron is captured by the target ion. When the energy of the incident electron is sufficiently high, a deeply bound electron in the core may be lifted to an outer shell in a dielectronic capture event. The resulting $A^{(q-1)+}$ ion may be so highly excited that it can emit three or more electrons so that net double or a higher multiple ionization is observed

$$e + A^{q+} \rightarrow [A^{(q-1)+}]^{**} \rightarrow A^{(q+n)+} + (n+1)e . \qquad (2.52)$$

For $n = 2$ in such a reaction the whole process has been called resonant-excitation-triple-autoionization (RETA) and for $n = 3$ resonant-excitation-quadruple-autoionization (REQA). The multiplicity of individual decay paths makes it impossible to invent new names for every single possibility for a cascade process.

2.2.3 Summation of Cross Section Contributions

In the previous two sections different contributions to single and multiple ionization have been discussed. For single ionization of an ion A^{q+} basically three different

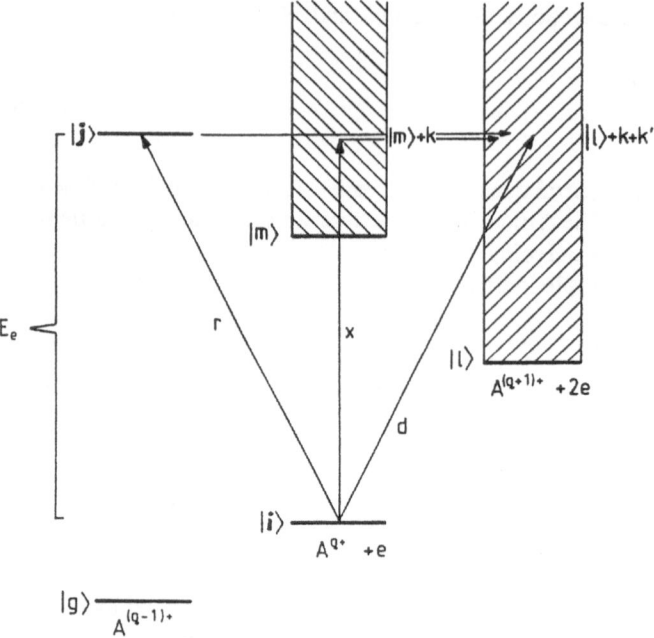

Fig. 2.12. Schematic illustration of processes contributing to net single ionization. The paths are: resonant recombination (r) plus subsequent two-electron emission, excitation (x) plus subsequent (single) autoionization, and direct ionization (d).

groups of mechanisms were found: direct, excitation-autoionization and resonant contributions. Figure 2.12 shows an energy level diagram illustrating the different kinds of transitions involved in ionization of an ion A^{q+}. The energies of the ground states $|g\rangle, |i\rangle$ and $|l\rangle$ of the ions $A^{(q-1)+}$, A^{q+} and $A^{(q+1)+}$ are indicated, respectively. When an electron with energy E_e collides with the ion A^{q+} direct ionization may lead to the ion $A^{(q+1)+}$ in its ground state plus two free electrons with certain energies k and k' (arrow d). The same products may be formed by direct excitation (arrow x) leading to the excited autoionizing state $|m\rangle$ of the ion A^{q+} plus an electron with energy k and an autoionization forming $A^{(q+1)+}(l) + k + k'$. This same result could be obtained when the quasi-bound state $|j\rangle$ of the ion $A^{(q-1)+}$ is formed by resonant dielectronic capture (arrow r) which then decays by two successive autoionization processes $|j\rangle \rightarrow |m\rangle + k \rightarrow |l\rangle + k + k'$. The indirect processes contribute through the interaction between the different continua.

In an experiment observing the production of $A^{(q+1)+}$ ions from parent A^{q+} ions one cannot distinguish between the three different paths. Hence one has to add transition amplitudes when the cross section is to be calculated. This can lead to interference effects. However, as mentioned above, there is very limited evidence for the presence of interference effects in electron-ion collisions so far. With continuing refinement of ionization measurements, it may well become necessary to include interference in the calculations. With few exceptions the total single ionization cross sections $\sigma_{q,q+1}^{(tot)}$ calculated so far have just been obtained by a summation over the cross sections for all partial contributions

$$\sigma_{q,q+1}^{(tot)}(i) = \sum_m \sigma_{q,q+1}^{(ion)}(i \to f)B_f^r + \sum_m \sigma_{q,q}^{(exc)}(i \to m)B_m^a$$

$$+ \sum_j \sigma_{q,q-1}^{(cap)}(i \to j)\left[B_j^{2a} + \sum_{m,s} B_j^a(j \to m)B_m^a(m \to s)\right]. \qquad (2.53)$$

Here $\sigma_{q,q+1}^{(ion)}(i \to f)$ is the direct ionization cross section from the initial state i to a particular level f within the singly ionized configurations. In case f is an autoionizing state populated by inner-shell ionization a branching ratio B_f^r has to be considered which describes the radiative stabilization of f. Non-radiative decay, that is, emission of one or more electrons, would lead to net double or multiple ionization. Thus already the first sum in (2.53) shows that even direct ionization is much more complex than suggested by the Lotz formula (2.27) which does not include the branching ratios B_f^r (which have to be set equal to 1 in case of non-autoionizing states f).

The second term in (2.53) contains contributions $\sigma_{q,q}^{(exc)}(i \to m)$ for direct non-resonant excitation from the initial state i to a particular level m within the core-excited configurations. The quantity B_m^a is the branching ratio for (a single) autoionization from level m. The third term involves contributions from resonant excitation (dielectronic capture) from the initial state i to a particular autoionizing level j within the core-excited configurations of the one-less charged ion. The branching from such states is given by sequential (single) Auger (or Coster-Kronig) processes via different intermediate autoionizing states m leading to non autoionizing levels s in the ion $A^{(q+1)+}$. In addition, simultaneous emission of two electrons from state j has to be considered by a branching ratio B_j^{2a}.

Equation (2.53) shows that calculation of cross sections $\sigma_{q,q+1} = \sigma_{q,q+1}^{(tot)}(i)$ involves a substantial amount of information on atomic structure, lifetimes of states and collision dynamics. In particular, for complex heavy ions, tracing all the possible reaction paths may well be beyond present computing capabilities. Qualitatively, the different contributions to net single ionization result in a cross section shape as indicated in Fig. 2.13. Real examples will be discussed later in this chapter.

2.3 Experimental Approaches

For the measurement of electron impact ionization cross sections σ or rate coefficients $\alpha = \int f(v)\sigma(v) \cdot v \, dv$, where $f(v)$ is the normalized Maxwell-Boltzmann distribution of electron velocities v, it is necessary to bring electrons and ions together in a collision volume and measure a signal due to ionization processes. A number of different approaches have been used for quantitative measurements which will be briefly reviewed in the following.

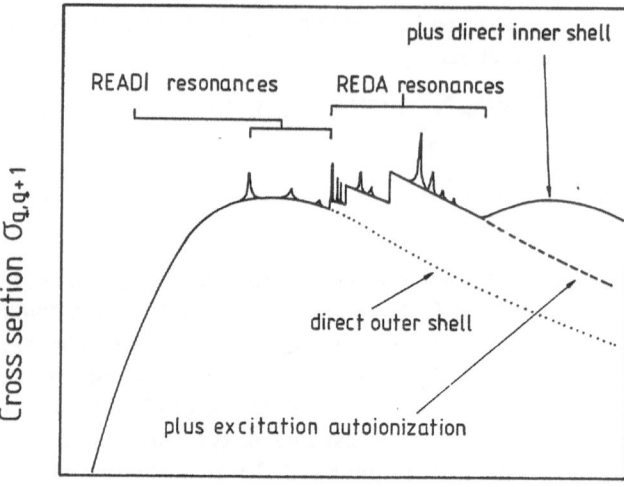

Fig. 2.13. Schematic cross section for electron impact single ionization of ions with different contributions from direct and indirect processes

2.3.1 Plasma Rate Measurements

In the introduction a simple model for the charge balance of ions and atoms in a plasma was discussed. By observing the time evolution of n_+, n_e and n_0 at a given temperature T the coefficients in the rate equation (2.1) can be determined by comparing the experiment with the solution of the model rate equations. Such techniques have been developed and used to determine rate coefficients for ionization, excitation and recombination of multiply charged ions. A more realistic description than (2.1) of the time evolution of a plasma is given by the following set of differential equations:

$$\frac{dN_q}{dt} = n_e N_{q-1} \alpha_{q-1} - n_e N_q (\alpha_q + \beta_q + n_e \gamma_q)$$
$$+ n_e N_{q+1} (\beta_{q+1} + n_e \gamma_{q+1}), \tag{2.54}$$

where N_q is the number density of ions in charge state q, n_e the electron density, α_q the ionization rate coefficient for an ion in charge state q, β_q the recombination rate coefficient, and γ_q the three-body recombination coefficient both for the initial ion charge state q.

When the electron density changes with time, as is the case, for example, in a θ-pinch, an additional term, $+N_q/n_e \cdot dn_e/dt$, has to be introduced which keeps the total concentration of the ions constant during the compression or expansion phases of the discharge. It has to be noted that (2.54) implies a low electron density so that any ion produced in an excited state relaxes by photon emission before the next collision with an electron (coronal model). Multiple ionization processes are not accounted for. Low electron density also implies that three-body recombination is negligible. Usually it is also assumed that recombination of ions and electrons can be neglected for the plasma conditions used in such experiments. A signature for the presence of ions in charge state q is the line radiation observed from a certain electronic transition in the ion A^{q+}. For a thin plasma, within the coronal model, the upper level is essentially populated by electron collisions from the ground state. The emission coefficient is given by

$$\varepsilon = h\nu n_e \chi N_q \, , \tag{2.55}$$

where $h\nu$ is the photon energy, χ the rate coefficient for excitation from the ground state and N_q the ground state population of A^{q+} ions. From a time-resolved observation of the line intensity $P_q = \varepsilon \cdot l$, where l equals the depth of the plasma along the line of sight, the normalized derivative $P_q^{-1} \cdot dP_q/dt$ is obtained experimentally. This can be compared with the same quantity calculated in the frame of the model

$$\frac{1}{P_q}\frac{dP_q}{dt} = \frac{1}{n_e}\frac{dn_e}{dt} + \frac{1}{\chi}\frac{d\chi}{dT}\frac{dT}{dt} + \frac{1}{l}\frac{dl}{dt} + \frac{1}{N_q}\frac{dN_q}{dt} \, . \tag{2.56}$$

The last term is inserted from (2.54).

Careful observation of the electron density n_e and temperature T are necessary to characterize the plasma. When the line radiation from ions A^{q+} occurs during a phase when the plasma parameters do not change much, the first three terms do not contribute significantly and the time history of the observed line basically reflects the time history of the respective ion. In particular, when dT/dt is small it is not necessary to know χ precisely. In essentially all cases, a set of rate coefficients based upon theory or semi-empirical formulae is put into the set of (2.56) and the equations are solved. The time-dependent spectral line intensities that result from this are compared with the observed line intensities. The coefficients are then iteratively adjusted until suitable agreement with observations is obtained. Ionization rate coefficients determined in this way are subject to large uncertainties. Quotations by authors are typically 20 % to 40 %. The averaging over the Maxwell-Boltzmann energy distribution causes any structure in the cross section, and the physical information contained therein, to be lost. Nevertheless, the method is important in that it has provided the only available information on ions in high ionization stages. For more details the reader is referred to original literature [2.4, 28–33].

2.3.2 Trapped-Ion Methods

These techniques are based on electron impact ionization of ions held in a trap. *Baker* and *Hasted* [2.5] obtained information on ionization of ions by the observation of sequential ionization in an ion source within which the target ions were trapped by the space charge of an intense electron beam. By applying suitable external electric potentials additional to the "internal" space charge potential it was possible to obtain trapping times of the order of one second and measure relative ionization functions [2.34]. Further development was carried out by *Hasted* and *Awad* [2.35] who used a cylindrically symmetric hollow electron beam for ion trapping and primary ionization (target preparation) and a concentric variable energy electron beam to further

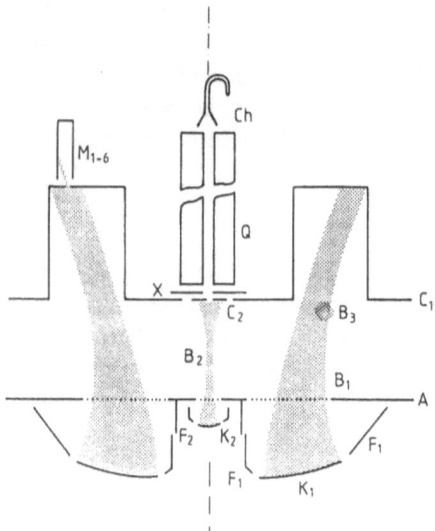

Fig. 2.14. Diagram of the concentric beam device [2.36] used by Hasted and coworkers in collision studies. K_1: toroidal cathode; K_2 spherical cathode; F_1, F_2: shaping electrodes; A: anode; B_1: hollow electron beam; B_2: axial electron beam; B_3: atomic beam; C_1, C_2: collecting electrodes; X: extraction electrode; Q: quadrupole mass filter; Ch: channel electron multiplier; M_{1-6}: current monitoring electrodes. The dot-dashed line is the cylindrical symmetry axis

ionize the ion targets (see Fig. 2.14). This device has been employed for the study of electron impact ionization of positive ions [2.35–37] as well as for electron impact dissociation of molecular ions [2.38] and for electron-ion recombination [2.39]. After calibrating the apparatus to crossed beams data, uncertainties of measured ionization cross sections were estimated to be in the vicinity of 30 %.

Perhaps the most productive use of traps for electron impact ionization studies is represented in the work of *Donets* et al. [2.40–42]. Using an EBIS trap (see Chap. 1) and analyzing time histories of the charge state evolution of trapped ions, they have deduced high energy (keV range) cross sections for ionization of ions over a range of charge states up to very highly charged, for example, Ar^{17+}, Kr^{33+}, Xe^{47+}.

Usually, recombination processes are neglected in ion traps; this may not be a good approximation when the electron energy hits a recombination resonance. Recently, *Ali* et al. [2.43] have demonstrated the substantial effect of dielectronic recombination on the charge state distribution of ions extracted from an EBIS. Outside the resonance region, however, the following equations may still provide a realistic description of the time history of ion charge states

$$\frac{1}{J}\frac{dN_q}{dt} = \sum_{i=0}^{q-1} N_i \sigma_{i,q} - N_q \sum_{i=q+1}^{Z} \sigma_{q,i} . \tag{2.57}$$

The electron flux density is denoted by J, N_q is the number density of ions in charge state q [as in (2.54)] and $\sigma_{i,j}$ is the cross section for $(j-i)$-fold ionization of an ion A^{i+} by a single electron impact.

By trapping ions for a given time t and subsequent charge state analysis of the extracted ions one can obtain $N_q(t)$. These experimental quantities can be compared with solutions to (2.57) using estimates of the single and multiple ionization cross

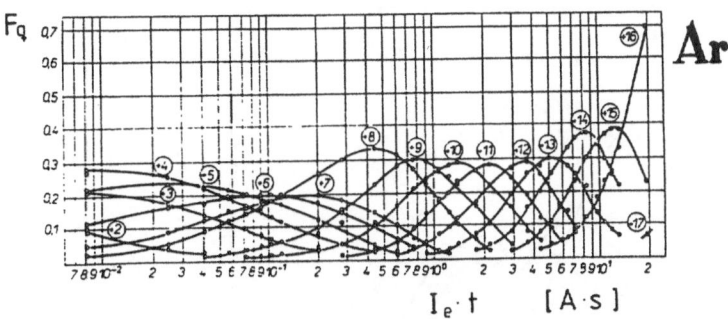

Fig. 2.15. Normalized charge state distributions $F_q = N_q / \sum_j N_j$ of Ar ions extracted from an EBIS [2.41] as a function of the product of electron current I_e and trapping time t. The solid lines are solutions to (2.58) with adjusted cross sections

sections. As in the case of the plasma measurements the cross sections $\sigma_{i,j}$ are adjusted for optimum agreement between the measured and calculated time histories of N_q. Although efforts have been made to deduce information on multiple ionization cross sections (2.57) generally have to be restricted to single ionization processes. Otherwise the fit programs do not yield stable solutions. Thus, (2.57) are reduced to the simple form

$$\frac{1}{J}\frac{dN_q}{dt} = N_{q-1}\sigma_{q-1,q} - N_q\sigma_{q,q+1} . \tag{2.58}$$

Figure 2.15 shows measured and calculated normalized charge state distributions $F_q = N_q(t)/\sum N_j$ of Ar ions as a function of time (multiplied by the electron beam current I_e).

In spite of the simplicity of the model description the ionization cross sections obtained are in remarkable agreement with crossed beams results and theoretical predictions for highly charged light ions. It must be doubted, however, that modeling the time history of charge states for heavy atoms without accounting for multiple ionization can yield real cross sections. In a case study assuming 700 eV electrons impinging on a sample of Xe atoms the time history of charge state fractions F_q was calculated [2.44] using alternatively (2.57) and (2.58). For the calculations experimental cross sections for single and multiple ionization were plugged into the equations. A comparison of the solutions of (2.57) and (2.58) with and without accounting for multiple ionization by a single electron impact shows substantial differences in the obtained fractions F_q (see Fig. 2.16). Hence, the analysis of experimental fractions F_q, which are determined by both single and multiple ionization processes, using a single ionization model, must lead to serious deficiencies in the evaluated cross sections.

2.3.3 Ion Channeling

Although plasma methods and the trapped-ion techniques have provided cross section information on quite highly charged ions the ultimate goal to study ionization

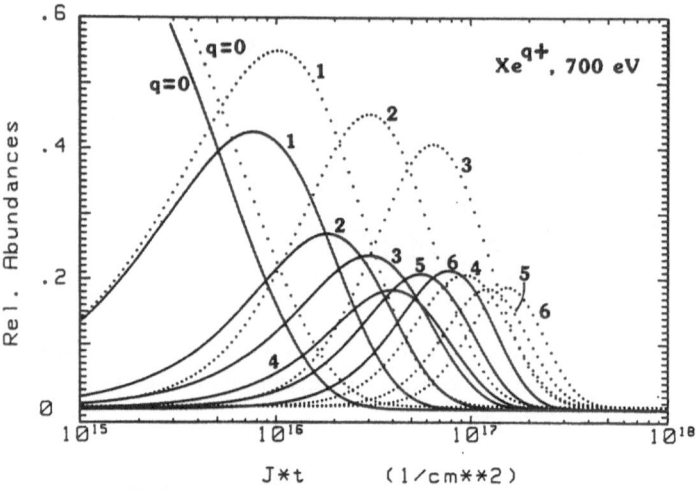

Fig. 2.16. Relative abundances F_q of Xe^{q+} ions ($q = 0, \ldots, 6$) as a function of the product of electron flux density and ion containment time at 700 eV electron energy [2.44]. The solid lines were calculated including multiple ionization (2.57), the dotted lines are based on single ionization only (2.58)

also of very heavy atoms to very high charge states, up to U^{91+}, cannot presently be reached by these methods. For the time being the only way to produce species such as U^{91+} is to accelerate ions to velocities comparable to the orbital velocities of the deeply bound core electrons and pass them through a stripper target. For efficient production of U^{91+} this technique requires an ion energy of the order of about 100 GeV!

Since the ionization cross sections for such highly charged ions are small (about 10^{-24} cm^2 for U^{91+}) it is difficult to obtain sufficient event rates for a measurement and a very dense electron target becomes desirable. One approach toward this goal is the use of the electron cloud in a channel of a single crystal, for example, of silicon [2.45, 46]. In the single crystal the atoms are arranged in a periodic structure with "channels" along which there are no nuclei. Ions traveling in these channels make only large impact parameter collisions with the distant nuclei of the crystal material and thus do not acquire enough energy to strip off their tightly bound electrons. Since the electron density in a crystal channel can be calculated theoretically the analysis of ion charge states emerging from a channel allows, in principle, extraction of electron impact ionization cross sections.

The experimental verification of channeling conditions, the unavoidable presence of ion-nucleus collisions in the crystal, the presence of impurities and defects in the crystal, the dependence on calculated electron densities, the velocity distribution of electrons in the crystal channel, the competition of different charge-changing electron-ion collisions in the channel and the verification of single collision conditions make the interpretation of the measured quantities difficult and leads to large uncertainties in the measured cross sections. It must be doubted whether the quality of such cross section measurements is sufficient to test theoretical predictions of electron impact ionization of highly charged ions.

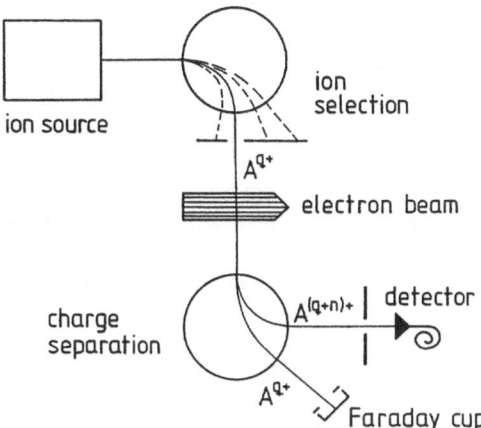

2.3.4 Crossed-Beams Method

The technique of intersecting electron and ion beams has provided the most accurate and detailed information on electron impact ionization of ions. The general features and associated problems have been discussed and reviewed in great detail by a number of authors [2.47, 48 and references therein].

The notion of colliding beams is conceptually simple and is represented in the diagram Fig. 2.17. Beams of electrons and ions are generated and then caused to intersect at some angle θ. Incident and product particles are separated, and their intensities measured with suitable instrumentation. The ion signal R (number of ions formed per second) resulting from ionizing collisions between electrons and target ions can be obtained from the relationship

$$R = \int n_i n_e v_r \sigma \, dV , \qquad (2.59)$$

where n_i and n_e are the position dependent number densities of target ions and electrons, respectively, v_r is the relative velocity of the colliding particles, and σ is the cross section. The integration extends over the collision volume V.

a) **The Static Case.** For electron and ion beams parallel in themselves and colliding at an angle $\theta = 90°$ the following relations can be applied

$$v_r = (v_e^2 + v_i^2)^{1/2} \qquad (2.60)$$

where v_e and v_i are the laboratory velocities of electrons and ions, respectively,

$$R = \sigma \cdot I_e \cdot I_i \cdot M/F \qquad (2.61)$$

where I_e and I_i are the electrical currents in the electron and the ion beam, respectively, and the quantities M and F are given by

$$M = v_r/(v_e v_i q e^2) , \qquad (2.62)$$

39

$$F = \frac{\int i_e(z)\,dz \int i_i(z)\,dz}{\int i_e(z)\,i_i(z)\,dz}.$$ (2.63)

The charge of a target ion is qe and e is the charge on an electron. The quantity F is called form factor. It describes the overlap of the interacting beams. The averaged line densities $i_e(z)$ and $i_i(z)$ of electrons and ions characterize the vertical intensity profiles of the two beams (z is the coordinate in vertical direction, that is, perpendicular to both beams). In particular, $i_e(z)dz$ and $i_i(z)dz$ are the differential currents of electrons and ions, respectively, flowing between z and $z + dz$. These currents can be measured behind a narrow slit which has to be moved across both beams in order to get the complete profiles. From the measured distributions the integrals can be evaluated numerically to obtain the form factor F. When one beam is larger than the other, and has a uniform intensity and height h, then $F = h$.

b) The Dynamic Method. A procedure alternative to probing the beam intensity profiles by slits has been introduced by *Defrance* et al. [2.49] and further developed by *Müller* et al. [2.50]. It involves a movement of one beam through the other in the z-direction while keeping it parallel to its initial axis. The velocity of the movement is $u(z')$ where z' describes the distance between the two beams measured, for example, from center to center. It is assumed that electron and ion beam profiles do not change during one complete sweep spanning between positions where there is no overlap of the two beams. The sweep interval has to cover the range where both beams can overlap. Neither a constant electron current nor constant ion current or constant sweep velocity are required.

The assumption of constant beam profiles implies that $g(z) = i_e(z, t)/I_e(t)$ and $f(z, z') = i_i(z - z', t)/I_i(t)$ do not depend on time t (or relative displacement z').

In the experiment the z'-axis is divided into discrete intervals of length Δz at positions z_j. Ionized ions are counted during the movement in intervals from $z_j - \Delta z/2$ to $z_j + \Delta z/2$ ($j = 1, 2, \ldots$). The accumulated number of counts in the jth interval

$$N(z_j) = \int_{z_j - \Delta z/2}^{z_j + \Delta z/2} R(z')\,dz' = R(z_j) \cdot \Delta z/u(z_j)$$ (2.64)

is stored in channel j of a multichannel analyzer. Here $R(z_j)$ is the average signal rate in the interval $z_j - \Delta z/2$ to $z_j + \Delta z/2$. At the same time the parent ion charge Q_i, the electron charge Q_e and the time τ are measured for that same interval

$$Q_i(z_j) = \int_{z_j - \Delta z/2}^{z_j + \Delta z/2} I_i(z')/u(z')\,dz' = I_i(z_j) \cdot \Delta z/u(z_j),$$ (2.65)

$$Q_e(z_j) = \int_{z_j - \Delta z/2}^{z_j + \Delta z/2} I_e(z')/u(z')\,dz' = I_e(z_j) \cdot \Delta z/u(z_j),$$ (2.66)

$$\tau(z_j) = \int_{z_j - \Delta z/2}^{z_j + \Delta z/2} r/u(z')\,dz' = r \cdot \Delta z/u(z_j).$$ (2.67)

Here, r is a pulse rate from a fixed-frequency pulser. The quantities $Q_i(z_j)$, $Q_e(z_j)$ and $\tau(z_j)$ are stored in channels j of three additional spectra.

After completion of data accumulation, the quantities

$$S_j = N(z_j) \cdot \tau(z_j)/[Q_e(z_j) \cdot Q_i(z_j)] \tag{2.68}$$

are calculated, each of which directly yields the desired cross section provided the beam overlap at position z_j is known. By summing over all positions z_j, including complete separation of both beams at the beginning and at the end of the sweep (with overlap in between), one obtains

$$S = \sum_j S_j = \sum_j \sigma M r \int_{-\infty}^{\infty} \frac{i_e(z,z_j)}{I_e(z_j)} \frac{i_i(z-z_j,z_j)}{I_i(z_j)} \, dz \,. \tag{2.69}$$

In (2.69) the instantaneous velocity $u(z_j)$ and the time spent in interval j have dropped out of the equation. Making use of the assumption of constant beam profiles one obtains

$$S = \sigma M r \int_{-\infty}^{\infty} g(z) \sum_j f(z,z_j) \, dz$$

$$= \sigma M r \int_{-\infty}^{\infty} g(z) \cdot \frac{1}{\Delta z} \, dz = \frac{\sigma M r}{\Delta z} \,. \tag{2.70}$$

The relationships $\int i_e/I_e \, dz = 1$ and $\int i_i/I_i \, dz = 1$ are employed. Since the integration (or summation in the experiment) spans from no overlap over optimum overlap to no overlap of the two beams we have

$$\sum_j f(z,z_j) = \sum i_i(z-z_j,z_j)/I_i(z_j) = \Delta z^{-1} \,, \tag{2.71}$$

where Δz corresponds to the width of one channel of the multichannel analyzer spectrum. The quantity S is determined experimentally in the following way. First, a channel-by-channel normalization of the signal spectrum to the accumulated ion charge is carried out. The resulting spectrum shows a peak on top of a constant background. This background is assumed to be solely due to the ion charge accumulated during the measurement. After subtraction of this background the content of each channel j in the remaining spectrum is divided by the corresponding electron charge $Q_e(z_j)$ and multiplied by the corresponding dwell time $\tau(z_j)$. Integration of the spectrum yields the quantity S from which the cross section directly results:

$$\sigma = \frac{S \Delta z}{M r \varepsilon} \,, \tag{2.72}$$

where ε is the efficiency of the product ion detector. No additional information on form factors is required in this method.

An example of experimental spectra taken for one cross section measurement is shown in Fig. 2.18. Spectrum number I is the position-dependent signal of ionized (Na^{2+}) ions produced at an electron energy of 70 eV. When the two beams (elec-

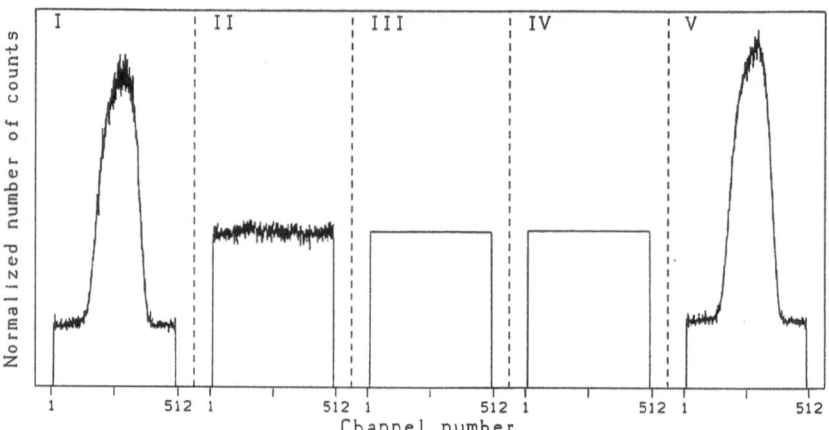

Fig. 2.18. Experimental spectra taken for one cross section measurement with the dynamic crossed beams method. The example is for 70 eV electrons ionizing Na$^+$ ions. Spectrum I is the number of Na^{2+} ions produced during a scan. As in all other spectra the abscissa is the position of the electron beam. Spectrum II is the ion charge, spectrum III the electron charge and spectrum IV the number of clock pulses. Spectrum V displays the numbers $S(z_j)$ (see 2.68). For convenience of display the spectra are normalized to the total number of counts. The measured cross section in this example is 7.34×10^{-18} cm^2 with a statistical uncertainty of 0.08×10^{-18} cm^2

trons and Na$^+$ ions) do not overlap one measures a flat background. The signal has its maximum at optimum overlap of the beams in the middle of a sweep. Spectrum number II is the position-dependent ion current measured by accumulating counts from a current-to-frequency converter. The spectrum shows slight changes of the ion current with time (position). The third spectrum (III) is the position-dependent electron current measured by accumulating counts from a second current-to-frequency converter. The fourth spectrum (IV) measures the inverse speed of the movement of the electron beam across the ion beam. This is accomplished by accumulating counts from a constant-frequency pulser as a function of the position of the electron gun. Spectrum V shows the quantities S_j obtained. The peak area is proportional to the ionization cross section.

As mentioned before, this technique does not require constant particle currents, nor constant speed of the movement of one beam through the other. This is demonstrated by two examples shown in Fig. 2.19 where the incident ion current was modulated (a) and the speed of the movement was arbitrarily changed during a sweep (b). When only the ion current is modulated the signal is modulated in the same way and the normalized spectrum V gives a restored distribution as expected for an "ordinary" measurement. When the scan speed is modulated the ion charge and signal counts accumulated per channel also change, however, the normalized spectrum V again has a normal shape. The cross sections obtained in these odd measurements (Fig. 2.19) agree with those obtained under good "normal" conditions (Fig. 2.18).

c) Electron-Energy Scans. Measurements of collision cross sections using interacting electron and ion beams have a reproducibility of ±0.5 % at best. This limit is

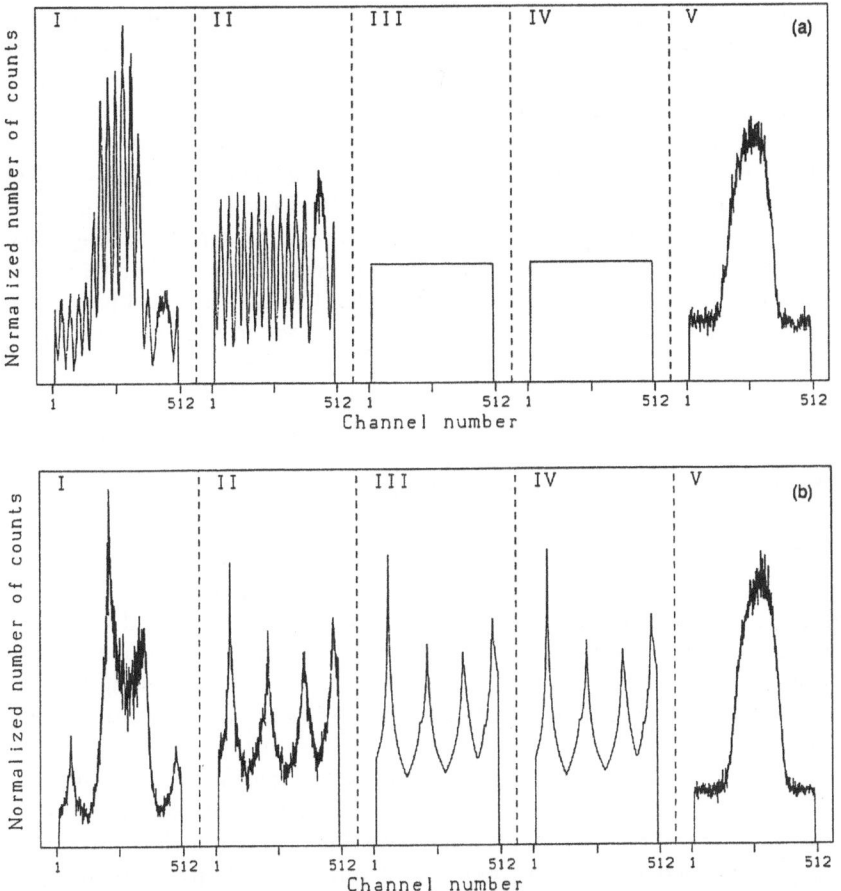

Fig. 2.19 a,b. Experimental spectra as in Fig. 2.18 for the same collision system: (a) The ion current was purposely modulated, hence the modulation in the signal and ion charge spectra. The resulting cross section is $(7.20 \pm 0.09) \times 10^{-18}$ cm^2 (b) The scan speed was purposely modulated, hence the modulation in the spectra for signal, ion charge, electron charge and time accumulated per channel (position interval). The resulting cross section is $(7.32 \pm 0.07) \times 10^{-18}$ cm^2

probably due to changes in the beam intensity profiles during a cross section measurement which cannot easily be accounted for by the techniques described above. Thus it is difficult to detect fine structures in total ionization cross sections caused by the presence of indirect ionization mechanisms.

Some of the problems with changing experimental conditions during a measurement can be avoided by performing measurements sufficiently fast and repeating them until the desired statistical precision is reached. Recently, such a concept has been successfully introduced to electron impact ionization of ions [2.51]. An intense electron beam is crossed with an ion beam at a position of optimum beam overlap. A computer sets the electron energy, allows for a signal accumulation time of typically only 3 ms and then sets the next energy. Pause times as short as 300 μs are possible to set a new electron energy and to read and reset the scalers accumulating the ionization signal, the ion and electron currents and counts from a constant-frequency pulse source. The contents of the scalers are stored in multichannel spectra. For a scan over a 40 eV energy interval typically 1024 channels (energies) are used with a spacing between two measurements of 0.039 eV, which is well below the energy resolution of an electron beam produced by a thermal cathode. One scan over 1024 energies takes about 3 s. By repeating scans many times one averages out possible fluctuations in the form factor, measurements of beam currents and counting rates, and in other sources of data scatter. It has been shown that this technique allows a precision which is at least a hundred times higher than that obtained previously by other methods. Fluctuations in measured cross sections could be reduced to the 0.01 % level [2.52].

2.4 Single Ionization Cross Section Data

Since 1961 a large body of data has been accumulated for ions in charge states up to 16 + by employing the crossed beams technique. It is not the purpose of this chapter to review all these measurements. The discussion will be concentrated on groups of experiments represented by specific examples. Cross section data published until 1985 have been compiled by *Tawara* and *Kato* [2.11]. Since then exciting new experiments have been carried out at different laboratories and an update of the data compilation is in order.

It has often been assumed that cross sections for multiple ionization are negligibly small compared to single ionization. Hence, for about two decades, most of the experiments on ionization of ions were directed to single ionization. It is now recognized that for the dominant cross section contributions, the differentiation between single and multiple ionization is somewhat artificial and that it is just the branching subsequent to the formation of excited states that determines whether a net single or net multiple ionization event is observed after an electron-ion collision. In this chapter single and multiple ionization will be treated in different sections. However, a more general notion of ionization will be applied in the discussion of some of the latest experiments.

2.4.1 Hydrogen-like Ions

Obviously, one of the most important classes of experiments in electron-ion ionization is that dealing with hydrogen-like ions. With only one target electron the H-like systems seem ideal for making comparisons between experiment and theory and to test the understanding of the direct ionization process. In spite of the relative simplicity of an H-like ion the free motion of two electrons and a nucleus within their mutual potential cannot be properly represented in ionization theories and experiments are needed to test and guide theory. Hence, it was no accident that He^+ was the object of the first crossed charged beams experiment carried out by *Dolder* et

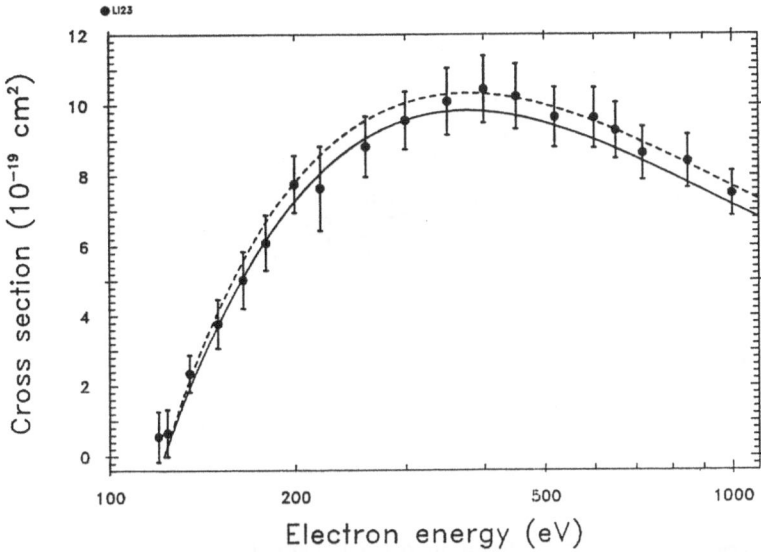

Fig. 2.20. Cross section for electron impact ionization of Li^{2+} ions. The solid points are the measurements by *Tinschert* et al. [2.53], the solid line is a distorted-wave exchange calculation by *Younger* [2.54] and the dashed line is the Lotz formula (2.27)

al. in 1961. Measurements for the next member of the hydrogen isoelectronic sequence, Li^{2+}, were not reported until very recently [2.53]. The reason for this delay is in the relative difficulty of producing intense beams of Li^{2+} ions necessary to measure a cross section in the range of 10^{-19} cm^2. The cross section data for Li^{2+} are shown in Fig. 2.20 together with the total experimental uncertainty. Also shown are cross sections calculated from the three-parameter Lotz formula [2.17] and the distorted-wave exchange approximation [2.54]. Theory and experiment are in good agreement. Total cross sections for direct single ionization of simple ions appear to be predicted with good accuracy by present day theory.

The early classical theoretical approach led to a scaling behaviour of electron impact ionization cross sections. This is confirmed by the Born approximation for direct single ionization. Hence, for H-like ions one may expect a universal curve when the cross section multiplied by the square of the ionization energy E_i is plotted as a function of the scaled electron energy E_e/E_i. Figure 2.21 displays scaled cross sections for ground state hydrogenic systems. In particular, cross sections are included obtained with the trapped-ion technique. The scaling appears to work quite well. All data points are grouped around a hypothetical universal curve. Much of the data scatter may be attributed to uncertainties in the EBIS measurements [2.41]; however, a closer look at the data shows that for small E_e/E_i the scaled cross sections increase with Z. The reason for this is the change of screening discussed above which has its highest influence on the ionization process at low electron energies due to a focusing of the incident electrons. This effect is also seen in the theoretical cross section curves calculated by *Rudge* and *Schwartz* [2.18] for atomic numbers $Z = 1$,

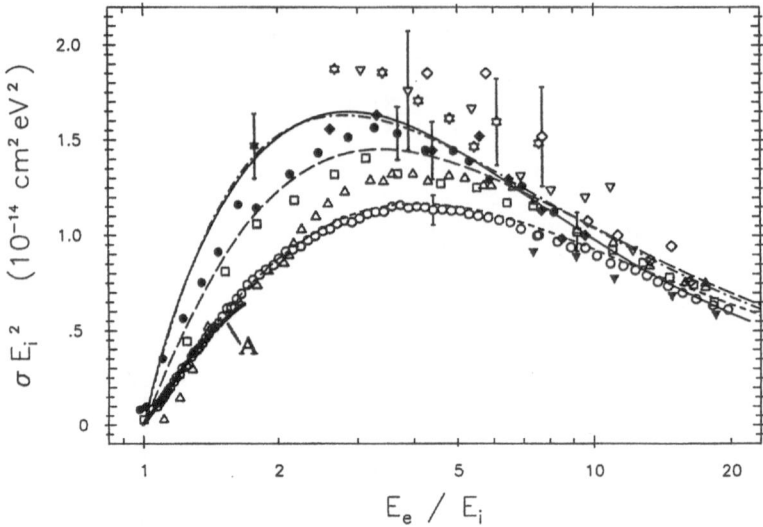

Fig. 2.21. Scaled cross sections σE_i^2 for electron impact ionization of the H-atom and H-like ions. Experimental data are: curve A for H(1s) [2.55], $\triangle$ for H(1s) [2.56], $\blacktriangledown$ for H(1s) [2.57], $\bigcirc$ for H(1s) [2.58], $\square$ for He$^+$ [2.59], $\bullet$ for Li^{2+} [2.53], $\Diamond$ for C^{5+}, $\triangledown$ for N^{6+}, $\blacklozenge$ for O^{7+}, $\star$ for Ne^{9+}, $\bigstar$ for Ar^{17+} [2.41]. Theoretical results are: short broken curve for H(1s), long broken curve for He$^+$, chain curve for a fictitious element with atomic number $Z = 128$ [2.18], full curve for the $Z = \infty$ limit [2.60]

$Z = 2$ and a fictitious $Z = 128$, which are also displayed in Fig. 2.21. For $Z = 1$ and $Z = 2$ there is almost perfect agreement between theory and experiment.

The full curve represents the theoretical cross section calculation of *Moores* et al. [2.60] for the $Z = \infty$ limit, the maximum of which coincides with the $Z = 128$ result of *Rudge* and *Schwartz* [2.18]. Minor discrepancies between theories at high ratios E_e/E_i where the Bethe limit is nearly reached are probably due to the special choice of parameters in scaling functions f used for fitting a limited number of theoretical cross sections. For $Z = 1$ and $Z = 2$, *Tinschert* et al. [2.53] calculated fit parameters from the theoretical cross sections of Rudge and Schwartz. For $Z = 128$ and $Z = \infty$ the parameters are given by the authors of the respective papers.

Figure 2.22 presents a collection of the most complete experimental data sets obtained so far for H, He$^+$ and Li^{2+}. The effect of increasing atomic number and ion charge state on the cross section shape becomes clearly visible. For sufficiently high ratios E_e/E_i all three data sets are merged by the scaling into a common functional dependence which is already predicted by the first-order Born approximation. In fact, measurements on the ionization of atomic hydrogen are sometimes calibrated against the Born or Bethe cross sections at high electron energies. This is not the case for the present H, He$^+$ and Li^{2+} absolute data. Comparison with the Lotz formula shows how well this scaling rule works for electron impact direct ionization.

The above discussions all lead to a consistent view of ionization processes of hydrogen-like ions. So, further experiments with higher atomic number would seem to be a waste of effort. It has been pointed out [2.53], however, that dramatic changes

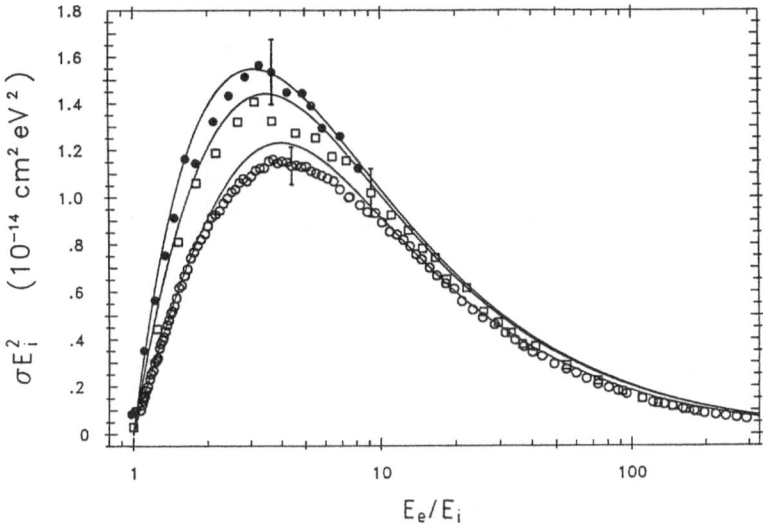

Fig. 2.22. Scaled cross sections for electron impact ionization of H (open circles [2.58]), He$^+$ (squares [2.59]) and Li^{2+} (filled circles [2.53]). Full curves show the predictions of the Lotz formula (2.27)

will occur in Fig. 2.21 when the atomic number increases to $Z = 50$ or even to $Z = 92$.

From Bohr's atomic model the electron binding energy for a hydrogen-like atom is $E_i = 13.6$ eV $\times Z^2$. For $Z = 92$ this would result in $E_i = 115$ keV and orbital velocities of the order of 2×10^{10} cm s^{-1}, suggesting relativistic effects both on the bound electron and the projectile electron which needs relativistic velocities to remove the bound electron. In fact the calculated binding energy of a 1s electron in an U^{91+} ion is $E_i = 131.8$ keV [2.61].

With respect to the relativistic energy of the ionizing electron one has to expect that the scaling fails. Thus the calculations for $Z = 128$ and $Z = \infty$ shown in Fig. 2.21 are fictitious in two ways: there is no atom with such atomic numbers and if it did exist, its cross section for electron impact ionization would look completely different from the non-relativistic calculation. For a guess of relativistic effects on ionization cross sections one could start, for example, from the Lotz formula, which agrees so well with the experiments for low-Z ions and introduce a relativistic correction. Such a correction factor was calculated by *Gryziński* [2.19] using classical methods. The factor is

$$R = \left(\frac{\tau + 2}{\varepsilon + 2} \right) \left(\frac{\varepsilon + 1}{\tau + 1} \right)^2 \left(\frac{(\tau + \varepsilon)(\varepsilon + 2)(\tau + 1)^2}{\varepsilon(\varepsilon + 2)(\tau + 1)^2 + \tau(\tau + 2)} \right)^{3/2} , \tag{2.73}$$

where $\varepsilon = E_e/(mc^2)$ and $\tau = E_i/(mc^2)$ are the projectile electron energy and the target binding energy, respectively, in electron rest energy units. For $Z > 3$ the ionization cross section can be written in the form

$$\sigma = 4.5 \times 10^{-14} \frac{R}{E_e \cdot E_i} \ln \left(\frac{E_e}{E_i} \right) \text{ cm}^2\text{eV}^2 . \tag{2.74}$$

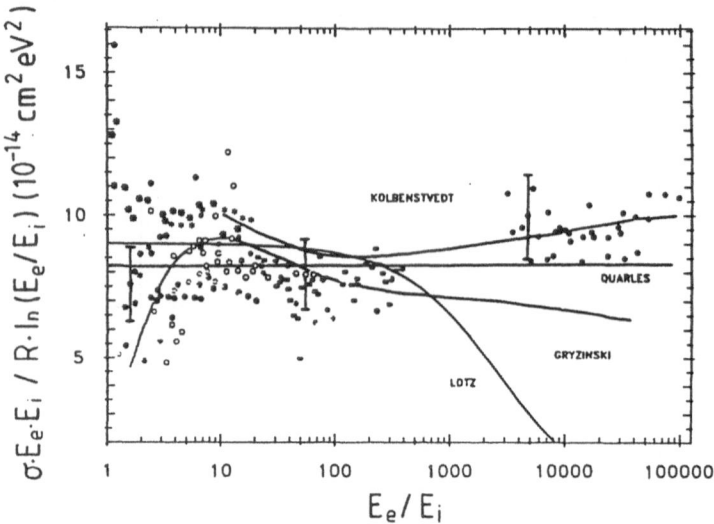

Fig. 2.23. Plot of scaled electron impact K-shell ionization cross sections of atoms with Z ranging from 3 to 83 [2.62]. For the assignments of symbols and the references to the original literature see [2.62]. The curve denoted QUARLES was calculated from (2.74)

This formula has been successfully applied before to scale inner-shell ionization cross sections of heavy atoms. Figure 2.23 shows a comparison of (2.74) with experimental cross sections including the range of relativistic energies. The agreement is so remarkable that (2.74) was used by *Tinschert* et al. [2.53] to estimate changes in the scaling behaviour of ionization cross sections for H-like atoms with high atomic number Z. Their calculations for $Z = 3, 10, 30, 50, 70$ and 92 are compared in Fig. 2.24 with an experiment [2.45] that was performed by employing the channeling technique. The experimental cross section together with its uncertainty bar had to be multiplied by 1/3 to bring it on scale. Also shown in Fig. 2.24 is a sophisticated relativistic calculation by *Pindzola* et al. [2.63] which supports the predictions made on the basis of (2.74). The fact that the experiment is far from all theoretical expectations may indicate an interesting new phenomenon. The uncertainties inherent in the channeling method, however, rather suggest that the experiment gàve a wrong result. In any case it is necessary to perform careful experiments possibly by using the crossed beams technique in a heavy ion storage ring in order to test the new theoretical and experimental results and to provide a basis for a better understanding of relativistic effects in the electron impact ionization of ions in general and, in particular, the scaling behaviour for hydrogenic species.

2.4.2 Helium-like Ions

Two-electron atoms and ions provide targets for ionization experiments next in complexity (or simplicity) to H-like systems. There is only one shell populated with two equivalent electrons, so that an adequate description by theory for direct single ion-

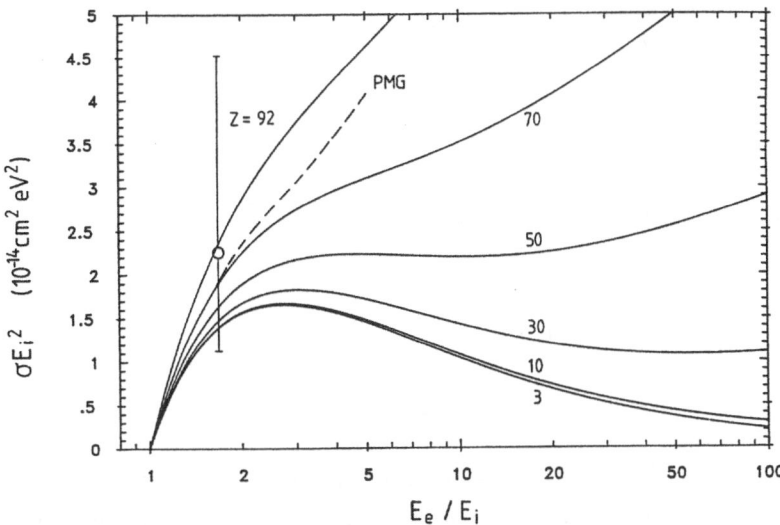

Fig. 2.24. Scaled cross sections for electron impact ionization of hydrogen-like ions. Full curves are from (2.74) with the atomic number indicated. The dashed line (PMG) was calculated by *Pindzola* et al. [2.63] for $Z = 92$. The data point is 1/3 of the experimental value found by *Claytor* et al. [2.45]. The error bars were also multiplied by 1/3 in order to fit into the plot

ization and even by the Lotz formula can be expected. Crossed beams measurements on multiply charged He-like ions are difficult because the two 1s electrons are tightly bound which leads to small cross sections, and because of the difficulty of obtaining intense ion beams. Thus the available data for multiply charged ions are sparse and have large associated uncertainties for the species investigated: B^{3+}, C^{4+}, N^{5+} [2.64], O^{6+} [2.65], Ne^{8+}, and Ar^{16+} [2.41]. Within the experimental uncertainties the data are well represented by the distorted-wave theory of *Younger* [2.66] and also the Lotz formula (2.27). One complication of the experiments with multiply charged He-like ions is the necessity to use "hot" ion sources which produce not only the desired ions in the ground state but also in the 2^3S and 2^1S metastable states. Usually, it is not easily possible to assess the fraction of such ions in the parent ion beam and hence the interpretation of experimental results is difficult [2.65]. Such difficulties can be avoided in experiments with Li^+ ions which can be produced by surface ionization. The necessary emitter temperatures are too low for significant production of ions in metastable states. It is no surprise therefore that several experiments have dealt with the ionization of Li^+ ions [2.67–69, 52]. Experimental results for single ionization are shown in Fig. 2.25.

The experimental cross sections obtained by different groups are in good agreement. Younger's distorted-wave-exchange calculation [2.66] (not shown in Fig. 2.25) is slightly above the measured data but still in good agreement. The cross section shape is typical for direct single ionization; it is essentially a smooth function of energy following roughly the prediction of the Lotz formula (2.27). However, although electrons from only one shell can be ionized and inner-shell effects cannot

49

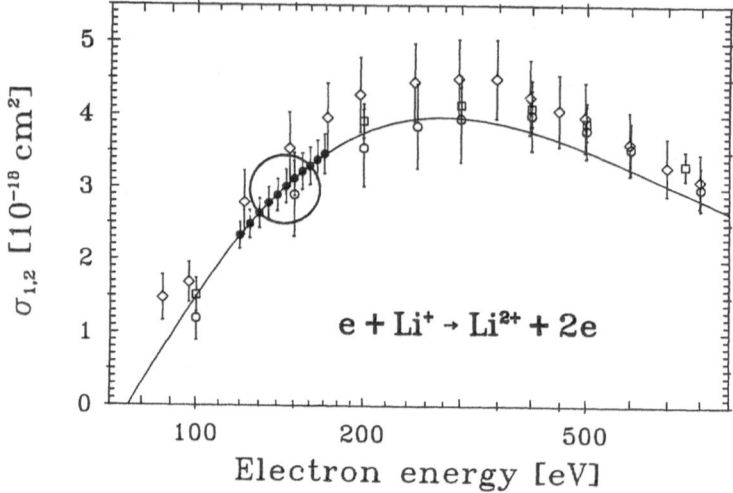

Fig. 2.25. Measured cross sections for electron impact ionization of Li$^+$ ions. The experimental data are: diamonds [2.67], open circles [2.68], open squares [2.69], solid points [2.52]. The solid line is a fit to the solid points based on a formula given by *Younger* [2.54] and represents the cross section $\sigma_{1,2}^d$ for direct single ionization

contribute to the total cross section, it is still possible to produce Li^{2+} from Li$^+$ via indirect processes. Such processes are depicted in Figs. 2.8 and 2.9. Recently, *Müller* et al. [2.52] performed an experiment to search for indirect ionization in e + Li$^+$ collisions. Relative cross sections were measured by the energy scanning technique with a precision (0.01 % statistical uncertainty) unprecedented in electron-ion crossed beams experiments. Thus it was possible to measure effects of indirect channels on the ionization cross section of Li$^+$.

In order to visualize fine details in the total single ionization cross section the absolute data were first fitted with a four-parameter scaling formula [2.54] suitable to represent direct single ionization. The result is the solid line in Fig. 2.25. It represents the direct ionization "background" and was therefore subtracted from the normalized scan data. The subtracted data are shown in Fig. 2.26 on a scale roughly expanded by a factor 10^3 as compared to Fig. 2.25. The displayed features represent only about 0.3 % of the total cross section; however, due to the extremely high precision of this measurement one can clearly see two dips in the cross section at about 137 eV and 140 eV and a step-like increase above 146 eV with resonance features at higher energies. The dips are at electron energies were resonant intermediate states 2s^{2}2p ^{2}P and 2s2p^2 ^{2}D can be populated by trielectronic capture, such as

$$e + Li^+(1s^2) \rightarrow Li(2s^2 2p) \rightarrow Li^{2+}(1s) + 2e \ . \tag{2.75}$$

When the triply excited intermediate Li atom decays by auto-double-ionization the whole process leads to the production of Li^{2+} ions. Thus, the resonant channel cannot be distinguished from direct ionization channels and interference becomes

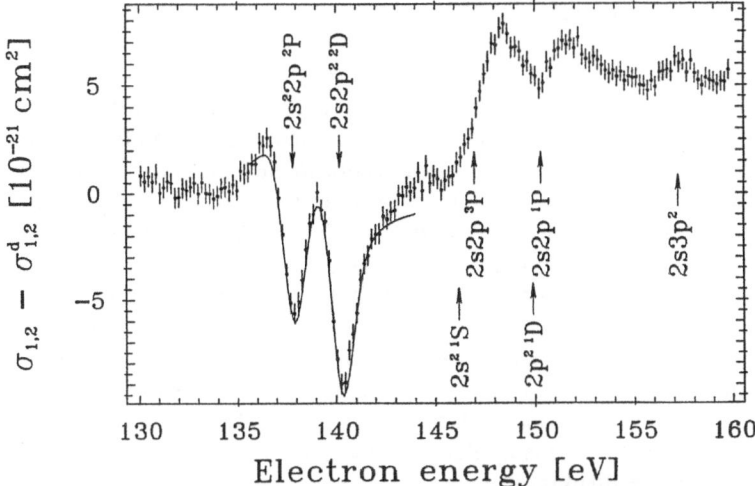

Fig. 2.26. Details of the Li+ electron impact ionization cross section which was measured with 0.01 % statistics by *Müller* et al. [2.52]. The points were obtained by subtracting the solid line in Fig. 2.25 from energy scan data. A number of energy levels are marked by arrows. The corresponding cross section structures indicate complex ionization channels like resonant-double-exitation-auto-double-ionization

possible. Indeed, the observed features clearly indicate the presence of destructive interference. The solid line in Fig. 2.26 was obtained by fitting the observed dips with two line profiles using Fano's q-parameter formula [2.70, 71]

$$\sigma(E_e) = \sigma_a[(q + \varepsilon)^2/(1 + \varepsilon^2)] + \sigma_b , \tag{2.76}$$

where σ_a and σ_b are the resonant and non-resonant portions of the cross section, $\varepsilon = 2(E_e - E_{res})/\Gamma$ and q is the line profile index. The quantities E_{res} and Γ are the same as in (2.36) and (2.37). Since in a real experiment the energy resolution is limited, (2.76) was folded with a Gaussian profile of variable width in order to obtain a realistic test function. From the fitted function the resonance positions 137.7 ± 0.1 and 140.1 ± 0.1 eV were obtained, respectively. These energies are only about 0.4 eV above calculated values for the $2s^2 2p\ ^2P$ and $2s2p^2\ ^2D$ resonances [2.72]. Comparison of the observed ^2P resonance feature (magnitude nearly 1×10^{-20} cm^2 with the strength calculated for the pure resonance case (as obtained by employing formulae similar to (2.37) and (2.45)) reveals a strong enhancement (roughly a factor 10) of the cross section's deviation from a smooth curve. This enhancement has to be attributed to the interference of the resonant with the partial kp direct ionization channel.

Just above the threshold energy for $1s^2 \rightarrow 2s^2$ double excitation at 146.2 eV a cross section increase by about 7.5×10^{-21} cm^2 is seen in Fig. 2.26. Additional thresholds for a number of non-resonant two electron excitations to configurations 2l2l′ (two-electron states) are indicated. On top of the direct double excitation contribution

$$e + Li^+(1s^2) \rightarrow Li^+(2l2l') + e \rightarrow Li^{2+}(1s) + 2e \,, \qquad (2.77)$$

several resonance-like features are found. These may be due to $2l3l'3l''$ or $2l3l'4l''$ intermediate configurations which can decay via two sequential Auger processes and thus contribute to the ionization channel. This experiment established for the first time contributions of double-excitation-autoionization to the total ionization of ions and provided quantitative data for these high-order processes. The ultimate precision accomplished by the use of the electron energy scanning technique made it possible to measure and distinguish contributions of less than 0.3 % of the total cross section.

Similar experiments were carried out earlier with He atoms; dips in the relative cross sections with a magnitude of nearly 1 % of the total ionization were also observed in He at energies resonant with the $2s^2 2p$ ^2P and $2s2p^2$ ^2D states [2.73]. These measurements are shown in Fig. 2.27. At slightly higher energies structure (not shown in Fig. 2.27) was found with a magnitude of only 0.03 % of the total cross section [2.73]. This is probably due to the double-excitation-autoionization channel in $e + He$ $(1s^2)$ collisions.

A third experiment in this category was carried out with H^- ions by *Peart* and *Dolder* [2.74, 75]. In the detachment cross section they found two dips at electron energies resonant with $2s^2 2p$ ^2P and $2s2p^2$ ^2D triply excited states of the H^{2-} ion. The experimental results are shown in Fig. 2.28. The big feature at about 14 eV makes up for a contribution of about 2×10^{-15} cm^2, that is, roughly one half of the total detachment cross section. The second feature at 17.2 eV is considerably smaller and above that energy additional structure in the cross section can hardly be recognized at the given level of accuracy.

These results give an impression of the relative sizes of the trielectronic-capture-auto-double-ionization and double-excitation-autoionization contributions along the He isoelectronic sequence. The ^2P feature observed with a given energy resolution

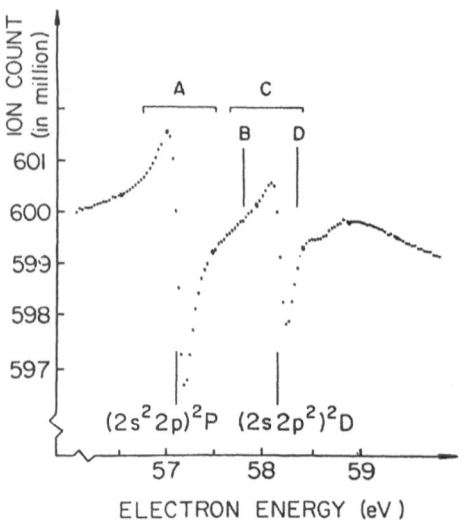

Fig. 2.27. Formation of He$^+$ by electron impact on He. The electron energy and ion count scales are both suppressed in order to visualize the fine details [2.73]

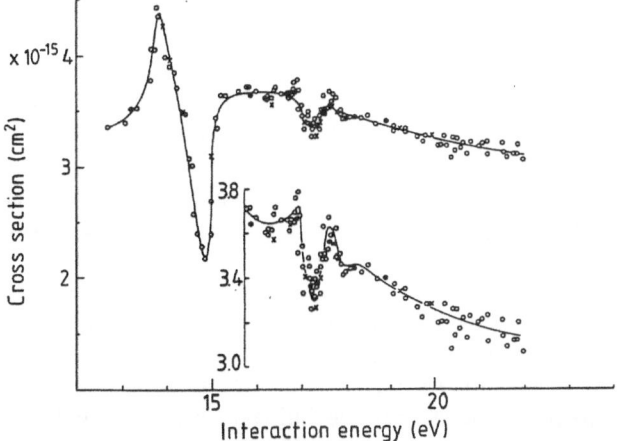

Fig. 2.28. Structure in the electron detachment cross section for $e + H^-$ collisions [2.75]. The dips in the cross section are due to resonances with H^- configurations. The inset is a blowup of the second cross section feature

ΔE_e of the experiment is 50 % of $\sigma_{-1,0}$ for H^- ($\Delta E_e = 0.25$ eV est.), only 1 % of $\sigma_{0,1}$ for He ($\Delta E_e = 0.1$ eV) and 0.3 % of $\sigma_{1,2}$ for Li^+ ($\Delta E_e = 1.2$ eV). The relative size of this feature drops dramatically from H^- to He but when considering the differences in energy resolution it appears to decrease more slowly between He and Li^+. The ^2D feature increases in size compared to the ^2P feature from 15 to 20 % for H^-, to nearly 50 % for He and over 100 % for Li^+. This trend suggests a dominance of the ^2D over the ^2P resonance features for ions in higher charge states.

The double-excitation-autoionization contribution is not visible in the detachment of H^-. It makes up for about 0.03 % of $\sigma_{0,1}$ for He and already 0.25 % for Li^+, suggesting an increasing importance with increasing atomic number. These comparisons show that high precision measurements with multiply charged He-like ions are desirable. The use of powerful state-of-the-art electron cyclotron ion sources (ECR) will hopefully make these experiments feasible in the near future.

The examples and the discussion show that already with simple two-electron targets interesting indirect ionization processes can be observed. This is even more true for the three-electron species.

2.4.3 Lithium-like Ions

The lithium isoelectronic sequence is the most extensively studied series both experimentally and theoretically. Experiments benefit from the absence of metastable ions in the parent beams so that well-defined collision systems can be investigated. The electronic structure is simple enough to facilitate fairly detailed theoretical calculations and to sort out individual ionization mechanisms. On the other hand, the presence of two inner-shell electrons makes the system complex enough to allow a variety of indirect ionization mechanisms which can be studied in great detail provided the experimental technique is sensitive enough to distinguish small features due to inner-shell processes, on top of a dominant contribution from direct ionization of the comparatively weakly-bound 2s electron.

Ionization rate measurements from plasma observations have been performed on Li-like ions [2.29, 31]. However, due to limited accuracy and due to the wide distribution of electron energies in a plasma, it is difficult, if not impossible, to isolate specific contributions to ionization. Similar arguments hold for data obtained by modeling the ion charge state evolution in an electron beam ion source (EBIS) [2.41].

Crossed beams approaches, in principle, allow one to resolve fine details in the energy dependence of cross sections; however, such experiments have been suffering from low event rates due to the low particle densities in electron and ion beams. Thus, in the first crossed beams experiment on Li-like C^{3+} and N^{4+} ions with statistical uncertainties of typically 5 to 10 % (two standard deviations) *Crandall* et al. [2.76] could hardly see contributions from indirect effects. With improved experimental conditions, however, they were soon able to clearly demonstrate the presence of indirect ionization mechanisms in Li-like ions of the elements C, N and O [2.77]. The two step mechanism observed in these experiments was that of excitation-autoionization such as

$$e + N^{4+}(1s^2 2s) \rightarrow N^{4+}(1s2s2l) + e , \tag{2.78a}$$

$$N^{4+}(1s2s2l) \rightarrow N^{5+}(1s^2) + 2e . \tag{2.78b}$$

The statistical uncertainty in these experiments was as low as 1.5 % for some of the measured data points. At this level of accuracy the sharp onset of electron impact excitation of ions at threshold produces a clearly distinguishable step in the cross section function. Measurements on the Li-isoelectronic sequence were then extended to Be^+ [2.78] and B^{2+} [2.79] ions providing data with similar precision. The experimental cross sections obtained for B^{2+} are displayed in Fig. 2.29 together with theoretical results. Theory (with a correction factor 0.90 for direct ionization) and experiment are consistent and the step-like increase of the cross section near 200 eV due to excitation-autoionization contributions is clearly established. New improved measurements were carried out by *Crandall* et al. on O^{5+} [2.79]. From the observation of small deviations from the expected energy dependence of the ionization cross section they concluded a possible influence of READI resonances on the measured ionization cross section. Subsequent to this, a serious attempt was undertaken at Oak Ridge to measure READI for O^{5+} ions [2.81]. In the experiment the capability of the ECR ion source was employed to produce intense beams of O^{5+} ions and an energy scanning technique was used suitable to detect small features in the cross section. After several days of data accumulation a statistical uncertainty of about 0.5 % (1 standard deviation) was accomplished which was not sufficient to establish the presence of READI resonances. Nevertheless, it was possible to resolve individual terms in an autoionizing configuration by measuring net electron impact ionization of an ion. Figure 2.30 shows the cross sections obtained near the excitation-autoionization threshold. Included are theoretical results for electron im-

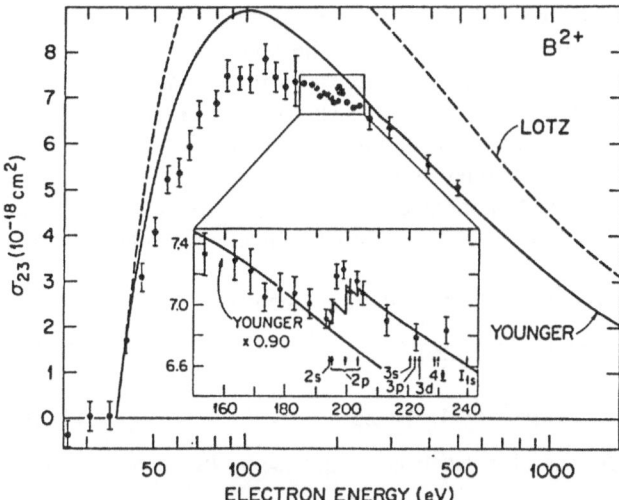

Fig. 2.29. Electron impact ionization cross sections of B^{2+} ions. The solid points are experimental data of *Crandall* et al. [2.79], the solid line is the distorted-wave prediction of *Younger* [2.80, 54], the dashed curve is the Lotz cross section from (2.27). The inset shows the energy where excitation-autoionization should contribute. Within the inset, the Younger theory of direct ionization has been multiplied by 0.90 and the arrows show the energies for inner-shell excitation of a 1s electron to the nl orbitals indicated. The upper curve in the inset adds a six-state close-coupling calculation of excitation-autoionization contributions to the direct ionization (from [2.79])

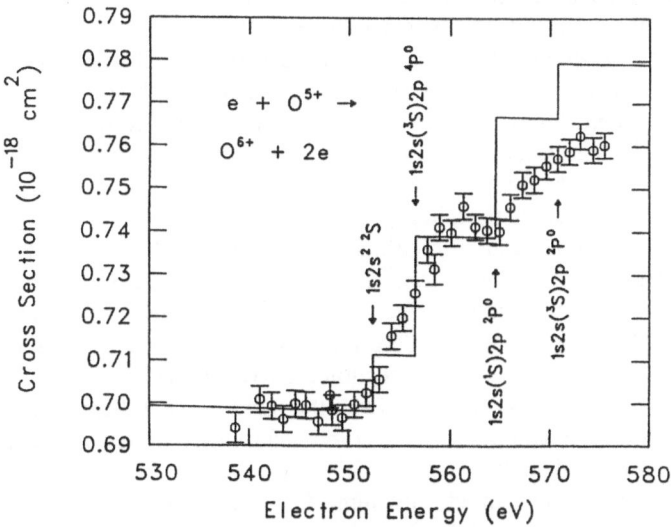

Fig. 2.30. Improved cross section measurements for electron impact ionization of O^{5+} ions [2.81]. The solid line is a six-state close-coupling calculation by *Henry* [2.79]

pact excitation as calculated in a six state close-coupling approximation by Henry, which are added to an extrapolated straight line fit to the experimental direct ionization cross sections just below the excitation-autoionization threshold. The agreement is excellent for the first two levels [$1s2s^2\,^2S$ and $1s2s(^3S)2p\,^4P^0$] suggesting a close to 100 % branching ratio for decay of the excited states by autoionization. Significant discrepancies exist for the higher two levels taken into account in the calculations [$1s2s(^1S)2p\,^2P^0$ and $1s2s(^3S)2p\,^2P^0$], indicating possible inaccuracies in the calculated energy levels or excitation cross sections, or a less than unity branching ratio for autoionization.

The possibility of observing READI contributions such as

$$e + A^{q+}(1s^2 2s) \rightarrow [A^{(q-1)+}(1s2s2p2l)]^{***} , \tag{2.79a}$$

$$[A^{(q-1)+}(1s2s2p2l)]^{***} \rightarrow A^{(q+1)+}(1s^2) + 2e \tag{2.79b}$$

was studied theoretically for the Li isoelectronic sequence by *Pindzola* and *Griffin* [2.82].

The energies of the $1s2s2p2l$ $A^{(q-1)+}$ Be-like configurations are less than the energies of the $1s2s^2$ and $1s2s2p$ (Li-like) configurations which are energetically the lowest autoionizing configurations of the Li-like A^{q+} parent ions. Contributions to the single ionization cross section can be made only if the $1s2s2p2l$ configurations decay by double Auger processes, that is, simultaneous emission of two electrons. The first explicit calculations on READI processes were performed for the $1s2s^2 2p$ ^3P term [2.82]. The relative contribution at resonance energy to single ionization of O^{5+} was found to be between 0.08 and 0.65 % for an assumed energy resolution of 2 eV. This result was not promising since the expected resonances sit on a "background" of direct ionization which is about 100 to 1000 times higher than a resonance peak. However, by employing a fast energy scanning technique it was possible to find READI contributions first in the ionization of C^{3+} [2.83] and then also in B^{2+}, N^{4+} and O^{5+} ions [2.84]. An example for these measurements is given in Fig. 2.31 which shows results for the ionization of B^{2+} ions in an energy range roughly corresponding to that of the inset in Fig. 2.29. Distinct from earlier data the new experiment shows structure in the excitation-autoionization step which is due to different terms in the $1s2s2l$ configurations and to resonance processes involving two-electron emission subsequent to dielectronic capture. The two little peaks at about 168 eV and 174 eV are due to READI processes. By subtracting a straight line fit to the observed direct ionization cross section the READI peaks can be isolated. Figure 2.32 shows these peaks obtained from the measurement displayed in Fig. 2.31. Packets of 5 original data points were combined to improve the statistics on each displayed experimental cross section.

The peak structures have been fitted with three Gaussian profiles which are shown by the dashed curves. The sum of the Gaussian curves is the solid line. On the basis of peak energies obtained from the fits specific terms in the $1s2s2l2l'$ Be-like configurations can be associated with the observed structures as indicated in the figure. The Gaussian fits also provide information about the resonance strengths S of the different READI contributions to the total ionization cross section. These numbers

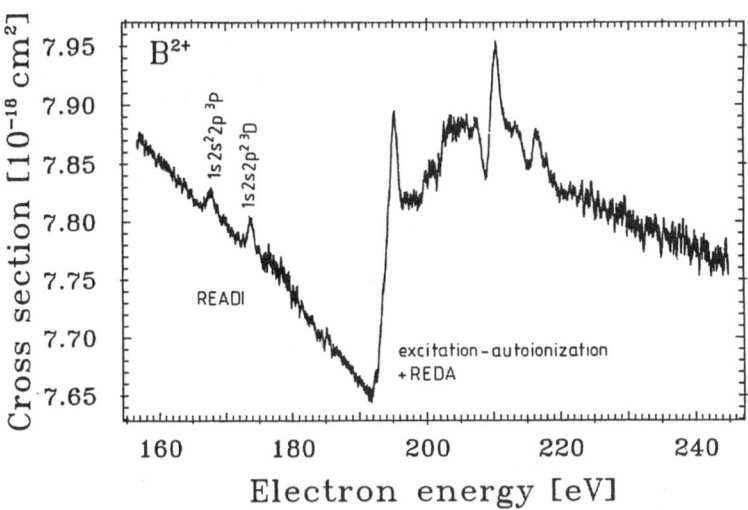

Fig. 2.31. Cross sections for electron impact ionization of B^{2+} ions obtained by the energy scan technique [2.84]. (Compare inset of Fig. 2.29)

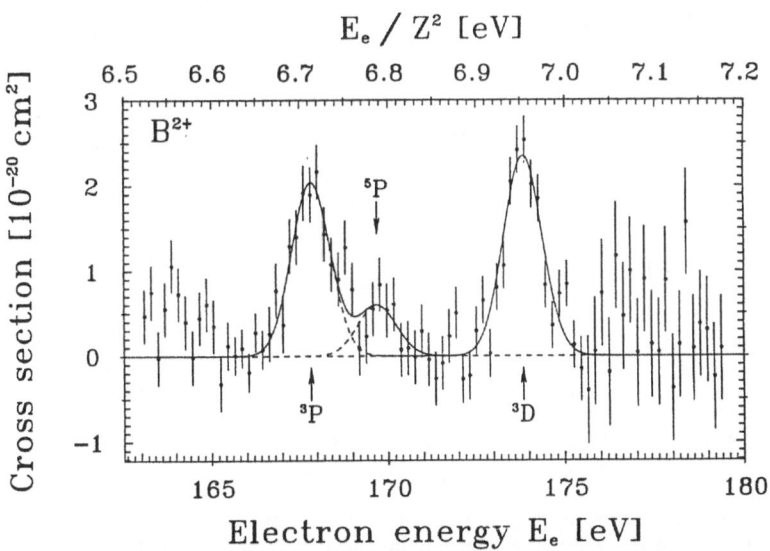

Fig. 2.32. READI contributions to the ionization of B^{2+} ions [2.84]. Direct ionization has been subtracted. The READI peaks are represented by Gaussian distributions. They correspond to Be-like states (in B^+) $1s2s^2 2p\ ^3P$, $1s2s2p^2\ ^5P$ and $1s2s2p^2\ ^3D$

can be compared to the predictions of *Pindzola* and *Griffin* [2.82] and will serve in guiding theory towards a proper description of double Auger processes. Figure 2.33 presents a comparison of experimental values of S for the $1s2s^2 2p\ ^3P$ READI resonance with the range theoretically predicted by Pindzola and Griffin. The atomic

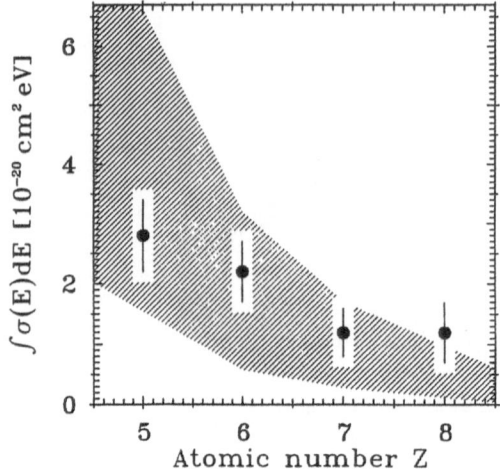

Fig. 2.33. Resonance strengths $S = \int \sigma(E)\,dE$ for the $1s2s^2 2p\ ^3P$ READI contributions to ionization of Li-like ions. The solid points are experimental data [2.83, 84]. The theoretical limits for $S(Z)$ [2.82] are indicated by the hatched area

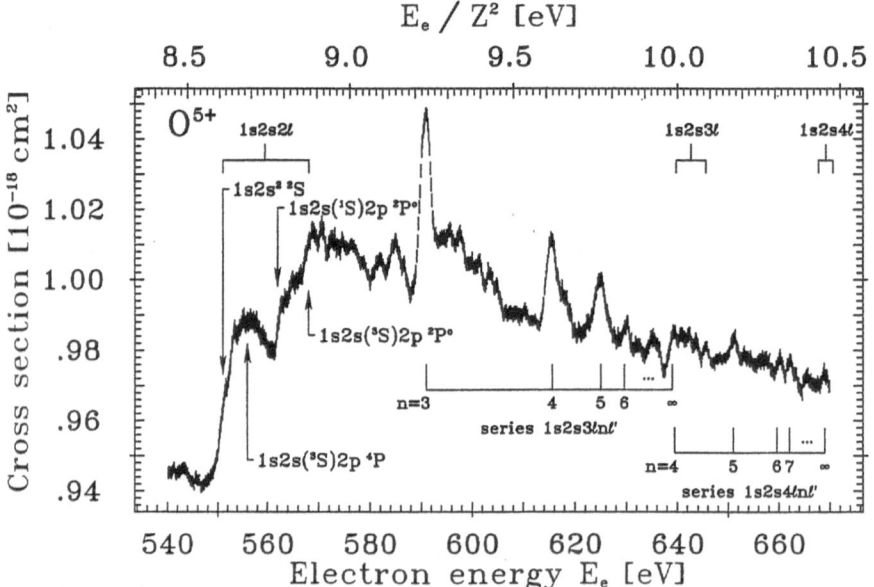

Fig. 2.34. Cross sections for electron impact ionization of O^{5+} ions obtained by the energy scan technique [2.84]. (compare Fig. 2.30). Assignments of intermediate excited states to the cross section features are given

numbers are between $Z = 5$ and $Z = 8$. The uncertainty of the theory is visualized by shading the area inside which the values of S are predicted (of course, at discrete atomic numbers). The solid points with error bars represent the experimental data obtained from Gaussian fits as the ones displayed in Fig. 2.32.

As expected, the cross section features above the excitation-autoionization threshold are qualitatively the same for all members of the Li isoelectronic sequence and therefore can be discussed on the basis of one example. For comparison with Fig. 2.30 a scan measurement for O^{5+} parent ions is displayed in Fig. 2.34 in the energy range at and above the excitation-autoionization threshold. Indirect cross section contributions are more prominent for higher-Z members of the sequence as evidenced by comparing Figs. 2.34 and 2.31. Intermediate highly excited states have been assigned to the observed cross section features. Besides the excitation-autoionization steps there are a number of resonance contributions whose size relative to the READI contributions suggests a different mechanism. Indeed, the peaks have to be assigned to REDA processes (see Fig. 2.6). It is interesting to note the sizable contributions from a Rydberg series of intermediate excited states with configurations $1s2s3lnl'$ ($n = 3, 4, 5 \ldots$) which are populated from the $1s^2 2s$ ground state by $\Delta n = 2$ excitation and capture of the incident electron. In a measurement on Li-like N^{4+} even resonant $\Delta n = 3$ excitation could be experimentally observed [2.84].

Lithium-like ions with one electron outside a closed shell provide rich structure in the ionization cross section. Although only three target electrons can take part in the ionization process a complete description of all possible indirect contributions to the total cross section is a very formidable task for theory now that such detailed experimental results are available as the ones displayed in Figs. 2.31–2.34. A first detailed theoretical description of resonant and non-resonant inner-shell excitation has been published by *Tayal* and *Henry* for C^{3+} ions [2.139]. The results are in surprising agreement with the experimental data.

2.4.4 Sodium-like Ions

As the number of electrons in the outermost shell increases also the ionization cross section increases. Indirect contributions are based on inner-shell processes and thus do not change their magnitude much when outer electrons are attached. Hence, it becomes experimentally more difficult to identify indirect ionization mechanisms in isoelectronic sequences with several electrons in the outer shell. Cross sections are then predominantly determined by direct ionization. Thus, the next interesting isoelectronic sequence after the Li-like is the Na-like sequence. In the ground state there is one 3s electron outside filled K- and L-shells, a situation promising significant ionization contributions from intermediate autoionizing states.

While the first crossed beams experiment with a sodium-like ion, Mg^+, by *Martin* et al. [2.85] in 1968, did not have the necessary precision to see indirect ionization effects, more recent measurements by *Crandall* et al. [2.86] clearly showed the presence of excitation-autoionization features in Mg^+, Al^{2+} and Si^{3+} ions. As an example Fig. 2.35 presents the experimental data available for Al^{2+}. The positions and orbital promotions associated with inner-shell excitation are indicated. Distorted-wave calculations for direct ionization of the 3s or one of the 2p electrons and for excitation-autoionization (added to the direct contribution) are also shown. While the direct part of the cross section appears to be adequately described theory fails to predict the correct magnitude for the indirect contributions and does not give the specific detailed shape of the cross section either. It was noted [2.88] that agreement with the experiment is substantially improved when monopole excitations $2p \rightarrow 3p$

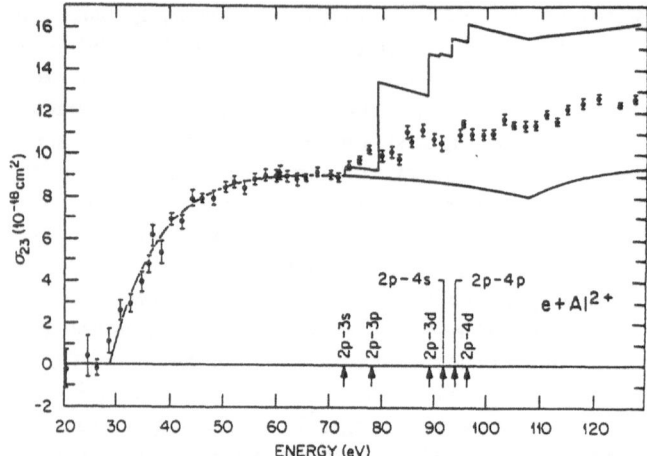

Fig. 2.35. Electron impact ionization of Al^{2+} near threshold. The solid points are crossed beams data [2.86], the solid curve is a distorted-wave calculation of direct ionization [2.87] normalized to the experiment at 70 eV by multiplying the theoretical result by 0.65. The distorted-wave excitation of *Griffin* et al. [2.88] is added to the direct ionization results with arrows indicating excitation thresholds

are left out of the calculation. Without further reasoning, however, such an exclusion is not plausible.

Henry and *Msezane* [2.89] have carried out coupled-state calculations to further investigate the observed features of indirect ionization in Na-like ions. They performed model calculations of resonances ($2p^53s3pnl$) for $e + Al^{2+}$ in a 3-state close-coupling approximation (see Fig. 2.36). These resonances decay predominantly by double-autoionization resulting in contributions to the ionization cross section (REDA, see Fig. 2.6). The curve denoted Gaussian average is a convolution of the resonances with a 2 eV Gaussian simulating the experimental energy spread. The lower part of Fig. 2.36 shows a comparison of the model calculations with measured cross sections for Al^{2+} ionization. The lower solid curve represents direct ionization. Added to this is the distorted wave calculation from Fig. 2.35 (dashed curve). The solid curve is a 2-state close-coupling calculation for excitation-autoionization including only $2p \to 3l$ excitations and folded with a 2 eV Gaussian energy spread. The long-dashed curve between 73 and 80 eV is the model result including the resonances from the upper part of Fig. 2.36. The coupled-state theory still overestimates the $2p \to 3l$ excitation component, but the estimated effect of resonances between 73 and 80 eV, just below this excitation threshold, matches the observed cross section in shape. This "agreement" of theory and experiment was the only evidence until recently for REDA resonances in the sodium isoelectronic sequence. Strong contributions to ionization of sodium-like Fe^{15+} from REDA had been estimated before by *LaGattuta* and *Hahn* [2.90] who initialized the search for resonances in the ionization of ions by their calculations. It was quite a step in the development of experimental techniques that led to the first measurement of electron impact ionization of Fe^{15+} ions by *Gregory* et al. [2.91] six years after the appearance of the

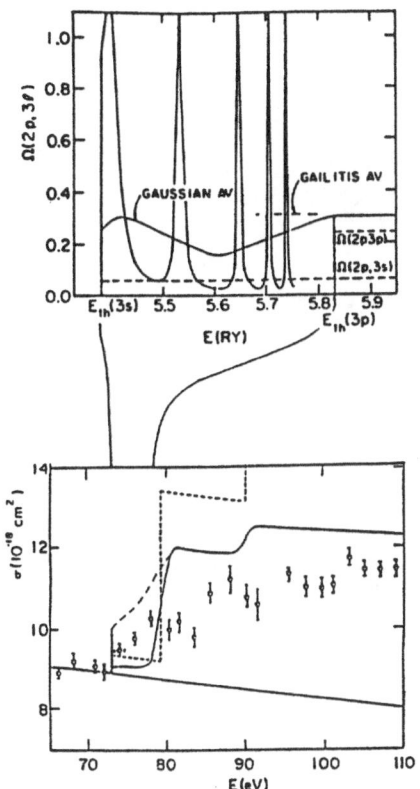

Fig. 2.36. Electron impact ionization of Al^{2+} ions. Upper figure: Close-coupling model calculations of resonances [2.89]. Lower figure: Lower solid curve is normalized direct ionization from Fig. 2.35. Dashed curve is distorted-wave excitation [2.88] added to this. Upper solid curve is close-coupling calculation [2.89] added to direct ionization. Long-dashed curve is estimated close-coupling result including both excitation-autoionization and resonances like those of the upper figure averaged with a 2 eV Gaussian. Open circles represent measurements [2.86]

theoretical predictions. Their results are shown in Fig. 2.37. The dashed line is the Lotz prediction (2.27) for direct ionization; added to this is a distorted-wave calculation of excitation-autoionization contributions (lower solid line) and added on top the hatched area indicates the predicted REDA contribution from [2.90]. The large error bars on the experimental data do not allow observation of resonances in the cross section. It can rather be ruled out that the resonance contribution near 770 eV is as big as predicted.

Meanwhile a close-coupling calculation for Fe^{15+} has been carried out [2.93] and the new theoretical result is compared with the experimental data in Fig. 2.38. The resonance contributions came out smaller in the close-coupling calculations and fairly good agreement with the experimental data is obtained. Still more recently a sophisticated close-coupling calculation was performed for Fe^{15+} with an immense computational effort [2.95]. These calculations provide a new challenge to experimentalists to further improve the experimental conditions and make the observation of REDA resonances in the electron impact ionization of sodium-like Fe^{15+} feasible.

One step towards this goal was made by employing the fast energy scanning technique to the ionization of sodium-like Mg^+ ions [2.96]. The result of this measurement is shown in Fig. 2.39. The energy resolution in this experiment was 0.3

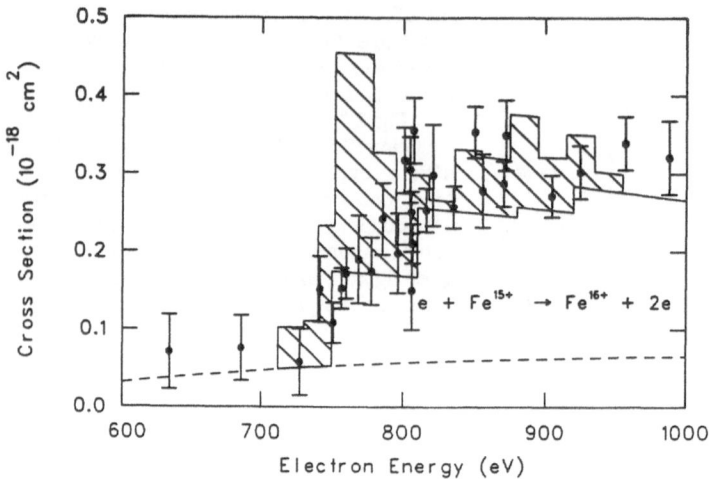

Fig. 2.37. Electron impact ionization of Fe^{15+} ions. Dashed curve: Lotz prediction (2.27) for direct ionization. Lower solid line: distorted-wave excitation-autoionization contributions [2.92] added to the direct cross section. The hatched region indicates the predicted cross section enhancement due to REDA processes [2.90]. The solid points are the measurements of *Gregory* et al. [2.91]

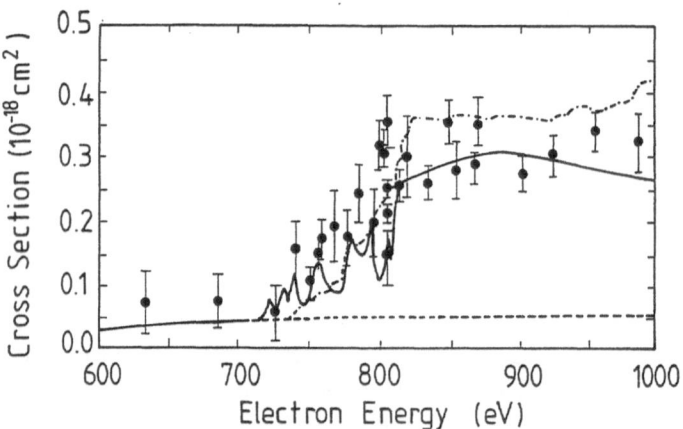

Fig. 2.38. Electron impact ionization of Fe^{15+} ions. The dashed curve is direct ionization from (2.27), the dot-dashed curve is excitation-autoionization calculated in a distorted-wave approximation [2.94] added to the direct part. The solid curve is a twelve-state close-coupling calculation by *Tayal* and *Henry* [2.93] which has been added to the direct part

eV as evidenced by the width of the peak observed at 56 eV. A wealth of structure is observed above the excitation-autoionization threshold near 50 eV. The dominant resonance series below 55 eV appears to be mainly due to $2p^53s3pnl$ intermediate excited states of the Mg atom. The first three peaks were interpreted to be due to $n = 3$, $n = 4$ and $n = 5$ and the shoulder on the high energy side of the $n = 5$ peak could be due to the unresolved resonance peaks with $n > 6$.

62

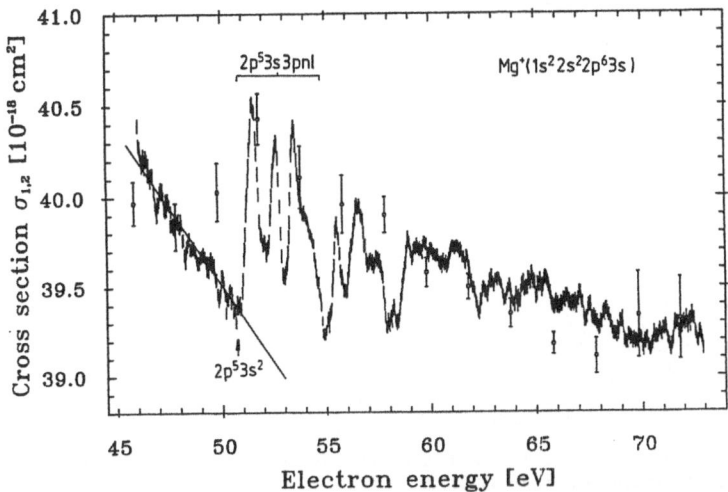

Fig. 2.39. Electron impact ionization of Mg$^+$ ions. Energy scan data [2.96] are compared to conventional measurements (open circles) [2.86]. The solid line respresents direct ionization. The 2p $\rightarrow$ 3s excitation threshold is indicated. The dominant resonances were identified to be due to the 2p^{5}3s3pnl Rydberg series of excited Mg

The complete theoretical analysis of the experimental cross section appears to be a monstrous task since there are so many excited Na-like and Mg-like intermediate states which can contribute to the observed net ionization. Already the 2p^{5}3s3p^2 configuration has 42 individual terms spread out over an energy range of nearly 7 eV. With respect to the multiplicity of contributing states it is a mild surprise that so much clear structure is still observable and not smeared out by a lumping of all possible resonances and excitation steps.

2.4.5 Heavy Alkali-like Ions

Beside the Li-like and Na-like isoelectronic sequences there are the other alkali-like sequences and groups of transition metals which are characterized by the presence of a single electron outside closed shells. It has been recognized by Peart and Dolder that excitation-autoionization is of increasing importance along the series Mg$^+$, Ca$^+$, Sr$^+$, and Ba$^+$ of alkali-like ions. Their results [2.97] are displayed in Fig. 2.40 together with the Be$^+$ data from [2.78]. While indirect contributions in the ionization of Mg$^+$ were not observed, excitation-autoionization was found to become the dominant ionization mechanism for the heavier elements in the series. The energy where the dominant excitation step occurs corresponds to transitions from the outermost p subshells to d subshells with $\Delta n = 0$, that is, 3p^{6}4s $\rightarrow$ 3p^{5}3d4s in the case of Ca$^+$, 4p^{6}5s $\rightarrow$ 4p^{5}4d5s in the case of Sr$^+$ and 5p^{6}5s $\rightarrow$ 5p^{5}5d5s in the case of Ba$^+$. In the ionization of Mg$^+$ (and of Be$^+$) there is no corresponding $\Delta n = 0$ transition possible since there is no d subshell for $n = 2$ (or 1, respectively) and hence only $\Delta n = 1$ excitations 2p^{6}3s $\rightarrow$ 2p^{5}3s3l can produce a (small) signal in the ionization

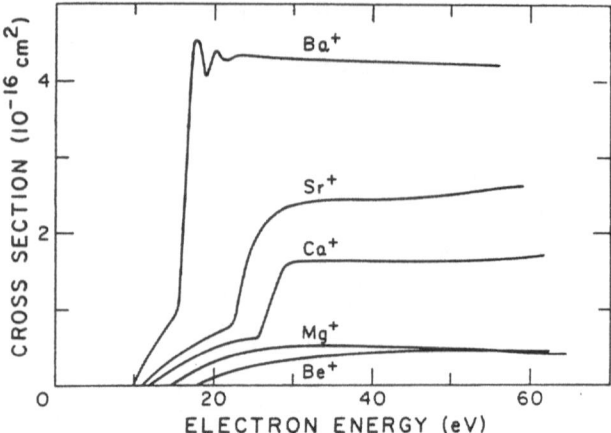

Fig. 2.40. Electron impact ionization of singly charged alkali-like ions. Lines are smooth curves drawn through crossed beams measurements [2.97, 78]

channel. Excitations within one shell are much more likely than those with $\Delta n \neq 0$ and, therefore, the $np^6 \rightarrow np^5 nd$ transitions (possible for $n > 3$) produce much bigger cross section contributions than the $np^6 (n+1)s \rightarrow np^5 (n+1)s(n+1)l$ transitions (possible for $n > 2$).

Measurements of cross sections for ionization of the three-times charged alkali-like series of ions Ti^{3+}, Zr^{3+}, and Hf^{3+} were carried out by *Falk* et al. [2.98]. These experiments yielded dramatic results. They showed an enhancement of ionization cross sections by factors like 10 or 15 by excitation-autoionization contributions. As an example, Fig. 2.41 shows a comparison of measured cross sections for Ti^{3+} ions along with calculated ones. The dashed curve toward the bottom represents the direct ionization cross section calculated using the Lotz formula (2.27). The solid curve was calculated using a distorted-wave dipole approximation for the excitation cross section [2.99] and that was added to the dashed curve. However, the calculations for excitation had to be reduced by a factor of 2.5 to get agreement with experiment, and it is the scaled theory which is plotted as the solid curve. The chain curve is a convolution of the electron energy distribution with the scaled prediction of theory. Although these curves are in quite good agreement with the experimental data, the correction factor of 2.5 points to the difficulties on the theory side to make useful predictions of cross section contributions. This is also emphasized by another example within the alkali-ion group. On the basis of distorted-wave calculations and close-coupling calculations [2.101, 102] structures in the ionization cross section for Ca^+ ions had been found which were not apparent in the measurements of *Peart* and *Dolder* [2.97]. Recently, new high resolution measurements on Ca^+ were performed [2.100]. These measurements (see Fig. 2.42) revealed much more detail in the cross sections than the first experiments. The big structure in the cross section predicted by theory near 32 eV is not there. However, a number of resonance features are resolved indicating the presence of REDA processes on top of

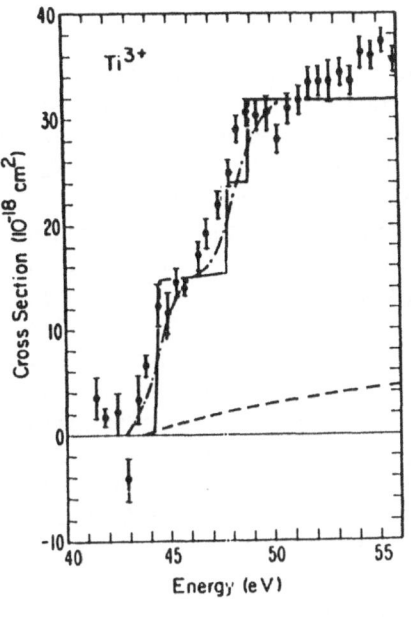

Fig. 2.41. Electron impact ionization of Ti^{3+} in the threshold region. The solid points are measurements by *Falk* et al. [2.98], the dashed line is direct ionization from (2.27), the solid line is a distorted-wave dipole approximation [2.99] for the excitation cross section which was reduced by a factor of 2.5 and then added to the direct part. The dot-dashed line results from the solid line by convolution with a 2 eV FWHM Gaussian

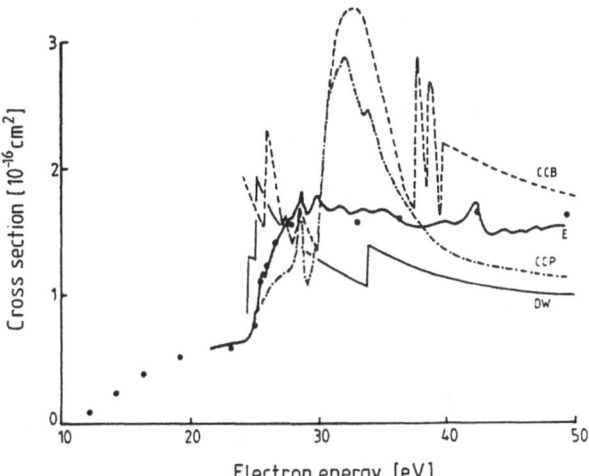

Fig. 2.42. Electron impact ionization of Ca^+ ions. Curve E is a measurement [2.100] with energy-resolved electrons which was normalized to the solid points representing earlier measurements [2.97]. Curves CCB, CCP, and DW are, respectively, the results of close-coupling calculations by *Burke* et al. [2.101] and *Pindzola* et al. [2.102] and a distorted-wave calculation by *Griffin* et al. [2.103]

the excitation-autoionization contributions. Experiments with a comparable energy resolution were also performed for Ba^+ [2.104]. The data obtained by Peart and Dolder are in agreement with unpublished energy scan data on Ca^+ and Ba^+ of the Gießen group. Apparently, new theoretical effort will be necessary to understand the experimental results on potassium-like and other heavy alkali-like ions.

2.4.6 Magnesium-like Ions

In investigations of Na-like ions (ground state configuration $1s^2 2s^2 2p^6 3s$) two remarkable difficulties had been met. In the experiments a background increasing with the ion charge made it impossible to perform cross section measurements beyond Si^{3+}. Only for Na-like Fe^{15+} was this background again low enough to facilitate the measurements shown in Figs. 2.37 and 2.38. The second difficulty, as mentioned earlier, was in understanding why the agreement of theory and experiment could be subtantially improved when monopole excitations $2p \rightarrow 3p$ were left out of the calculations. In subsequent studies of the Mg-like sequence it was possible to shed light on these issues [2.105–108]. It was found that Na-like $2p^5 3s 3p$ excited levels are metastable against both radiative decay and autoionization. Lifetimes are apparently sufficient for excited ions to survive the flight path from the ion source to the electron-ion interaction region. An ion decaying in the region between the mass filter and the crossing region and the post-collision analyzer (see Fig. 2.17) produces a background signal. Thus, background count rates (of S^{6+}) up to $10^4 s^{-1}$ per particle-nA could be observed for 40 keV incident S^{5+} ions [2.105]. The possible occurrence of metastable ions in the parent beam is a general problem in ionization studies. It has to be kept in mind when looking at experimental cross sections for Mg-like ions as well. Figure 2.43 shows cross sections for electron impact ionization of Mg-like S^{4+} ions. The solid points are measurements by *Howald* et al. [2.105]. The dashed curve is the direct ionization cross section, the solid curves include excitation-autoionization contributions from the $2p^6 3s^2$ ground state configuration. The experimental direct ionization threshold is about 10 eV lower than

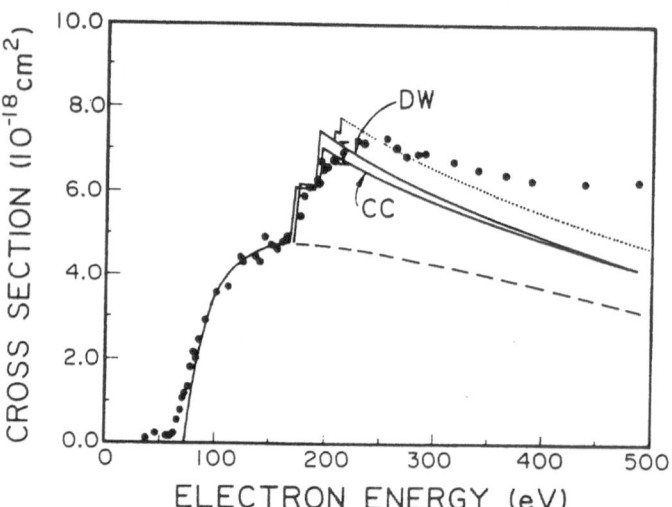

Fig. 2.43. Ionization cross sections for S^{4+} ions. Dashed curve: direct ionization distorted-wave results [2.109]; solid curve labeled CC: indirect ionization close-coupling results [2.107] added to the direct part; solid curve labeled DW: indirect ionization distorted-wave results [2.106] plus direct; dotted curve: additional $2p \rightarrow 4l$ excitation contributions added to the DW results [2.108]; full circles: experiment [2.105]

the spectroscopic value of 72.68 eV, indicating the presence of metastable ions in the parent beam. The calculations were therefore repeated, taking into account ionization from the $2p^6 3s3p$ excited configuration. Of the twelve states in this configuration, six are forbidden to decay. The ratio of extremely metastable states (millisecond lifetimes or longer) in the $2p^6 3s3p$ excited configuration to all states of the ground and excited state configurations is 0.46. Thus a substantial fraction of the parent ion beam may be in the metastable 3s3p states. Consequently, the theoretical calculations were also carried out for the metastable ions and the low-energy onset of the cross section was reproduced. The distinct excitation steps resulting for the ground state configuration are smeared out as seen in the experiment because of the increased number of possible intermediate excited states. One discrepancy, however, still remained, and that is an underestimation of the measured cross section by theory above about 240 eV. The explanation for this discrepancy is the following. Direct ionization out of the 2p subshell beginning at 238.3 eV leads to an autoionizing state when ground state parent ions are used. Hence, through the ionization-autoionization mechanisms a net double ionization is observed and 2p direct ionization does not appear in the single ionization channel. When starting with $2p^6 3s3p$ metastable ions, however, direct 2p ionization leads to the $2p^5 3s3p$ configuration in which 2/3 of all states are metastable against autoionization. Hence, the product ions cannot decay before their charge state is detected. In particular they do not have time enough to eject another electron. Therefore, it is reasonable to assume 2/3 of the direct ionization of the 2p subshell to contribute to the observation of net single ionization. The theory curve for metastable $2p^6 3s3p$ S^{4+} shown in Fig. 2.44 includes the 2p direct inner-shell ionization contribution. The resulting agreement with experiment is outstanding, though in

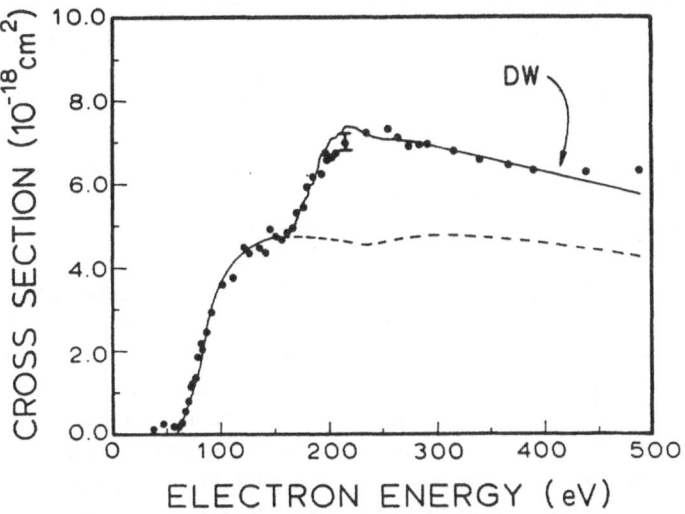

Fig. 2.44. Ionization cross sections for S^{4+} ions. Full circles: experiment [2.105]; theory curves are for the metastable $2p^6 3s3p$ parent ion configuration. Dashed curve: direct ionization distorted-wave results including the 2p subshell; solid curve labeled DW: indirect ionization distorted-wave results plus the direct contribution [2.106]

the light of the previous comparisons, for example, for Ti^{3+} and Ca^+ this agreement may appear somewhat fortuitous.

With the effect of the metastability of many of the $2p^5 3s3p$ states one can also understand the problems encountered in the studies of Na-like ions. The $2p^5 3s3p$ configuration with its metastability against autoionization is responsible for the background problem encountered in experiments with Na-like ions mentioned earlier in this chapter. Also due to the dominance of metastable levels in the $2p^5 3s3p$ configuration there is a reduced probability for this configuration to be seen in the excitation-autoionization channel in the ionization of Na-like ions; hence the missing $2p \rightarrow 3p$ monopole excitation strength in the ionization of $2p^6 3s$ Na-like ions.

This brief discussion shows that it is useful to look at ionization along certain isoelectronic sequences. It is also necessary to study the neighbouring sequences for a complete understanding of the various cross section contributions involved in ionization.

2.4.7 Isonuclear Sequences

Studies of isoelectronic sequences have mostly been carried out for ions with one or, at most, few electrons outside closed shells. The purpose of such studies is to gain insight into the fundamental processes contributing to electron impact ionization of ions and into their dependence on the charge of the ion. From an applied point of view it is often also desirable to know cross sections for an isonuclear sequence, that is, for ions of one element in different charge states. Such data are needed, for instance, to model the ion charge state development with time in a plasma or with distance from the plasma edge. For such purposes a number of isonuclear sequences have been studied in quite some detail. One of these studies is on Fe atoms and ions in charge states up to $q = 15$ by experiments [2.110–113, 91] and for all charge states by theoretical calculations [2.114].

The Fe atom in its ground state has a configuration Ar $3d^6 4s^2$. The participation of an open 3d shell in excitation-autoionization processes leads to numerous intermediate states taking part in collisions involving the low charge states of Fe. Thus, theory has to be restricted to average-configuration calculations in order to limit computational effort. An additional complication is that many of the ions in the iron isonuclear sequence have metastable states which are present in the experiments with unknown fractions. Inclusion of ionization from metastable states generally leads to satisfactory agreement of theory with the available experiments. Figure 2.45 presents the experimental data available for electron impact single ionization of Fe^{q+} with q ranging from 0 to 15. The figure clearly shows the increasing threshold energy for ionization of Fe^{q+} and the decrease of the cross sections for increasing ion charge state q. In the present range of q the cross section maxima change within 4 orders of magnitude. This also throws some light on the experimental difficulties which are met in a crossed beams experiment on ionization of Fe^{15+} ions where the cross sections are small and where any ion source based on electron impact ionization comes to its limits of efficient production of an ion beam.

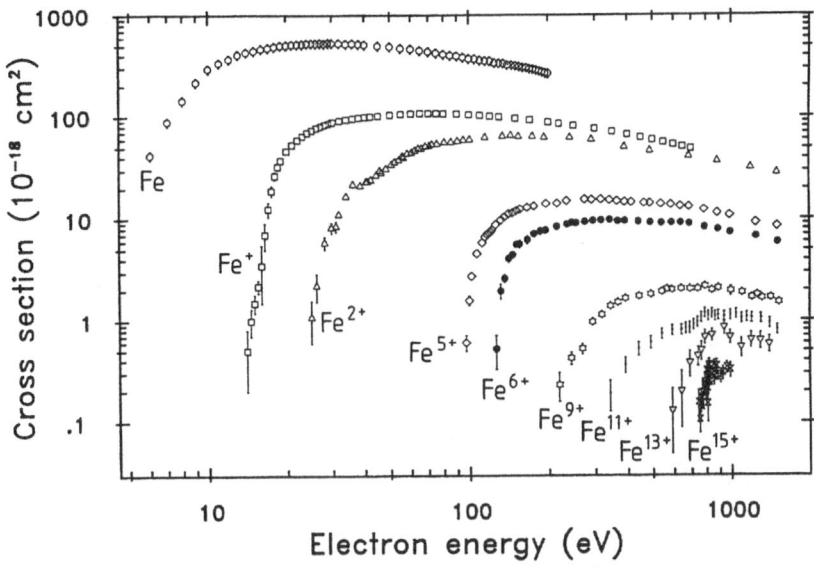

Fig. 2.45. Electron impact single ionization cross sections for Fe^{q+} ($q = 0, \ldots, 15$) [2.110–113, 91]

2.4.8 Ionization of Heavy Ions

It had been noted in experiments with heavy ions that ionization of the 4d subshell in I^+ and Xe^+ ions is characterized by a pronounced peak in the cross section at around 100 eV [2.115]. This peak, both in shape and size, resembles the 4d photoionization cross section of Xe atoms. In photoionization this resonance-like feature has been called a 'giant resonance'. Much of the theoretical and experimental work done on this subject has concentrated on the term dependence of core-excited states of the form d^9f. The strong exchange potential occurring between the f electron and the d^9 core results in an effective double-well potential and properties associated with the bound nf and continuum kf states of such configurations depend critically on the degree to which the orbital penetrates into the inner well in which the bound core orbitals reside. The term 'giant resonance' has come to be applied to almost any excited orbital which exhibits pronounced term dependence, even if the strict conditions for resonance behaviour, that is, an increase of the phase shift by π and the addition of a node to the orbital, are not satisfied.

In a calculation on the 4d ionization in Cs^+ ions *Younger* [2.116] produced the first real identification (in terms of the strict conditions mentioned above) of a genuine giant resonance appearing in the scattered electron channel in the ionization of an ion. The Cs^+ ion is Xe-like, that is, it has an outer shell configuration $4d^{10}5s^25p^6$. A local approximation of the potential for f waves in the $4d^{10}kf$ configuration (where k denotes a continuum orbital) in Cs^+ is given in Fig. 2.46. Clearly, there is a potential barrier separating the inner and outer wells. The transfer of continuum orbital density from the outer well to the inner potential well with increasing electron energy is reflected in the phase shift of the partial kf wave (see inset of Fig. 2.46). As the

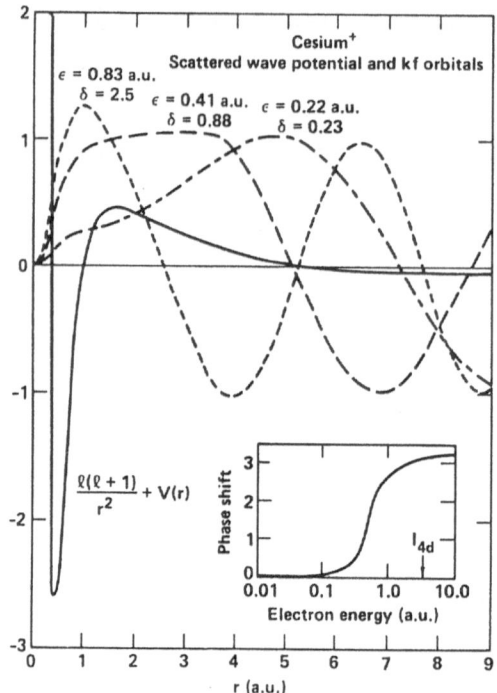

Fig. 2.46. Scattering potential for the incident and scattered channels of Cs$^+$ in atomic units (solid line). The dashed curves correspond to kf partial waves computed in this potential, illustrating the formation of an additional node as the scattering resonance is traversed. The inset shows the phase shift as a function of energy (from [2.117])

shape resonance is traversed the phase shift of the partial wave increases by π and the innermost portion of the continuum orbital gains a node. Figure 2.47 presents the theoretical cross section for ionization of a 4d electron in Cs$^+$. The solid curve is a distorted-wave Born exchange approximation using term dependent Hartree-Fock ejected f waves and including ground state correlations such as $d^{10}+d^8f^2$. The calculated cross section shows a strong giant resonance structure close to 100 eV electron energy corresponding to an energy range for the scattered electron as indicated in the inset of Fig. 2.46 (the ionization energy for a 4d electron is approximately 92 eV).

There are no experiments where the direct single 4d ionization cross section of Cs$^+$ ions is measured. However, theory can be compared to data for net double ionization of Cs$^+$ ions because 4d ionization leads to an inner-shell excited state which can be assumed to decay by autoionization with probability 1, and thus leads to a net double ionization of the ion. In addition to the ionization-autoionization contribution the experiment includes direct double ionization for which there are no cross section calculations available. Fig. 2.47 shows the measurements for Cs$^+$ double ionization by *Müller* et al. [2.51] and *Hertling* et al. [2.118]. The experimental cross sections also show the pronounced giant resonance feature. Apparently, there is a cross section contribution below the threshold of 4d ionization indicating the presence of direct double ionization in the experiment. Considering the complexity of the ionization process the agreement between theory and experiment is truly remarkable.

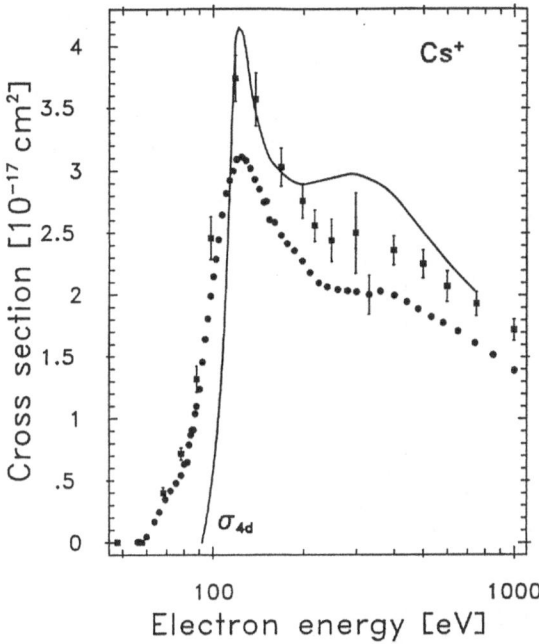

Fig. 2.47. Electron impact ionization of Cs^+ ions. The experimental data points are for double ionization $e+Cs^+ \rightarrow Cs^{3+}+3e$ (squares [2.118], solid circles [2.51]). The solid line is Younger's calculation of the 4d inner-shell direct ionization [2.116]

Younger [2.119] has studied the systematics of giant resonance effects along isonuclear and isoelectronic sequences. For Xe-like ions (and atoms) Xe, Cs^+, Ba^{2+}, La^{3+} he found a shift of the 4d scattering resonance towards lower incident electron energies as the nuclear charge increases. Thus the giant resonance in Xe is found to be quite broad and to occur above about 1.5 times the threshold energy for 4d ionization. For Cs^+ the feature has become narrower and its position is closer to the threshold. For Ba^{2+} the giant resonance only causes a shoulder in the cross section just above the 4d threshold and there is no resonance structure visible in the 4d ionization of La^{3+} ions.

The influence of outer shell excitation on 4d giant resonance effects was also investigated. Fig. 2.48 shows calculated 4d ionization cross sections for the $5s^2 5p^6 5d^2$ ground state of La^+ and the $5s^2 5p^6 5d4f$ and $5s^2 5p^6 4f^2$ excited states of the same ion. For the ground state the theoretical cross section shows a sharp resonance just above threshold. The excitation of one or both of the 5d electrons to N-shell localized 4f orbitals adds to the amplitude of the potential barrier separating the inner and outer wells which define the scattering wave potential. As the amplitude of the barrier increases, the resonance moves to higher incident electron energy and broadens.

As in the case of Cs^+ the ionization of a 4d electron in La^+ leads to autoionizing states. Therefore, the net result of the direct single 4d ionization is a multiple ionization of the La^+ ion. Comparison of the quoted calculations with experiments thus

71

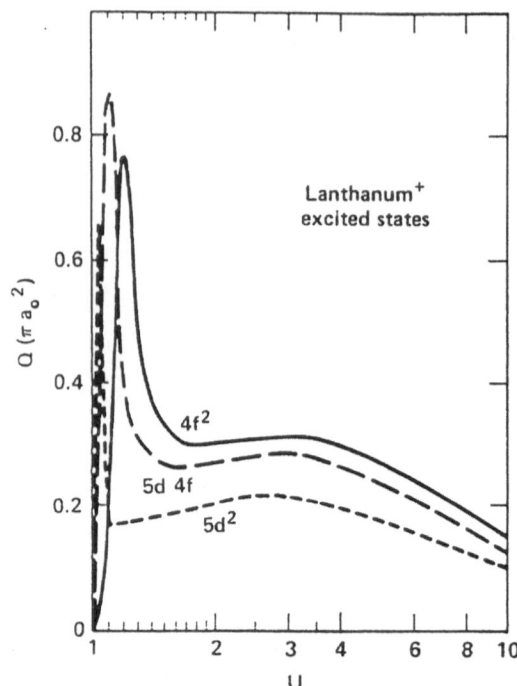

Fig. 2.48. Cross sections for the electron impact ionization of the $5s^2\,5p^6\,5d^2$ ground state and the $5s^2\,5p^6\,5d4f$ and $5s^2\,5p^6\,4f^2$ excited states of La$^+$. All curves were computed in a single configuration distorted-wave Born-exchange approximation using term-dependent ejected f waves (from [2.119])

requires additional considerations about branching ratios leading to the final charge state of the ion. The presence of direct multiple ionization contributions in the experimental cross sections further adds to the complication. This discussion shows the interplay of single and multiple ionization processes and the necessity of a 'unified picture' of ionization phenomena. For practical reasons, the following section will deal with multiple ionization of ions per se.

2.5 Multiple Ionization Cross Section Data

2.5.1 Helium-like Ions

Two-electron ions are the simplest targets for double ionization. Experimental data are available for He atoms [2.120], H$^-$ [2.121, 122] and Li$^+$ [2.123] ions. For sufficiently high energy of the incident electron it has been stated that a "sudden approximation" should be applicable with the argument that both outgoing electrons in a single ionization are fast and leave the residual ion before the other bound electron can adjust to the new effective charge. Consequently, this electron remains in an orbital which is not an eigenstate, and in the process of relaxation there is a finite probability for a transition into the continuum [see (2.47)]. Hence, the ratio of cross sections for single to double ionization can be calculated using the shake-off for-

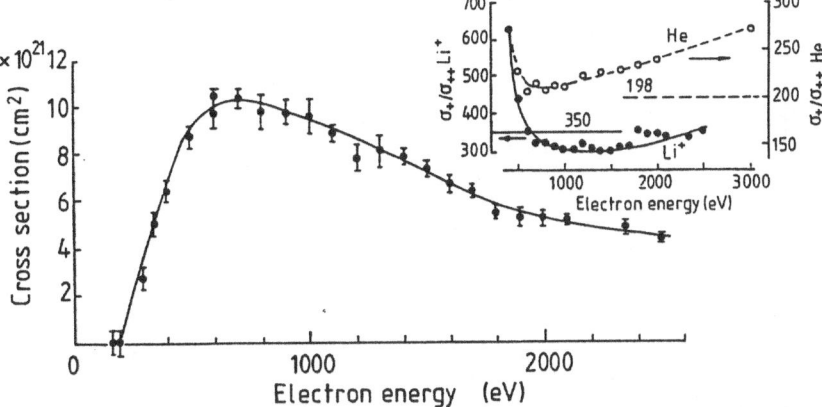

Fig. 2.49. Cross section σ_{++} for double ionization of Li^+ ions [2.123]. The inset shows ratios of cross sections for single (σ_+) and double (σ_{++}) ionization of He and Li^+. The solid and dashed lines are results of "sudden approximations" [2.124]

malism and the result should be the same as for ionization by high energy photons. Unlike photoionization, however, there is always a possibility that the distribution of outgoing electron energies will also have slow electrons, even if the incident electron energy is much higher than the ionization energy. With this fact in mind one can never expect completely satisfactory results from the shake-off theory for electron impact ionization.

Figure 2.49 shows the experimental cross section $\sigma_{1,3}$ for double ionization of Li^+ ions [2.123]. The maximum of $\sigma_{1,3}$ is about 10^{-20} cm^2 which is about 400 times lower than the maximum of $\sigma_{1,2}$ (see Fig. 2.25). The ratio of the two cross sections $\sigma_{1,2}$ (denoted σ_+) and $\sigma_{1,3}$ (denoted σ_{++}) for single and double ionization of Li^+, respectively, is plotted in the inset of Fig. 2.49. The shake-off model would predict a ratio of 350. At energies well above 500 eV the experimental ratio is quite close to this prediction indicating that the shake-off concept takes care of the dominant double ionization mechanisms.

2.5.2 The Argon Isonuclear Sequence

Ions with more than 2 electrons can also be triply ionized; however, the cross sections for direct multiple ionization decrease rapidly with increasing number of electrons removed, and also with increasing ion charge state. For the Ne isonuclear sequence these dependences were systematically studied [2.125]. With the charge state of the parent ion increasing from 1 to 4 the double ionization cross sections drop from 1.1×10^{-18} cm^2 to about 2.2×10^{-20} cm^2 at maximum. Staying with Ne^+ and looking at single, double and triple ionization the cross section maxima change from 3.5×10^{-17} cm^2 to 2.5×10^{-20} cm^2. These numbers seem to confirm the often used assumption that multiple ionization processes are unlikely and can be neglected compared to single ionization. However, as we go to more complex ions, things change

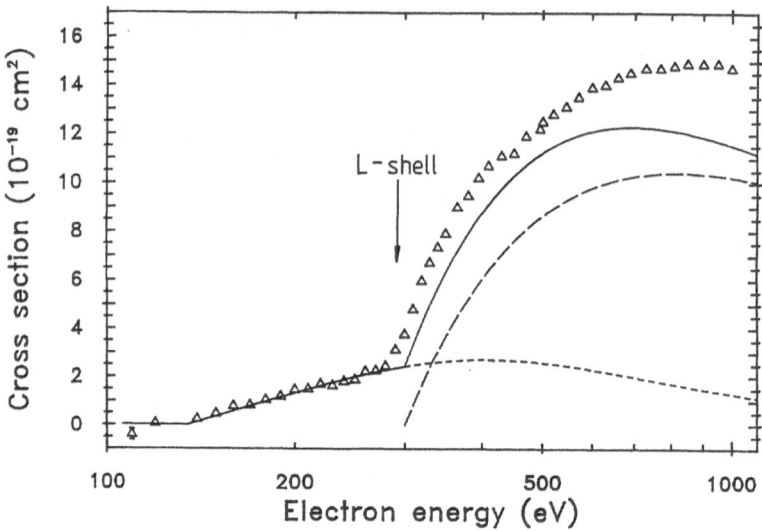

Fig. 2.50. Cross section for double ionization of Ar^{3+} ions [2.127]. The short broken curve is Gryziński's estimate (2.30) of direct double ionization scaled by a factor 0.19. The long broken curve is direct inner-shell ionization from (2.27) followed by autoionization. The solid curve is the sum of the two calculated contributions

due to the possibility of inner-shell ionization-autoionization processes. These effects were first demonstrated for ions in the Ar isonuclear sequence [2.126].

Figure 2.50 shows the experimental cross section for double ionization of Ar^{3+} ions [2.127]. At about 300 eV, that is, close to the threshold for L-shell ionization, the cross section rapidly increases and reaches a magnitude characteristic of direct single L-shell ionization. Since there is no general quantum theory available for direct multiple ionization, the classical approach of *Gryziński* [2.19] was employed to obtain a cross section estimate. After multiplying the Gryziński cross section by a factor 0.19 the short broken curve results which fits the experimental data below 300 eV, giving some confidence in the prediction above 300 eV. At the threshold for direct single ionization of an electron from the 2p subshell a new cross section contribution has its onset. This contribution can be represented by the Lotz formula (2.27) for the partial 2p ionization cross section (long broken curve in Fig. 2.50). Here a branching ratio near unity is assumed for Auger processes [2.126], leading to the final net double ionization. The sum of the direct and the ionization-autoionization contributions is the solid curve in Fig. 2.50 which is in quite good agreement with the experimental data. Similar calculations for Ar ions in other charge states show that the inner-shell contribution stems from a direct single L-shell ionization with subsequent autoionization. Since the branching ratio is close to unity, independent of the incident electron energy, the cross section can be described reasonably well by using the Lotz formula.

For demonstration of cross section dependences on the ion charge state Fig. 2.51 shows experimental results [2.127] for double ionization of all Ar^{q+} ions with q be-

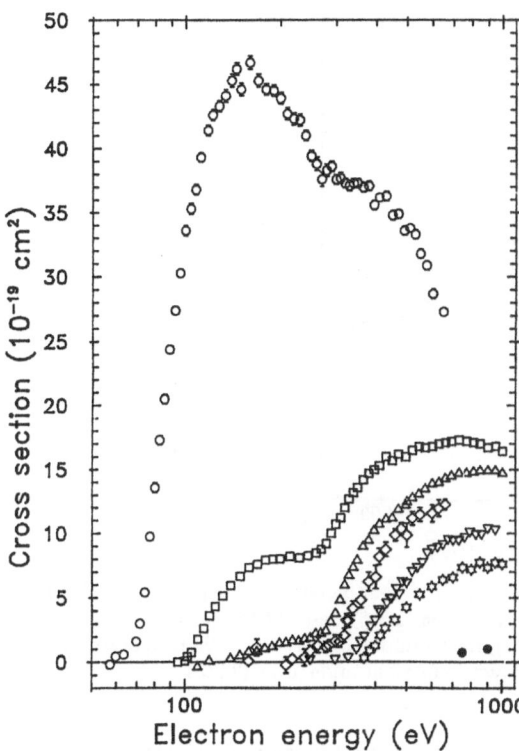

Fig. 2.51. Cross sections for double ionization of Ar^{q+} ions [2.127, 128]: open circles ($q = 1$), squares ($q = 2$), triangles ($q = 3$), diamonds ($q = 4$), inverted triangles ($q = 5$), stars ($q = 6$), solid circles ($q = 7$).

tween 1 and 7. All cross sections have an additional feature above 250 to 300 eV which is due to L-shell ionization. Below the L ionization threshold double ionization is only possible if two outer electrons are removed in a direct process (there are clear indications from other measurements that also L-shell excitation or even resonances can contribute to the measured cross sections, however, these effects are small).

For Ar^+, the direct double ionization has a sizable cross section which exceeds the L-shell contribution by a factor of three while, for Ar^{5+}, this process is insignificant compared with the inner-shell ionization contribution which is only slightly decreased. For Ar^{7+} the inner-shell contribution should be completely gone provided the parent ion beam is in its ground state. The L-shell ionization in a $2s^2 2p^6 3s$ configuration does not produce an autoionizing state and hence the observed cross section drops by about one order of magnitude between $q = 6$ and $q = 7$. Between $q = 1$ and $q = 6$ the maximum absolute contribution from L-shell ionization-autoionization decreases only by about a factor 2 which can be explained by the increasing binding energy of the L-shell electrons. Along the same sequence of ion charge states the maximum contribution from direct double ionization goes down about two orders of magnitude.

Comparisons of inner-shell versus direct contributions to multiple ionization of Ar^{q+} ions give similar results for triple and quadruple ionization [2.126]. The di-

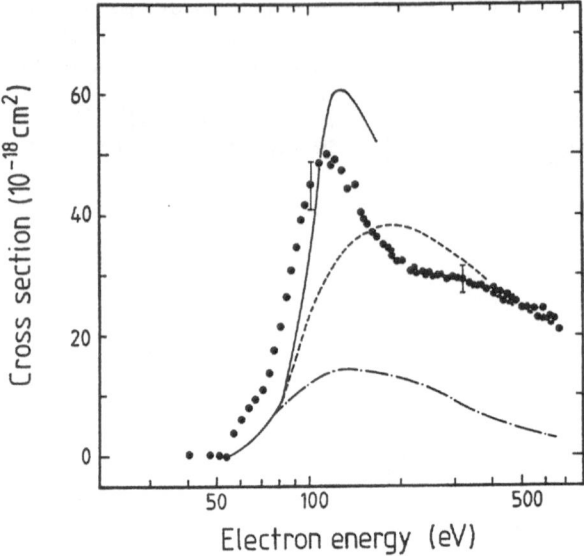

Fig. 2.52. Cross sections for double ionization of Xe^{1+} ions. Solid points: experiment [2.129, 130]; chain curve: estimate of the direct contribution (from (2.30)); dashed line: sum of the direct contribution and a Lotz cross section (from (2.27)) for 4d subshell ionization; solid line: distorted-wave calculation of *Pindzola* et al. [2.131] plus direct double ionization (from (2.30))

rect processes become very unlikely with increasing number of electrons removed in a single collision. Thus the inner-shell contributions can dominate the multiple ionization cross sections.

2.5.3 The Xenon Isonuclear Sequence

The results on multiple ionization of lighter ions discussed above indicate an increasing relative importance of these processes when the number of electrons in the ion increases. This trend continues in heavier ions and is manifest in measurements on the Xe isonuclear sequence [2.129, 130]. The cross section for double ionization of Xe^+ ions reaches up to 5×10^{-17} cm^2 which is due to a giant resonance in the 4d ionization. In Fig. 2.52 the measured data are compared with theoretical cross sections. The dot-dashed curve is the Gryziński direct double ionization estimate (2.30), the dashed curve is the sum of this direct contribution and the 4d ionization-autoionization cross section calculated from the Lotz formula and assuming unit branching ratio. The solid curve is a sum of the direct double ionization cross section and a 4d ionization cross section calculated by *Pindzola* et al. [2.131]. This calculation is a distorted-wave approach including ground state correlations and term dependent Hartree-Fock calculations of the ejected-electron continuum orbitals. These calculations are similar in spirit to those of *Younger* [2.116]. They also represent the unusual shape of the 4d contribution to double ionization of Xe^+ reasonably well as did Younger's calculations for Cs^+ ions.

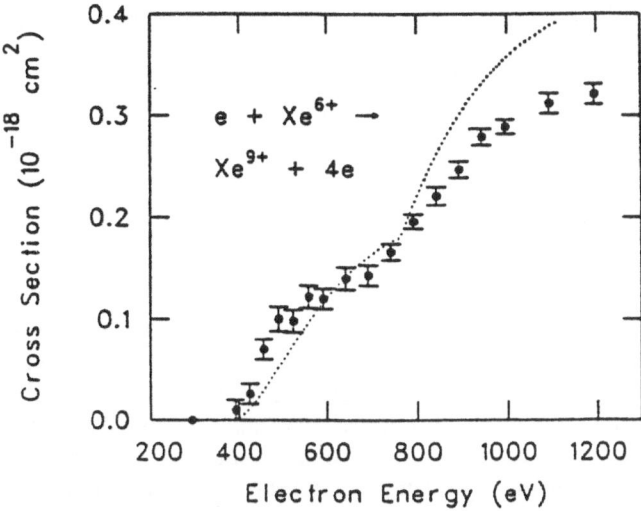

Fig. 2.53. Cross sections for triple ionization of Xe^{6+} ions [2.132]. The dotted curve is the sum of contributions from direct double ionization [2.19] and direct ionization of a single, inner-shell electron [2.17] followed by autoionization

While the double ionization cross section maxima for Ne^{q+} ions ($q = 1, \ldots, 4$) decrease by two orders of magnitude between $q = 1$ and $q = 4$, the same data for Xe^{q+} ($q = 1, \ldots, 4$) remain within a factor of 5. Even triple and quadruple ionization of Xe ions in low charge states are in the range 1 to $10 \times 10^{-18} \, cm^2$. These numbers clearly indicate that multiple ionization of heavy ions cannot simply be neglected compared to single ionization. The influence of multiple ionization of Xe atoms on the charge state evolution, for example, in an EBIS ion source [2.44] is demonstrated by Fig. 2.16. Ratios of cross sections for multiple versus single ionization can reach up to a magnitude of 0.5 to nearly 1 at certain energies and charge states of the Xe^{q+} parent ions.

The reason for the high multiple ionization cross sections is not only due to the many possibilities of efficient inner-shell ionization processes in many-electron systems; direct multiple ionization is also enhanced when there are more equivalent electrons. This was demonstrated by *Howald* et al. [2.132] in the triple ionization of Xe^{6+} ions. Their measured cross sections, which are still of $10^{-18} \, cm^2$ magnitude, are shown in Fig. 2.53. By a careful analysis of inner-shell binding energies and possible excitation and decay processes *Howald* et al. could make the statement that triple ionization of Xe^{6+} in a single-electron process (with subsequent autoionizations) can only occur above 670 eV. The dominant contribution to triple ionization is expected to come from direct ionization of a single 3d inner-shell electron, a process which has its onset at 762 eV. As seen from Fig. 2.53 the observed cross section has its onset already near 374 eV which is the minimum energy to remove the three least bound electrons. Between this threshold and 670 eV the measured cross section can only be due to multiple electron processes, such as direct double ionization followed by autoionization. Above 762 eV the cross section exhibits a change in slope which is due to the 3d contribution. An extrapolation of the low energy portion of the cross section to higher energies shows that multiple electron ionization and the direct inner-shell ionization-autoionization processes are of comparable importance in the triple ionization of Xe^{6+}.

2.6 A Unified Picture of Ionization

Our discussions have clearly shown that direct single ionization of an inner shell electron can result either in net single or in net multiple ionization depending on the branching ratios of the decay processes. Rather than looking at n-fold ionization of ions along isonuclear or other sequences it is necessary to study all single and multiple ionization processes of a given ion in order to obtain a more complete picture. Here we consider features occurring in net single, double and triple ionization cross sections which can often be traced back to one intermediate multiply excited state branching into different decay channels.

2.6.1 Single and Double Ionization of Sb and Bi Ions

Bismuth is the heaviest stable element. Looking at the trends in the ratio of multiple versus single ionization Bi^{q+} ions are expected to have extremely high cross sections for net double ionization. In a study of single and multiple ionization of Bi^{q+} ions ($q = 1, 2, 3$) it was noted [2.133] that, for energies greater than 200 eV, the cross section $\sigma_{2,3}$ was larger than $\sigma_{1,2}$, that is, single ionization of Bi^{2+} is easier at sufficiently high electron energy than of Bi^+ although the ionization energy of Bi^{2+} (25.56 eV) is about 9 eV higher than that of Bi^+. This result is illustrated in Fig. 2.54. Both cross sections are quite large at maximum. At an energy of 1000 eV $\sigma_{2,3}$ exceeds $\sigma_{1,2}$ by about 20 %. The reason for this unexpected behaviour can be found by looking at the cross sections for double ionization of Bi^{q+} ions ($q = 1, 2, 3$). Experimental results are shown in Fig. 2.55. The cross section $\sigma_{1,3}$ for Bi^{1+} reaches a magnitude [2.133] 7.5×10^{-17} cm^2 and this is the largest cross section ever observed for multiple ionization of an ion. As the ion charge increases from $q = 1$ to $q = 2$ double ionization goes down by a factor 3. A further increase to $q = 3$ leads to a cross section $\sigma_{3,5}$ only slightly smaller than $\sigma_{2,4}$ at energies well above the respective thresholds. This behaviour indicates the loss of an important cross section contribution when the ion charge state becomes greater than 1.

A closer look at the ground state configuration $5d^{10}6s^26p^2$ and the energy levels of inner shell excited states suggests a dominant contribution to double ionization of Bi^+ by 5d ionization-autoionization. The dot-dashed curve in Fig. 2.55 is a calculation for direct single ionization of a 5d electron by using the Lotz formula (2.27) for this partial contribution. Unit branching ratio for autoionization is assumed. Direct outer shell double ionization is estimated by the dashed line calculated on the basis of Gryziński's binary encounter approximation (2.30). The sum of both contri-

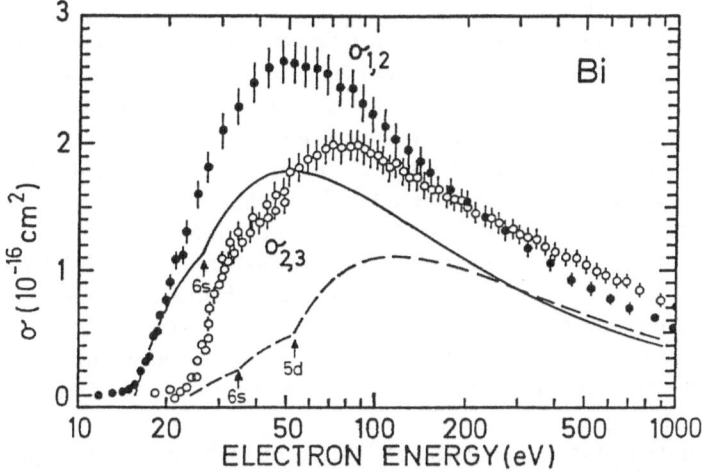

Fig. 2.54. Cross section for single ionization of bismuth ions. Solid circles: experiment for Bi^{1+}, open circles: experiment for Bi^{2+} [2.133]. The curves are summed Lotz cross sections from (2.27) for direct ionization of the 6p and 6s subshells of Bi^{1+} (solid line) and of the 6p, 6s and 5d subshells of Bi^{2+} (dashed line)

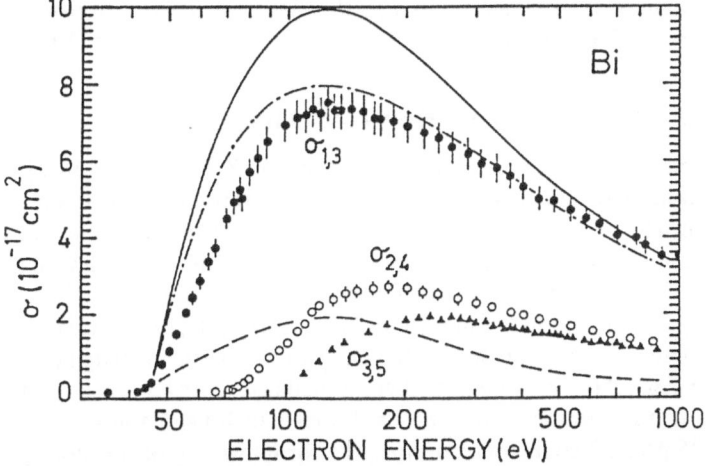

Fig. 2.55. Cross sections for double ionization of bismuth ions. The experimental data [2.133] are for Bi^+ (solid circles), Bi^{2+} (open circles) and Bi^{3+} (solid triangles). The dashed curve is the direct double ionization estimate (2.30); the chain curve is the direct single ionization Lotz calculation from (2.27) for the 5d subshell of Bi^{1+}, the solid curve is the sum of direct double and 5d ionization-autoionization contributions

butions to net double ionization is the solid line. Considering the uncertainty in the binary encounter approximation this comparison clearly shows the dominance of the 5d contribution to $\sigma_{1,3}$. As one goes from $q = 1$ to $q = 2$ the 5d vacancy goes from autoionizing to bound and hence double ionization cannot result. With this important contribution missing the cross sections $\sigma_{2,4}$ and $\sigma_{3,5}$ are considerably smaller than $\sigma_{1,3}$. However, the cross section for 5d ionization does not change much with the charge state of the ion and the collision strength has to be seen somewhere else. Since the 5d vacancy produced in a parent Bi^{2+} ion can only decay by photoemission processes, a 5d ionization will contribute to net single ionization. There the addition of such a big cross section contribution results in a considerable enhancement so that it is possible to have $\sigma_{2,3} > \sigma_{1,2}$. The curves in Fig. 2.54, which are based on simple Lotz cross section calculations, illustrate this situation. In Bi^+ ions single ionization can only occur when one 6p or one 6s electron is removed (ionization of a 5d electron leads to net double ionization because of subsequent autoionization). The structure in the solid curve reflects the two possible contributions. In Bi^{2+}, also direct single ionization of a 5d electron adds to $\sigma_{2,3}$ and the sum of the three contributions from the 6p, the 6s and the 5d subshells can produce a cross section $\sigma_{2,3}$ (dashed curve in Fig. 2.54) with $\sigma_{2,3} > \sigma_{1,2}$.

This discussion shows the interplay between single and multiple ionization processes. For a complete understanding of ionization phenomena, at least in heavy ions, it is necessary to look at all the single and multiple ionization channels and to consider the branching ratios of decays from intermediate excited states.

A situation very similar to the one in Bi^{q+} ions was observed for Sb^{q+} ions [2.133] which have homologous configurations. Also other heavy ions show the effect that $\sigma_{q+1,q+2}$ may be larger than $\sigma_{q,q+1}$ [2.129, 134–136]. This is for instance also the case for $q = 3$ in Xe^{q+} where $\sigma_{4,3} > \sigma_{3,4}$ for energies greater than 200 eV.

2.6.2 Single and Multiple Ionization of Heavy Metal Ions

Heavy atoms and ions, especially those with atomic numbers $Z = 54$ to $Z = 58$, offer an attractive opportunity to study photoabsorption and electron scattering processes in a region where the atomic structure is dominated by electron-electron interactions. Some of the dramatic effects involving electrons of the closed 4d subshell have already been discussed. Here we will deal with fine details revealed by the fast energy scan technique in systematic studies of many ions in the Xe, Cs, Ba, La and Ce isonuclear sequences where up to net four-fold ionization (for Ba^+) was investigated.

As an example Fig. 2.56 shows measured and calculated cross sections $\sigma_{3,4}$ for electron impact single ionization of Xe-like La^{3+} ions which have a ground state configuration $4d^{10}5s^25p^6$. The theoretical threshold for ionization of this configuration is between 49 and 52 eV. The rise in the experimental cross section at 30 eV appears to come from a metastable component, possibly low-lying triplet terms of the odd parity $5p^55d$ configuration (thresholds 20–33 eV), but most likely low-lying terms of the even parity $5p^54f$ configuration (thresholds 26–33 eV).

In Fig. 2.56 configuration-average distorted-wave calculations for ionization from both the $5p^6$ ground configuration and the $5p^54f$ excited configuration of La^{3+} are compared with the experimental crossed beams results [2.137]. The dashed line is for direct single ionization of the $5p^6$ ground configuration. The solid line is the sum of the direct plus excitation-autoionization contributions from the $5p^6$ configuration. The dot-dashed curve is the sum of direct ionization (dotted line) plus

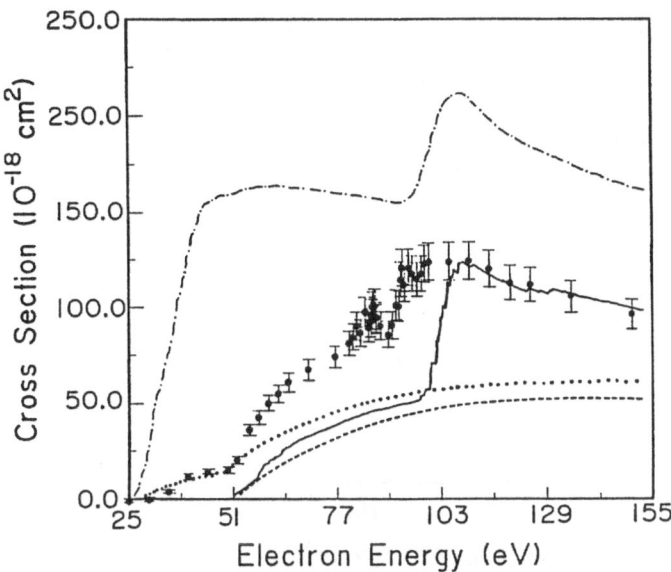

Fig. 2.56. Threshold region of the cross section for single ionization of La^{3+} ions. The solid circles are experimental data for which the technique described in Sect. 2.4.4 was employed. The solid line is a distorted-wave calculation of the sum of direct single ionization (dashed line) plus excitation-autoionization contributions from the $5s^2 5p^6$ ground configuration. The dot-dashed curve is the sum of direct ionization (dotted line) plus excitation-autoionization contributions as calculated for the $5s^2 5p^5 4f$ excited configuration of La^{3+} (from [2.137])

excitation-autoionization contributions as calculated for the $5s^2 5p^5 4f$ excited configuration. Resonant contributions are not included in the calculations. Comparing with the theoretical results the experimental measurements appear to be a mixture of approximately 95 % ground and 5 % excited configuration contributions. The rapid rise of the $5p^5 4f$ excited configuration cross section from 25 to 50 eV is mainly due to the $5p \rightarrow 5f, 5p \rightarrow 6l$ $(l = 1, 2)$, $5s \rightarrow 5d$ excitation-autoionization contributions, which are absent in the $5p^6$ ground configuration cross section. The small rise in both the ground and excited configuration cross sections from 50 to 70 eV is due to $5s \rightarrow nl$ $(n > 6)$ excitation-autoionization contributions. Above 95 eV, $4d \rightarrow 4f$ excitation-autoionization contributions lead to a rapid rise in the cross sections. The step-like character of the $5p^6$ ground configuration cross section is due to the energy onset of the 20 different levels of the $4d^9 5s^2 5p^6 4f$ core-excited configuration. The steps have blended into a smooth curve for the $5p^5 4f$ excited configuration cross section due to the presence of 626 different levels in the $4d^9 5s^2 5p^5 4f^2$ core-excited configuration.

At energies above 110 eV the theoretical prediction agrees remarkably well with the experimental cross section. In the range from 50 eV to 100 eV theory apparently misses important cross section contributions. In particular, it does not reproduce the onset of the excitation feature in the experiment above 90 eV nor the structure observed between 80 and 90 eV. Theory predicts no energy onsets due to excitation-

81

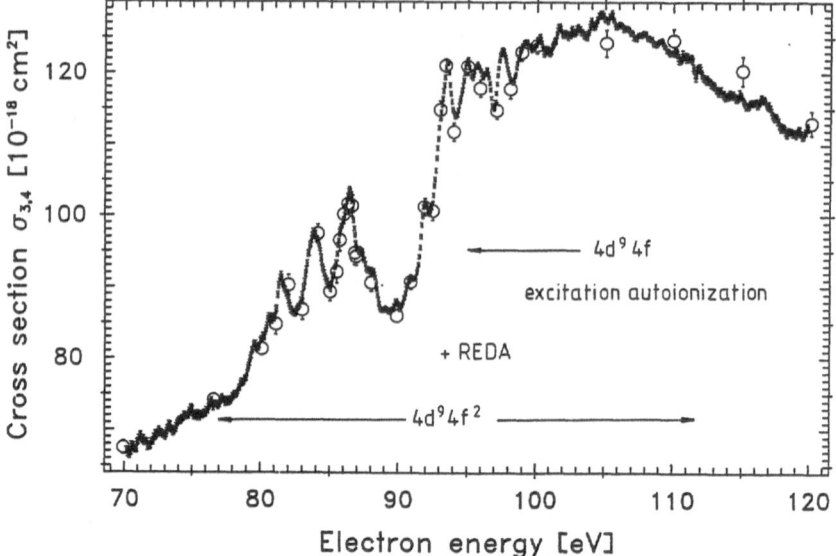

Fig. 2.57. Energy scan measurement [2.51] of cross sections for single ionization of La^{3+} ions. The open circles are data from Fig. 2.56. Ranges are indicated for contributions from excitation-autoionization via the 4d^{9}4f excited configuration in La^{3+} and from REDA processes via the 4d^{9}4f^2 doubly excited configuration in La^{2+}

autoionization contributions in the 70–95 eV energy range. The 5p ionization energy is about 50 eV, so all 5p processes (neglecting multi-electron excitations) are too low in energy. The same reasoning holds for 5s excitations, since the 5s ionization threshold is 71 eV. The 4d ionization thresholds are in the range 137–140 eV, and lead to net double ionization of the original target. The 4d excitation thresholds start at about 95 eV for the 5p^{5}4f excited configuration as seen in Fig. 2.56. This leaves only the possiblity of REDA processes with dielectronic capture into a 4d^{9}5s^{2}5p^{6}4fnl configuration of La^{2+} and subsequent double-autoionization. With respect to the observed cross section features the energetics are correct: 4d^{9}5s^{2}5p^{6}4f^2 has thresholds from 79–111 eV above the 5p^6 ground configuration.

A closer look at the cross section in the energy range 70–120 eV using the fast energy scan technique reveals a wealth of fine details. The result of such a measurement [2.137] with La^{3+} ions is displayed in Fig. 2.57. Resonances with a width as narrow as 0.4 eV are found. Due to the complexity of the electron configurations involved it is not readily possible to assign doubly excited states of La^{2+} ions to the observed features; however, one cannot think of any process other than REDA to account for such narrow peaks as those observed here. The dominant resonances make up about 20 % of the measured cross section indicating that REDA may not be neglected in the ionization of heavy ions in low charge states.

Resonance features are also found in cross sections for net double ionization. The most prominent one observed so far [2.137] was found in $\sigma_{2,4}$ for La^{2+}. Already the conventional cross section measurement shows this feature. Figure 2.58

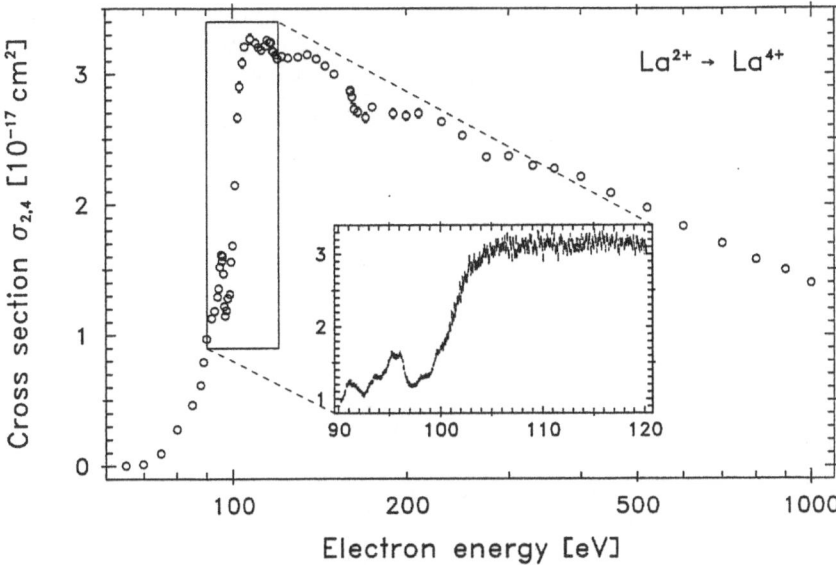

Fig. 2.58. Cross section for double ionization of La^{2+} ions. The inset shows energy scan data in the resonance region (from [2.137])

displays the measured cross section for double ionization of La^{2+} ions up to 1000 eV. Just below 100 eV the cross section shows a peak with a height which is about 30 % of the total cross section $\sigma_{2,4}$. The scan measurement which is represented in the inset, clearly exhibits that peak and additional resonance structure. Apparently, a RETA resonance (see (2.52)) is observed here and it is amazing how big a contribution this can give to net double ionization. Again, energy considerations lead to the conclusion that the resonances are due to 4d excited states produced by dielectronic capture. The recombined (La^+) state must then lose three electrons via autoionization. The sequence of autoionizations must include multiple-electron emission such as autoionization plus auto-double-ionization, auto-double-ionization plus autoionization, or auto-triple-ionization, since the $4d^9 5s^2 5p^6 5d^3$ configuration in La^+ (at 95 eV) cannot sequentially triple autoionize. Even though it lies above the double ionization threshold of La^{2+}, two sequential single autoionization processes lead to bound states (in La^{3+}), resulting in net single ionization of the target ion La^{2+}. Of course, the latter is a possible decay path and hence the same resonant state should be found also in the cross section for net single ionization of La^{2+}. The total cross section $\sigma_{2,3}$ is about 4×10^{-16} cm^2 and therefore only a detailed scan measurement can reveal small features due to REDA processes. The result of the scan of $\sigma_{2,3}$ is shown in Fig. 2.59 together with the scan result on $\sigma_{2,4}$ in the same energy range [2.51, 137]. The structures observed in $\sigma_{2,3}$ and $\sigma_{2,4}$ sit on top of a smooth "background" of direct ionization contributions. In the cross section $\sigma_{2,3}$ this background is going down with energy, in $\sigma_{2,4}$ it is rising. With this in mind one can see that the resonance features in the two cross sections perfectly line up in energy. Af-

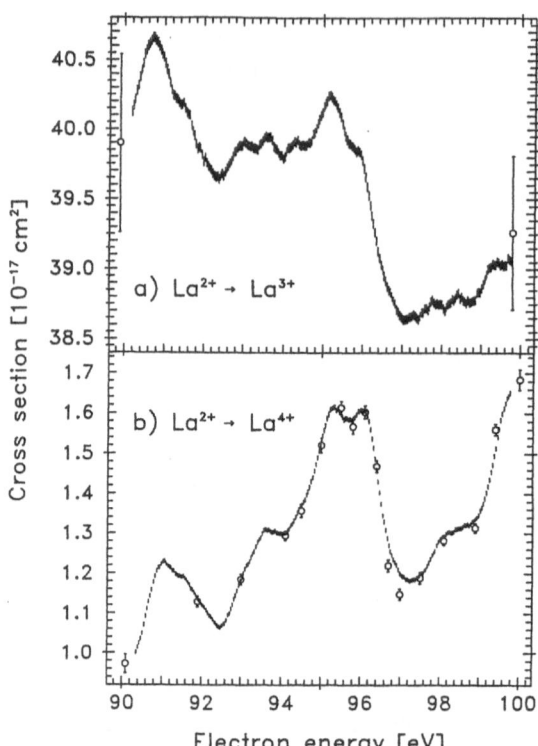

Fig. 2.59 a,b. Comparison of energy scan measurements [2.137] of (a) single and (b) double ionization of La²⁺ ions. The open circles are experimental data taken with the dynamic crossed beams technique

ter dielectronic capture of the incident electron into a multiply excited intermediate state, probably with a configuration $4d^9 5s^2 5p^6 5d^3$, either two or three electrons can be emitted resulting in net single or net double ionization of the parent ion, respectively. Emission of only one electron would be observed in an electron scattering channel.

Resonances have been found even in net triple or quadruple ionization of other heavy metal ions [2.51]. As an example Fig. 2.60 shows features in cross sections of Ba⁺ ions [2.138]. Here, the background from direct multiple ionization processes was determined from the total cross sections measured below the resonance features. A straight line was fitted to these data and then subtracted from the total experimental cross section. What remains is a contribution involving the 3d subshell. There are two dominant peaks in both cross sections which are due to dielectronic capture plus subsequent emission of four or five (!) electrons, respectively. In addition to the resonances there are contributions from 3d excitation with subsequent emission of three or four electrons, respectively, and probably also 3d ionization with subsequent emission of two or three electrons, respectively, as evidenced by an increased "background" level at energies above the resonances. All these processes can produce the net three-fold and net four-fold ionization observed in the experiments. It is particularly interesting to note the different decay probabilities for the intermediate resonant 3d-excited states. Branching into net three-fold ionization yields a resonant

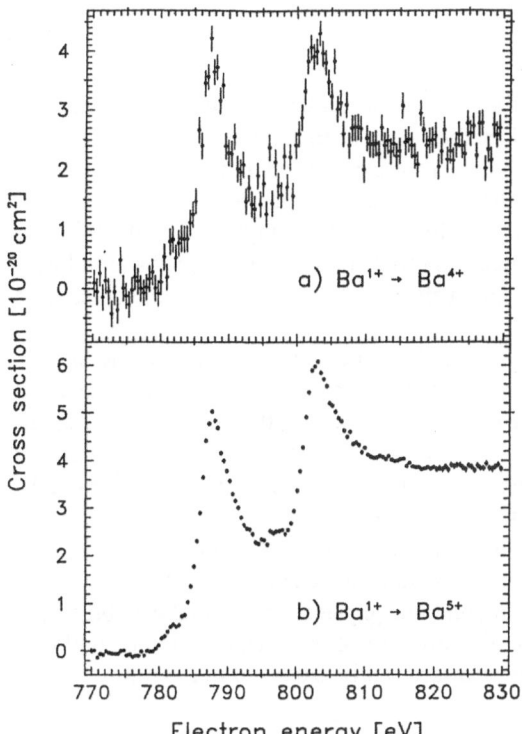

Fig. 2.60 a,b. Comparison of energy scan measurements [2.138] of net triple (**a**) and net quadruple (**b**) ionization of Ba^+ ions. Direct contributions to the cross section were represented by straight lines and subtracted from the measured total cross section in order to isolate the resonance features displayed here

cross section contribution of about 4×10^{-20} (for the peak at 787 eV) and branching into net four-fold ionization yields a cross section of about 5×10^{-20} cm². This means that the branching ratio for the emission of five electrons from the resonant intermediate excited state exceeds that for the emission of four electrons by a factor of roughly 1.3. When there is a vacancy in a deep shell the cascading Auger processes lead to a probability distribution of the final charge states. In the case of a 3d vacancy in an intermediate Ba atom the maximum of the probability distribution may be around charge state $q = 5$.

Since the non-resonant contributions to the production of Ba^{q+} from Ba^+ ions are much higher for lower q it becomes increasingly difficult to find the low q resonance contributions in experiments measuring the total cross section. This is also the reason why the statistical uncertainty of the data displayed in Fig. 2.60 is higher for four-fold compared to five-fold ionization: the "background" from non-resonant multiple ionization is a factor 7 higher in $\sigma_{1,4}$ than in $\sigma_{1,5}$.

2.7 Conclusions

Electron impact ionization of atoms and ions is a fundamental collision process. It is of very wide applied interest and has, therefore, been studied for nearly a century. The importance of electron-ion collisions in astrophysical and laboratory plasmas led to the first electron-ion crossed beams experiments in 1961. Since then much work has been carried out on ionization of ions including single and multiple ionization, excitation-autoionization and resonant contributions. Now a substantial body of experimental and theoretical data are available. Only few of these have been presented as examples in this chapter.

Theory has made important progress and it appears to be now possible to make reliable cross section predictions for many of the ionization processes in lighter ions involved in different fields of applied research. However, the complexity of the interplay of many different ionization contributions from different electron shells and myriads of intermediate excited states with term-to-term dependent branching ratios will never fully be accounted for by theory. For very heavy ions it is still not possible even to decide which processes give the dominant cross section contributions, and a complete calculational analysis of all channels leading to a net single (not to speak of multiple) ionization appear to be beyond our physical and mental capabilities.

The gap between theoretical predictability and experimental findings will keep research in the field lively and more surprises can be expected in the future.

Acknowledgements. I am indebted to many colleagues with whom I have worked on electron impact ionization experiments: R. Becker, G. Hofmann, K. Tinschert, G. H. Dunn, C. Achenbach, R. A. Phaneuf, D. C. Gregory, K. Huber, E. Salzborn, H. Klein, H. Winter, R. Frodl, P. Defrance, K. Rinn, F. W. Meyer, A. M. Howald, J. Peschina, N. Djurić, B. Weißbecker, M. Stenke, M. Wagner, L. J. Wang, and D. H. Crandall.

Particular recognition is also due to a number of theorists with whom I have been collaborating: M. S. Pindzola, S. M. Younger, D. C. Griffin, V. P. Shevelko, A. M. Urnov and L. A. Vainshtein.

Our work performed at E. Salzborn's laboratory in Gießen was supported by Deutsche Forschungsgemeinschaft, Bonn-Bad Godesberg, and Max-Planck-Institut für Plasmaforschung, Garching. The NATO collaborative research grant No. RG 86/0510 is gratefully acknowledged.

References

2.1   P. Lenard: Wied. Ann. **51**, 225 (1894); Ann. d. Phys. 15, 485 (1904)
2.2   J. J. Thomson: Phil. Mag. **23**, 449 (1912)
2.3   K. T. Dolder, M. F. A. Harrison, P. C. Thonemann: Proc. Roy. Soc. A **264**, 367 (1961)
2.4   E. Hinnov: J. Opt. Soc Am. **56**, 1179 (1966); J. Opt. Soc. Am. **57**, 1392 (1967)
2.5   F. A. Baker, J. B. Hasted: Phil. Trans. Roy. Soc. A **261**, 33 (1966)
2.6   R. A. Phaneuf: "Experiments on Electron-Impact Excitation and Ionization of Ions", in *Atomic Processes in Electron-Ion and Ion-Ion Collisions*, ed. by F. Brouillard (Plenum, New York 1986) pp. 117–156
2.7   G. H. Dunn: "Electron-Ion Ionization", in *Electron Impact Ionization*, ed. by T. D. Märk and G. H. Dunn (Springer, Wien, New York 1985) pp. 277–319

2.8   K. Dolder, B. Peart: Adv. At. Molec. Phys. **22**, 197 (1986)
2.9   E. Salzborn: "Electron-Impact Ionization of Ions", in *Physics of Ion-Ion and Electron-Ion Collisions*, ed. by F. Brouillard, J. W. McGowan (Plenum, New York 1983) pp. 239–277
2.10   D. H. Crandall: "Electron-Ion Collisions", in *Atomic Physics of Highly Ionized Atoms*, ed. by R. Marrus (Plenum, New York 1983) pp. 399–453
2.11   H. Tawara, T. Kato: At. Data Nucl. Data Tables **36**, 167 (1987)
2.12   M. S. Pindzola, D. C. Griffin, C. Bottcher: "Indirect Processes in Electron Impact Ionization of Atomic Ions", in *Electronic and Atomic Collisions*, ed. by H. B. Gilbody, W. R. Newell, F. H. Read, A. C. H. Smith (North Holland, Amsterdam 1988) pp. 129–145
2.13   S. M. Younger: "Quantum Theoretical Methods for Calculating Ionization Cross Sections", in *Electron Impact Ionization*, ed. by T. D. Märk, G. H. Dunn (Springer, Wien, New York 1985) pp. 277–319
2.14   Y.-K. Kim: "Theory of Electron-Atom Collisions", in *Physics of Ion-Ion and Electron-Ion Collisions*, ed. by F. Brouillard, J. W. McGowan (Plenum, New York 1983) pp. 101–165
2.15   H. Bethe: Ann. d. Phys. **5**, 325 (1930)
2.16   S. M. Younger, T. D. Märk: "Semi-Empirical and Semi-Classical Approximations for Electron Ionization", in *Electron Impact Ionization*, ed. by T. D. Märk, G. H. Dunn (Springer, Wien, New York 1985) pp. 24–41
2.17   W. Lotz: Z. Phys. **232**, 101 (1970); Z. Phys. **216**, 241 (1968); Z. Phys. **206**, 205 (1967)
2.18   M. R. H. Rudge, S. B. Schwartz: Proc. Phys. Soc. **88**, 563 (1966); Proc. Phys. Soc. **88**, 579 (1966)
2.19   M. Gryziński: Phys. Rev. **138**, A305 (1965)
2.20   A. Burgess, M. C. Chidichimo: Mon. Not. R. Astr. Soc. **203**, 1269 (1983)
2.21   G. H. Wannier: Phys. Rev. **90**, 817 (1953)
2.22   H. Klar: J. Phys. B **14**, 3255 (1981)
2.23   S. Cvejanović, F. H. Read: J. Phys. B **7**, 1841 (1974)
2.24   H. Kossmann, V. Schmidt, T. Andersen: Phys. Rev. Lett. **60**, 1266 (1988)
2.25   G. H. Wannier: Phys. Rev. **100**, 1180 (1955)
2.26   H. Klar, W. Schlecht: J. Phys. B **9**, 1699 (1976)
2.27   J. A. R. Samson, G. C. Angel: Phys. Rev. Lett. **61**, 1584 (1988)
2.28   H. J. Kunze, A. H. Gabriel, H. R. Griem: Phys. Rev. **165**, 267 (1968)
2.29   H. J. Kunze: Phys. Rev. A **3**, 937 (1971)
2.30   R. U. Datla, M. Blaha, H. J. Kunze: Phys. Rev. A **12**, 1076 (1975)
2.31   E. Källne, L.A. Jones: J. Phys. B **10**, 3637 (1977)
2.32   C. Breton, C. DeMichelis, M. Finkenthal, M. Mattioli: Phys. Rev. Lett. **41**, 110 (1978)
2.33   R. U. Datla, J. R. Roberts: Phys. Rev. A **28**, 2201 (1983)
2.34   P. A. Redhead: Can. J. Phys. **49**, 3059 (1971)
2.35   J. B. Hasted, G. L. Awad: J. Phys. B **5**, 1719 (1972)
2.36   J. B. Hasted: "Confinement of Ions for Collision studies", in *Physics of Ion-Ion and Electron-Ion Collisions*, ed. by F. Brouillard, J. W. McGowan (Plenum, New York 1983) pp. 461–500
2.37   M. Hamdan, K. Birkinshaw, J. B. Hasted: J. Phys. B **11**, 331 (1978)
2.38   D. Mathur, J. B. Hasted, S. U. Khan: J. Phys. B **12**, 2043 (1979)
2.39   D. Mathur, S. U. Khan, J. B. Hasted: J. Phys. B **11**, 3615 (1978)
2.40   E. D. Donets, A. I. Pikin: Zh. Tekh. Fiz. **45**, 2373 (1975) (English transl.: Sov. Phys. Tech. Phys. **20**, 1477 (1976))
2.41   E. D. Donets, V. P. Ovsyannikov: Zh. Eksp. Teor. Fiz. **80**, 916 (1981) (English transl.: Sov. Phys. JETP **53**, 466 (1981))
2.42   E. D. Donets: Phys. Scr. T3, 11 (1983)
2.43   R. Ali, C. P. Bhalla, C. L. Cocke, M. Stöckli; Phys. Rev. Lett. **64**, 633 (1990)
2.44   A. Müller: Phys. Lett. **113A**, 415 (1986)
2.45   N. Claytor, B. Feinberg, H. Gould, C. E. Bemis, J. G. del Campo, C. A. Ludemann, C. R. Vane: Phys. Rev. Lett. **61**, 2081 (1988)
2.46   S. Andriamonje, R. Anne, N. V. de Castro Faria, M. Chevallier, C. Cohen, J. Dural, M. J. Gaillard, R. Genre, M. Hage-Ali, R. Kirsch, A. L'Hoir, B. Farizon-Mazuy, J. Mory, J. Mouilin,

J. C. Poizat, Y. Quere, J. Remillieux, D. Schmaus, M. Toulemonde: Phys. Rev. Lett. **63**, 1930 (1989)

2.47   K. T. Dolder, B. Peart: Rep. Prog. Phys. **39**, 693 (1976)

2.48   P. Defrance: "Electron impact excitation and ionization of ions – experimental methods" in *Atomic Processes in Electron-Ion and Ion-Ion Collisions*, ed. by F. Brouillard (Plenum, New York 1986) pp. 157–183

2.49   P. Defrance, F. Brouillard, W. Claeys, G. Van Wassenhove: J. Phys. B **14**, 103 (1981)

2.50   A. Müller, K. Huber, K. Tinschert, R. Becker, E. Salzborn: J. Phys. B **14**, 2993 (1985); A. Müller, K. Tinschert, C. Achenbach, E. Salzborn, R. Becker: Nucl. Instr. Methods Phys. Res. B **24/25**, 369 (1987)

2.51   A. Müller, K. Tinschert, G. Hofmann, E. Salzborn, G. H. Dunn: Phys. Rev. Lett. **61**, 70 (1988)

2.52   A. Müller, G. Hofmann, B. Weißbecker, M. Stenke, K. Tinschert, M. Wagner, E. Salzborn: Phys. Rev. Lett. **63**, 758 (1989)

2.53   K. Tinschert, A. Müller, G. Hofmann, K. Huber, R. Becker, D. C. Gregory, E. Salzborn: J. Phys. B. **22**, 531 (1989)

2.54   S. M. Younger: J. Quant. Spectrosc. Radiat. Transfer **26**, 329 (1981)

2.55   J. W. McGowan, E. M. Clarke: Phys. Rev. **167**, 43 (1968)

2.56   W. L. Fite, R. T. Brackmann: Phys. Rev. **112**, 1141 (1958)

2.57   E. W. Rothe, L. L. Marino, R. H. Neynaber, S. M. Trujillo: Phys. Rev. **125**, 582 (1962)

2.58   M. B. Shah, D. S. Elliot, H. B. Gilbody: J. Phys. B **20**, 3501 (1987)

2.59   B. Peart, D. S. Walton, K. T. Dolder: J. Phys. B **2**, 1347 (1969)

2.60   D. L. Moores, L. B. Golden, D. H. Sampson: J. Phys. B **13**, 385 (1980)

2.61   W. R. Johnson, G. Soff: At. Data Nucl. Data Tables **33**, 405 (1985)

2.62   C. A. Quarles: Phys. Rev. A **13**, 1278 (1976)

2.63   M. S. Pindzola, D. L. Moores, D. C. Griffin: Phys. Rev. A **40**, 4941 (1989)

2.64   D. H. Crandall, R. A. Phaneuf, D. C. Gregory: Report ORNL/TM-7020, Oak Ridge National Laboratory, 1979

2.65   P. Defrance: J. de Phys. C1 **50**, 229 (1989)

2.66   S. M. Younger: Phys. Rev. A **22**, 1425 (1980)

2.67   W. C. Lineberger, J. W. Hooper, E. W. McDaniel: Phys. Rev. **141**, 151 (1966)

2.68   J. B. Wareing, K. T. Dolder: Proc. Phys. Soc. **91**, 887 (1967)

2.69   B. Peart, K. T. Dolder: J. Phys. B **1**, 872 (1968)

2.70   U. Fano: Phys. Rev. **124**, 1866 (1961)

2.71   U. Fano: J. W. Cooper: Phys. Rev. **137**, 1364 (1965)

2.72   R. L. Simons, H. P. Kelly, R. Bruch: Phys. Rev. A **19**, 682 (1979)

2.73   J. J. Quémenér, C. Paquet, P. Marmet: Phys. Rev. A **4**, 494 (1971)

2.74   D. S. Walton, B. Peart, K. Dolder: J. Phys. B **3**, L148 (1970)

2.75   B. Peart, K. T. Dolder: J. Phys. B **6**, 1497 (1973)

2.76   D. H. Crandall, R. A. Phaneuf, P. O. Taylor: Phys. Rev. A **18**, 1911 (1978)

2.77   D. H. Crandall, R. A. Phaneuf, B. E. Hasselquist, D. C. Gregory: J. Phys. B **12**, L249 (1979)

2.78   R. A. Falk, G. H. Dunn: Phys. Rev. A **27**, 754 (1983)

2.79   D. H. Crandall, R. A. Phaneuf, D. C. Gregory, A. M. Howald, D. W. Mueller, T. J. Morgan, G. H. Dunn, D. C. Griffin, R. J. W. Henry: Phys. Rev. A **34**, 1757 (1986)

2.80   S. M. Younger: Phys. Rev. A **22**, 111 (1980)

2.81   K. Rinn, D. C. Gregory, L. J. Wang, R. A. Phaneuf, A. Müller: Phys. Rev. A **36**, 595 (1987)

2.82   M. S. Pindzola, D. C. Griffin: Phys. Rev. A **36**, 2628 (1987)

2.83   A. Müller, G. Hofmann, K. Tinschert, E. Salzborn: Phys. Rev. Lett. **61**, 70 (1988)

2.84   G. Hofmann, A. Müller, K. Tinschert, E. Salzborn: Z. Phys. D – Atoms, Molecules and Clusters **16**, 113 (1990)

2.85   S. O. Martin, B. Peart, K. T. Dolder: J. Phys. B **1**, 537 (1968)

2.86   D. H. Crandall, R. A. Phaneuf, R. A. Falk, D. S. Belić, G. H. Dunn: Phys. Rev. A **25**, 143 (1982)

2.87   S. M. Younger: Phys. Rev. A **24**, 1272 (1981)

2.88   D. C. Griffin, C. Bottcher, M. S. Pindzola: Phys. Rev. A **25**, 154 (1982)

2.89   R. J. W. Henry, A. Z. Msezane: Phys. Rev. A **26**, 2545 (1982)

2.90 K. J. LaGattuta, Y. Hahn: Phys. Rev. A **24**, 2273 (1981)

2.91 D. C. Gregory, L.-J. Wang, F. W. Meyer, K. Rinn: Phys. Rev. A **35**, 3256 (1987)

2.92 A. L. Merts, J. B. Mann, W. D. Robb, N. H. Magee, Jr.: Los Alamos Scientific Laboratory Report No. LA-8267-MS, 1980 (unpublished), as reported in [2.90, 91]

2.93 S. S. Tayal, R. J. W. Henry (1987): reported by R. J. W. Henry, A. E. Kingston: Adv. At. Molec. Phys. **25**, 267 (1988)

2.94 D. C. Griffin, M. S. Pindzola, C. Bottcher: Phys. Rev. A **36**, 3642 (1987)

2.95 M. H. Chen, K. J. Reed, D. L. Moores: Phys. Rev. Lett. **64**, 1350 (1990)

2.96 A. Müller, G. Hofmann, K. Tinschert, B. Weißbecker, E. Salzborn: Z. Phys. D **15**, 145 (1990)

2.97 B. Peart, K. T. Dolder: J. Phys. B **8**, 56 (1975)

2.98 R. A. Falk, G. H. Dunn, D. C. Griffin, C. Bottcher, D. C. Gregory, D. H. Crandall, M. S. Pindzola: Phys. Rev. Lett. **47**, 494 (1981)

2.99 D. C. Griffin, C. Bottcher, M. S. Pindzola: Phys. Rev. A **25**, 1374 (1982)

2.100 B. Peart, J. R. A. Underwood, K. Dolder: J. Phys. B **22**, 2789 (1989)

2.101 P. G. Burke, A. E. Kingston, A. Thompson: J. Phys. B **16**, L385 (1983)

2.102 M. S. Pindzola, C. Bottcher, D. C. Griffin: J. Phys. B **20**, 3535 (1987)

2.103 D. C. Griffin, M. S. Pindzola, C. Bottcher: J. Phys. B **17**, 3183 (1984)

2.104 B. Peart, J. R. A. Underwood, K. Dolder: J. Phys. B **22**, 1679 (1989)

2.105 A. M. Howald, D. C. Gregory, F. W. Meyer, R. A. Phaneuf, A. Müller, N. Djurić, G. H. Dunn: Phys. Rev. A **33**, 3779 (1986)

2.106 M. S. Pindzola, D. C. Griffin, C. Bottcher: Phys. Rev. A **33**, 3787 (1986)

2.107 S. S. Tayal, R. J. W. Henry: Phys. Rev. A **33**, 3825 (1986)

2.108 M. S. Pindzola, D. C. Griffin, C. Bottcher: Comments At. Mol. Phys. **20**, 337 (1987)

2.109 S. M. Younger: in Atomic Data for Fusion, Vol. 7, 190 (1981), Controlled Fusion Atomic Data Center Newsletter, Oak Ridge National Laboratory (unpublished)

2.110 R. G. Montague, M. J. Diserens, M. F. A. Harrison: J. Phys. B **17**, 2085 (1984)

2.111 D. W. Mueller, T. J. Morgan, G. H. Dunn, D. C. Gregory, D. H. Crandall: Phys. Rev. A **31**, 2905 (1985)

2.112 D. C. Gregory, F. W. Meyer, A. Müller, P. Defrance: Phys. Rev. A **34**, 3657 (1986)

2.113 R. S. Freund, R. C. Wetzel, R. J. Shul, T. R. Hayes: Phys. Rev. A **41**, 3575 (1990)

2.114 M. S. Pindzola, D. C. Griffin, C. Bottcher, S. M. Younger, H. T. Hunter: Report ORNL/TM-10297, Oak Ridge National Laboratory, 1987

2.115 C. Achenbach, A. Müller, E. Salzborn, R. Becker: Phys. Rev. Lett. **50**, 2070 (1983)

2.116 S. M. Younger: Phys. Rev. Lett. **56**, 2618 (1986)

2.117 S. M. Younger: Phys. Rev. A **35**, 2841 (1987)

2.118 D. R. Hertling, R. K. Feeney, D. W. Hughes, W. E. Sayle: J. Appl. Phys. **53**, 5427 (1982)

2.119 S. M. Younger: Phys. Rev. A **35**, 4567 (1987)

2.120 K. Stephan, H. Helm, T. D. Märk: J. Chem. Phys. **73**, 3763 (1980)

2.121 B. Peart, D. S. Walton, K. T. Dolder: J. Phys. B **4**, 88 (1971)

2.122 P. Defrance, W. Claeys, F. Brouillard: J. Phys. B **15**, 3509 (1982)

2.123 B. Peart, K. T. Dolder: J. Phys. B **2**, 1169 (1969)

2.124 M. H. Mittleman: Phys. Rev. Lett. **16**, 498 (1966)

2.125 G. Hofmann, A. Müller, K. Tinschert, R. Sauer, E. Salzborn: to be published

2.126 A. Müller, R. Frodl: Phys. Rev. Lett. **44**, 29 (1980)

2.127 K. Tinschert, A. Müller, R. A. Phaneuf, G. Hofmann, E. Salzborn: J. Phys. B **22**, 1241 (1989)

2.128 A. Müller, K. Tinschert, C. Achenbach, R. Becker, E. Salzborn: J. Phys. B **18**, 3011 (1985)

2.129 A. Müller, C. Achenbach, E. Salzborn, R. Becker: J. Phys. B **17**, 1427 (1984)

2.130 C. Achenbach, A. Müller, E. Salzborn, R. Becker: Phys. Rev. Lett. **50**, 2070 (1983)

2.131 M. S. Pindzola, D. C. Griffin, C. Bottcher: Phys. Rev. A **27**, 2331 (1983) and J. Phys. B **16**, L355 (1983)

2.132 A. M. Howald, D. C. Gregory, R. A. Phaneuf, D. H. Crandall, M. S. Pindzola: Phys. Rev. Lett. **56**, 1675 (1986)

2.133 A. Müller, K. Tinschert, C. Achenbach, E. Salzborn, R. Becker, M. S. Pindzola: Phys. Rev. Lett. **54**, 414 (1985)

2.134 C. Achenbach, A. Müller, E. Salzborn, R. Becker: J. Phys. B **17**, 1405 (1984)

2.135 D. C. Griffin, C. Bottcher, M. S. Pindzola, S. M. Younger, D. C. Gregory, D.H. Crandall: Phys. Rev. A **29**, 1729 (1984)

2.136 M.S. Pindzola, D. C. Griffin, C. Bottcher, D. H. Crandall, D. C. Gregory: Phys. Rev. A **29**, 1749 (1984)

2.137 A. Müller, K. Tinschert, G. Hofmann, E. Salzborn, G. H. Dunn, S. M. Younger, M. S. Pindzola: Phys. Rev. A **40**, 3584 (1989)

2.138 G. Hofmann, A. Müller, B. Weißbecker, M. Stenke, K. Tinschert, E. Salzborn: Proceedings of the 5th International Conference on the *Physics of Highly-Charged Ions*, Gießen, 10.–14.9.1990. Z. Phys. D – Atoms, Molecules and Clusters, Suppl. Vol. ATOMIC PHYSICS OF HIGHLY-CHARGED IONS, ed. by E. Salzborn, P. H. Mokler and A. Müller (Springer, Berlin Heidelberg New York 1991)

2.139 S. Tayal, R. J. W. Henry: Phys. Rev. A **42**, 1831 (1990)

3. Ion-Neutral Reactions: Collision Spectrometry of Multiply Charged Ions at Low Energies

E. Y. Kamber[1] *and C. L. Cocke*[2]

With 21 Figures

In considering the topic of collisions between atomic and molecular particles at low energies, one must pay particular attention to the processes that involve the transfer of an electron from one of the colliding partners to the other, since frequently this is one of the most probable processes. Over the last thirty years, a great deal of interest has been focused on the physics of atomic collisions in an attempt to understand phenomena in nature and apply this science to practical situations. The investigation of electron capture by multiply charged ions is needed for such diverse purposes as the development of particle detectors, short wavelength gas laser systems and the physics of the upper atmosphere. Electron capture also plays an important role in the behaviour of high temperature plasmas containing impurities and is therefore important to the controlled thermonuclear fusion program. For example, electron capture into excited states of multiply charged ions followed by the radiative decay of these states is a well recognised plasma-cooling mechanism. The experimental physics of atomic collisions owes much to the many and varied contributions of John Hasted throughout his long scientific career. We shall be concerned with electron capture processes involving low energy multiply charged ions, a subject initiated experimentally by the pioneering work of John Hasted and his collaborators [3.1–5].

For low velocity collisions (the relative velocity of the colliding particles is smaller than the velocity of the bound target electrons), a simple physical picture of electron capture can be described in terms of the molecular orbitals of the quasi-molecule formed by the collisional partners. The electron capture is then a transition between electronic states of the transitory molecule formed temporarily during the collision. Therefore, the dominant mechanisms in the single electron capture processes are the diabatic transitions at pseudocrossings of the ingoing potential energy curve with the multitude of outgoing potential energy curves. The predicted cross sections depend critically on the internuclear distances R_x at which the adiabatic potential energy curves have minimum energy separation and on the magnitude of the corresponding dynamic coupling matrix elements.

The basic parameters governing the electron capture are the collision velocity and the ionization energy of the target. Other parameters of the colliding system influence the collision dynamics. These include i) the charge state of the projec-

Springer Series in Chemical Physics, Vol. 54

Deepak Mathur (ed.): Physics of Ion Impact Phenomena

© Springer-Verlag Berlin Heidelberg 1991

tile, which has the role of scaling the velocity regions in which the quasi-molecular and other theoretical methods are valid, and ii) the number of electrons involved in the collision system, which influences the collision dynamics by determining, at low collision energies, the complexity of the interaction paths for different reaction channels.

3.1 Dynamics of Collisions

When multiply charged ions collide with atoms, a variety of reaction channels are possible through which electron capture may take place. These include one- and multiple-electron capture, electron capture accompanied by target excitation or ionization, electron capture accompanied by autoionization of the projectile, etc. These processes may be summarised by the equation

$$A^{q+} + B \rightarrow A^{m+} + B^{n+} + (m+n-q)e + \Delta E ,\tag{3.1}$$

where q and m, n are the initial and final charge states of the projectile and target respectively, and ΔE represents the energy defect of the reaction channel involved. Moreover, the collision products may be formed in excited states. In a general consideration of state-selective electron capture processes, an important parameter is the total change in internal energy that takes place during an inelastic collision. This is also called the energy defect of the collision. The energy defect can either be positive (a decrease in internal energy with corresponding increase in kinetic energy) or negative (an increase in internal energy and a decrease in kinetic energy). The former is called an exoergic reaction and the latter an endoergic one. The energy defect can also be zero, as in the resonant electron capture processes. The study of energy defect analysis is known as translational energy spectrometry.

From the kinematic calculations, based on classical two-body dynamics, the translational energy of an ion undergoing inelastic scattering differs from the energy of the projectile ion E_0 by

$$E_{\text{inelastic}} - E_0 = \Delta E - \Delta K ,\tag{3.2}$$

where ΔK is the translational energy given to the target, and ΔE is the energy defect of the reaction. ΔE is calculated according to the formula

$$\Delta E = I(A^{m+}) - I(B) - E_j ,\tag{3.3}$$

where $I(A^{m+})$ is the ionization energy of the product ion A^{m+}, $I(B)$ the ionization energy of the target atom (assuming the target to be in its ground state and the captured electron to be the most loosely bound) and E_j the excitation energy of the jth level of projectile product ion A^{m+}.

The translational energy ΔK given to the target, is given by [3.6]

$$\Delta K = \frac{m_p}{m_p + M}(1 - \cos\theta)\left(\frac{2ME_0}{m_p + M} - \Delta E\right) + \frac{m_p(\Delta E)^2}{4ME_0}\cos\theta , \qquad (3.4)$$

where m_p and M are, respectively, the projectile and target masses, E_0 is the laboratory translational energy of the projectile, and θ is the scattering angle. For zero scattering angle, this energy reduces to

$$\Delta K = \frac{m_p(\Delta E)^2}{4ME_0} . \qquad (3.5)$$

Experimentally, studies of state-selective single electron capture by multiply charged ions from atomic and molecular targets have generally been carried out at keV energies and zero scattering angle. In this energy range, the translational energy given to the target is very small. Thus the energy defect in an inelastic scattering process is conveniently expressed in terms of the energy difference between the inelastically and elastically scattered ions, that is, (3.2) reduces to

$$E_{\text{inelastic}} - E_0 = \Delta E \qquad (3.6)$$

which can be readily determined from an energy spectrum of the scattered particles. However, at lower impact energies and for heavy projectile ions in collision with light targets, values of ΔK, calculated on the basis of zero and non-zero scattering angles, were found to be large enough to be observed as a shift for a particular reaction channel in the translational energy spectrum, indicating that it is necessary to take account of ΔK in the analysis of energy gain measurements.

Olson and *Kimura* [3.7] have pointed out the importance of angular scattering in collisions between slow multiply charged ions and atoms. They obtained a relation between the center-of-mass scattering angle (Θ_{CM}), the final charge states, center-of-mass collision energy and impact parameter by assuming zero interaction in the entrance channel and Coulomb scattering for the exit channel:

$$\Theta_{\text{CM}} = \frac{\pi}{2} - \cos^{-1}\left(\frac{\alpha}{\sqrt{\alpha^2 + 4}}\right) , \qquad (3.7)$$

where

$$\alpha = \frac{Z_a Z_b e^2}{E_{\text{CM}} b} \qquad (3.8)$$

and Z_a and Z_b are the charges of the product ions, b is the impact parameter and E_{CM} is CM collision energy. They [3.7] obtained a rough estimate of the maximum scattering angle by taking an impact parameter equal to the crossing radius. This rough estimate indicates that the instrumental acceptance angle for product ions following electron capture should be larger than the maximum laboratory scattering angle for the collision system under investigation to ensure 100 % transmission through the post-collision electrostatic analyser. This result pertains specifically to total cross section measurements.

3.2 Theoretical Aspects

In the low velocity region, when the collision velocity is small compared with the velocity of the captured electron, there are basically three different approaches to describe the electron capture process in this region: (i) numerical solution of the coupled channel equations with a molecular basis expansion of the total electron wavefunction; (ii) reduction of the problem to a multichannel model which allows analytical solution; and (iii) construction of decay models of the process. Each of these approaches has its own merits and disadvantages in describing state-selective electron capture in particular physical situations. The general theory and the most frequently used methods in the study of electron capture processes between multiply charged ions and atomic targets are discussed by *Janev* et al. [3.8], and in a number of other review articles [3.9–12]. Within this section we shall restrict ourselves to the main ideas of the theoretical methods, and emphasize mainly those aspects which are directly related to the electron capture processes presented here.

3.2.1 Landau-Zener Model

The earliest theoretical work on the crossing problem was done by *Landau* [3.13] and *Zener* [3.14] who, independently, derived the well-known formula for the transition probability which bears their names. Both derivations, however, were based on the classical picture of the nuclear motion within the impact parameter formulation and, consequently, they ignored the inherent quantal phase interference associated with the coupling classical trajectories within the crossing region. *Stueckelberg* [3.15] described the motion of the nuclei semi-classically by an expansion in powers of Planck's constant and, using the extension of the JWKB method, he derived a set of connection formulae which describe the effects of inelastic transitions. In this manner, Stueckelberg obtained a formula for the transition probability which included phase interference effects and which reduced in the appropriate limit to the Landau-Zener formula. Consider the transfer of an electron from state (p) of the atomic system (A) to state (q) of an atomic system (B) represented by [3.16]

$$(A + e)_p + B \rightarrow A + (B + e)_q .\tag{3.9}$$

In the absence of interaction, the left- and right-hand members of this equation can be imagined to form two quasi-molecules. Figure 3.1 shows the associated potential energy curves in the absence of interaction (the curves indicated by dashed lines) between the two systems, where the curves cross at an internuclear distance $R = R_x$. If there is an interaction, these curves usually do not cross provided that the interaction matrix element does not vanish. In the region of the pseudo-crossing point (R_x), the difference in the two potential energies is small, and a transition is possible, that is, electron capture may occur. If the number of outgoing reaction channels of the collision system interacting with the incoming reaction channel is not excessively large, and if the corresponding pseudo-crossing regions are mutually well separated, the electron capture process can be treated by the multichannel Landau-Zener (MCLZ)

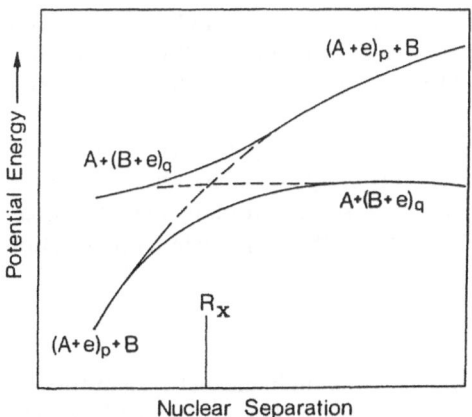

Fig. 3.1. Associated potential energy curves between the two systems of (3.9)

A+(B+e)$_q$

(A+e)$_p$+B

A+(B+e)$_q$

(A+e)$_p$+B

R$_x$

Potential Energy ⟶

Nuclear Separation

theory [3.17] by treating each crossing separately, making the analysis a two-state problem.

According to the Landau-Zener model [3.8], the single-crossing transition probability at each potential energy crossing R_x is given by

$$P_n = \exp\left[\frac{-2\pi H_{12}^2}{v_r \Delta F}\right] , \qquad (3.10)$$

where, in the straight-line trajectory approximation, the radial velocity at the crossing distance R_x is

$$v_r = v_0 \left[1 - \left(\frac{b}{R_x}\right)^2\right]^{\frac{1}{2}} , \qquad (3.11)$$

where v_0 is the relative velocity and b is the impact parameter. The quantity H_{12} in (3.10) is the matrix element and is approximately equal to half the splitting of the adiabatic curves at the curve-crossing, and ΔF is the difference in slopes of the diabatic potential curves at the curve-crossing, assuming a repulsive Coulombic interaction (U_e) in the outgoing channel and zero interaction (U_i) in the incoming channel (neglecting the small attractive polarisation force between the projectile and the target). Then it is possible to represent ΔF by

$$\Delta F = \frac{d}{dR}[U_e(R) - U_i(R)] = \frac{(q-1)}{R^2} . \qquad (3.12)$$

Olson and *Salop* [3.18] have developed an analytical formula for evaluating the reduced coupling matrix element H_{12} for application to the collision of fully stripped ions with hydrogen atoms, using the method discussed by *Bates* and *Moiseiwitch* [3.19]. Empirically, *Olson* and *Salop* [3.18] determined a functional expression for H_{12}, given by

$$H_{12}^{os} = 9.13q^{-\frac{1}{2}} \exp[-1.324\alpha R_x q^{-\frac{1}{2}}] , \qquad (3.13)$$

where $\alpha = (2I_t)^{\frac{1}{2}}$ and I_t is the ionization energy of the target atom (in atomic units). A reduction of H_{12} by 40 % has been proposed by *Kimura* et al. [3.20] to fit their experimental results, and *Taulbjerg* [3.21] has suggested a correction term to H_{12} accounting for projectiles that are only partially stripped.

Clearly, p is the probability that in traversing the crossing the system remains on the same potential curve, and $(1 - p)$ is the probability that a jump from one curve to another occurs, since the system actually traverses the crossing point R_x twice, on the incoming and outgoing passages. The single-crossing probability for electron capture at a given impact parameter is given by

$$P = 2p(1 - p) . \tag{3.14}$$

The extension of the Landau-Zener (LZ) model to a multichannel system (that is, several possible exit channels) has been carried out by *Salop* and *Olson* [3.22]. The probability P_n $(n = 1, 2, \ldots, N)$ for capture into the nth final state, assuming that there is no interference between different paths leading to a particular exit channel, is

$$P_n = p_1 p_2 \ldots p_n (1 - p_n)[1 + (p_{n+1} p_{n+2} \ldots p_N)^2$$
$$+ (p_{n+1} p_{n+2} \ldots p_{N-1})^2 (1 - p_N)^2 + (p_{n+1} p_{N-2})^2$$
$$+ (1 - p_{N-1})^2 + \ldots + p_{n+1}^2 (1 - p_{n+2})^2 + (1 - p_{n+1})^2] . \tag{3.15}$$

The cross section σ_n for capture into a particular state n is given by

$$\sigma_n = 2\pi \int_0^{R_x} P_n b \, db . \tag{3.16}$$

The predictive ability of the MCLZ model could be improved by calculating the coupling matrix H_{12} for individual reaction channels of a specific collision system. For the case of partially stripped projectiles, for which the subshell degeneracy is removed, we may use the *Taulbjerg* [3.21] expression for H_{12} to describe electron capture by partially stripped ions, which is given by

$$H_{12} = f_{nl} H_{12}^{os} , \tag{3.17}$$

where

$$f_{nl} = \frac{(-1)^{n+l-1}(2l + 1)^{\frac{1}{2}} \Gamma(n)}{[\Gamma(n + l + 1)\Gamma(n - 1)]^{\frac{1}{2}}} , \tag{3.18}$$

where n and l characterise the final orbital quantum numbers of the captured electron.

It should be noted that the validity of the MCLZ model is restricted to the adiabatic energy region. Furthermore, the MCLZ model presented here completely neglects the electron transition caused by the rotation of the internuclear axis. This may lead to an underestimation of the cross section in the low energy region by a factor of two, or even more. *Janev* et al. [3.23] have proposed a modification of the MCLZ model to include rotational mixing for the total transition probability for the collision of fully stripped ions with hydrogen atoms.

3.2.2 The Classical Over-Barrier Transition Model

The classical over-barrier transition model (CBM) [3.24, 25], originally proposed by *Bohr* and *Lindhard* [3.26], has been widely used for a qualitative explanation of the main features of the electron capture process. In this model capture is assumed to take place at the crossing of the diabatic potential curves at large internuclear distances provided that electron capture simultaneously becomes classically allowed; that is, the electron can overcome the barrier between the potential wells generated by the ionic charges of the target and the projectile. The success of this simple model stems in part from the fact that the resonant electron capture populates preferentially states with large quantum numbers n and l, yielding large cross sections for capture into the principal shell n.

According to this model, the transition of the active electron from the target to the projectile becomes classically allowed for internuclear distances smaller than R_{cbm}, given by

$$R_{cbm} \leq \frac{(2\sqrt{q}+1)}{I_t} , \tag{3.19}$$

with R_{cbm} being in atomic units, q the charge state of the projectile and I_t the ionization energy of the target in atomic units, assuming that the charge state of the projectile (q) is equal to the effective charge.

The effective principal quantum number n_{cbm} of the projectile having a diabatic crossing near R_{cbm} is given by

$$n_{cbm} = q \left(2I_t \left[\frac{1+(q-1)}{(2\sqrt{q}+1)} \right] \right)^{-\frac{1}{2}} . \tag{3.20}$$

The model predicts that the exit channel with the largest possible principal quantum number n satisfying the condition $n \leq n_{cbm}$ will be predominantly populated. This is equivalent to taking the exit channel with the largest possible crossing radius R_x satisfying $R_x \leq R_{cbm}$, where $R_x(au) = ((q-1)27.2/\Delta E(eV))$, neglecting polarization.

In addition, the CBM model has been used to predict the magnitude of the electron capture cross section

$$\sigma_t = A\pi R_x^2 . \tag{3.21}$$

The value of A is somewhat arbitrary, both $A = 1$ and $A = 0.5$ have been used by various authors. $A = 1$ corresponds to unit capture probability for all impact parameters $b \leq R_x$, which is analogous to the so-called absorbing-sphere model of *Olson* and *Salop* [3.18]. $A = 0.5$ corresponds to the maximum possible capture probability in the case of a pure two-state system, if the coupling is localised at the crossing and if interference effects are neglected. The CBM model is static in the sense that the collision velocity does not enter as a system parameter, and thus predicts a constant cross section in the velocity range where the model is valid, that is, for $0.1 < v < 0.6$ au.

Recently *Hvelplund* et al. [3.27] used an extended version of the CBM model for multiple-electron capture to calculate the energy gains (Q_n) and the capture radii (R_n) for collisions of multiply charged argon ions with Ne, Ar and Xe. This extended model treats a multiple-electron capture process as a consecutive progression of single electron captures. The energy gain and capture radii are given by

$$Q_n = \frac{[q - (2n - 1)]}{R_n} + Q_{n-1}, \quad \text{where} \quad n = 2, 3, \dots, \quad (3.22)$$

$$R_n = \left(\frac{1}{I_t}\right) [2(q - n + 1)^{\frac{1}{2}} n^{\frac{1}{2}} + n], \quad \text{where} \quad n = 1, 2, \dots. \quad (3.23)$$

Here n represents the number of electrons captured.

3.2.3 Reaction Window

In exoergic electron capture processes of low energy multiply charged ions with atoms or molecules, a wide range of final energy states become accessible and present a broad sampling of energy defects and avoided curve-crossing distances. Such curve-crossings appear frequently in these processes because the interaction of the entrance channel is dominated by a relatively weak polarization attraction at large internuclear separations, while the interaction in the exit channels is dominated by Coulombic repulsion. The important avoided curve-crossings between the adiabatic potential energy curves associated with the entrance channel and various exit channels are those which occur at moderate internuclear separations, depending on the initial projectile charge state. This is well understood when considering the gradual decrease of quasimolecular state-coupling with internuclear separation. This intermediate range of separations is commonly referred to as the reaction window for the electron capture process [3.20, 21, 28]. The position of the reaction window for any collision system depends mainly on the collision energy of the projectile. When the collision energy is reduced the adiabaticity at inner crossings will become increasingly pronounced, while the transition probability at a distant crossing becomes larger. Therefore, the reaction window shifts towards larger internuclear separations if the collision energy is reduced and vice versa.

The multichannel Landau-Zener model and the classical over-barrier transition model have been used by several authors [3.20, 29–34] to interpret the final-state populations by calculating the location of the crossing radii where the probability of electron capture is large, in terms of the energy gain values of the final states.

3.3 Experimental Techniques

The study of low-energy state-selective electron capture processes has shown spectacular growth in recent years. This interest has been stimulated by instrumental developments including high-energy resolution spectrometers, position-sensitive detectors based on multichannel plate electron multipliers, recoil ion sources, electron beam ion sources, electron cyclotron resonance ion sources, and other experimental tools. In addition to translational energy spectrometry, different experimental methods may be applied in order to identify the individual states of the products of electron capture processes.

a) Photon measurements determine specific atomic states with high resolution. However, ground and metastable states are not detectable. Furthermore, large primary ion currents are necessary, because only a small part of the total space angle is covered by the detection unit. Although the energy resolution is good, the interpretation of the observed results is not simple. Accurate wavelengths of transitions and their probabilities (branching ratios) for electron captured ions have to be known. This technique has been applied by different authors [3.35, 36].

b) By measuring the kinetic energy of the ejected electrons, autoionizing states of the atomic particles as well as states of the quasimolecule formed during the collision may be investigated. The process of transfer ionization (double-electron capture into the autoionizing states) has been studied by this method [3.37–39].

c) In the translational energy spectrometry technique [3.40–50], an ion beam is scattered from a gas or crossed beam target and is detected after travelling over a collimated path at a certain polar scattering angle. The kinetic energies before and after collision are measured, so that the entrance and exit channels of the electron capture process can be inferred. The abundance of the individual peaks in the observed energy spectra can be measured so as to obtain either the total cross section or the differential cross section. The length of the post collision collimation must ensure that an angular resolution of better than $0.3°$ is obtained for impact energies in the range of ≤ 1 keV. A range of scattering angles between $0°$ and $10°$ is covered either by traversing the detecting system, together with its energy analyzer, in angle or by electrical deflection. By means of conventional electrostatic momentum analysis, an energy resolution of 1 % can be obtained. The advantages of this technique at low impact energy are high detection efficiencies and straightforward interpretation of experimental data. In order to ensure unambiguous assignment of the observed reaction channels, the absolute scale of the energy in the translational energy spectra must be obtained. Measurements of the translational energy gain/loss are usually calibrated with respect to the elastic channel, which is the measured source of experimental uncertainty. Errors in the translational energy scale, as calibrated against the unscattered ions and due to stray fields or surface charges, should always be less than ± 0.5 eV. A number of excellent reviews of the experimental techniques and results related to the study of electron capture processes are now available [3.51–55].

3.4 Discussion of Translational Energy Spectrometry

3.4.1 State-Selective Single Electron Capture Processes by Ground and Metastable Doubly Charged Ions

Ion beams from sources usually contain a mixture of ground state and metastable excited states. It is very important to know the fractional populations of the long-lived excited states in an ion beam, at what rates each one of these is destroyed, and how effective each one is in transferring its charge to a target in order to obtain detailed knowledge of the collision phenomenon. When an ion beam consists of two components, the populations of the states and the scattering cross section for each state are, in principle, possible to determine by using an attenuation method if the cross sections are appreciably different from each other. However, if the cross sections are almost identical, the attenuation method becomes useless. If the ion beam consists of more than two components, the analysis of the attenuation curve will be difficult.

Recently, different techniques have been used for determining the fractional population of metastable and highly-excited states of doubly charged ion beams. *Varga* and *Winter* [3.56] have determined the metastable fraction of singly and doubly charged noble-gas ion beams by investigation of the secondary electron yield resulting from the impact of these ion beams on a clean polycrystalline tungsten surface. *Matsumoto* et al. [3.57], *Brazuk* and *Winter* [3.58], and *Huber* and *Kahlert* [3.59] have used state-selective optical and translational attenuation methods, respectively, for detecting low-excited metastable states. On the other hand, *Nakamura* et al. [3.60, 61] have succeeded in determining the fractional populations of low-lying metastable states in Ar^{2+} and Kr^{2+} ion beams by measuring the energy changing spectra for excitation and de-excitation processes among these states. Very recently *Kamber* et al. [3.62] have reported the detection of metastable and long-lived highly excited Ar^{2+} ions by means of state-selective charge-stripping processes in collisions of Ar^{2+} ions with He and Ar by using the translational energy spectrometry technique.

The process of single electron capture by doubly charged ions shows less complication, because the two collision products have equal positive charge and the interaction between them is Coulombically repulsive, which ensures that there is a suitable avoided crossing even though the energy defect is large. Translational energy spectrometry has been extensively used to study state-selective electron capture by doubly charged ions in collision with rare-gas atoms. Figure 3.2 shows the translational energy gain spectrum for the Ar^{2+}-He collision system at 6 keV impact energy [3.63]. Three peaks are clearly resolved and seen, due to the following reaction channels:

$$Ar^{2+}(3p^4\ ^3P) + He\ \rightarrow\ Ar^+(3p^5\ ^2P) + He^+ + 3.04\,eV \qquad I\alpha X$$

$$Ar^{2+}(3p^4\ ^1D) + He\ \rightarrow\ Ar^+(3p^5\ ^2P) + He^+ + 4.78\,eV \qquad II\alpha X$$

$$Ar^{2+}(3p^4\ ^1S) + He\ \rightarrow\ Ar^+(3p^5\ ^2P) + He^+ + 7.16\,eV \qquad III\alpha X$$

The energy defect, ΔE, corresponding to each of the observed reaction channels is represented by a vertical line in the translational energy spectrum. The observed reaction channels are labelled according to Hasted's notation [3.64]: Roman numerals I, II and III represent the ground state and metastable states of the incident ion beam, respectively; $\alpha, \beta, \gamma, \dots$ represent the ground and successively higher excited states of the projectile product, respectively; X represents the ground state of the product target.

In the above collision system $II\alpha X$ is the dominant reaction channel. It corresponds to the capture from metastable state (^1D) of Ar^{2+} into the ground state (^2P)

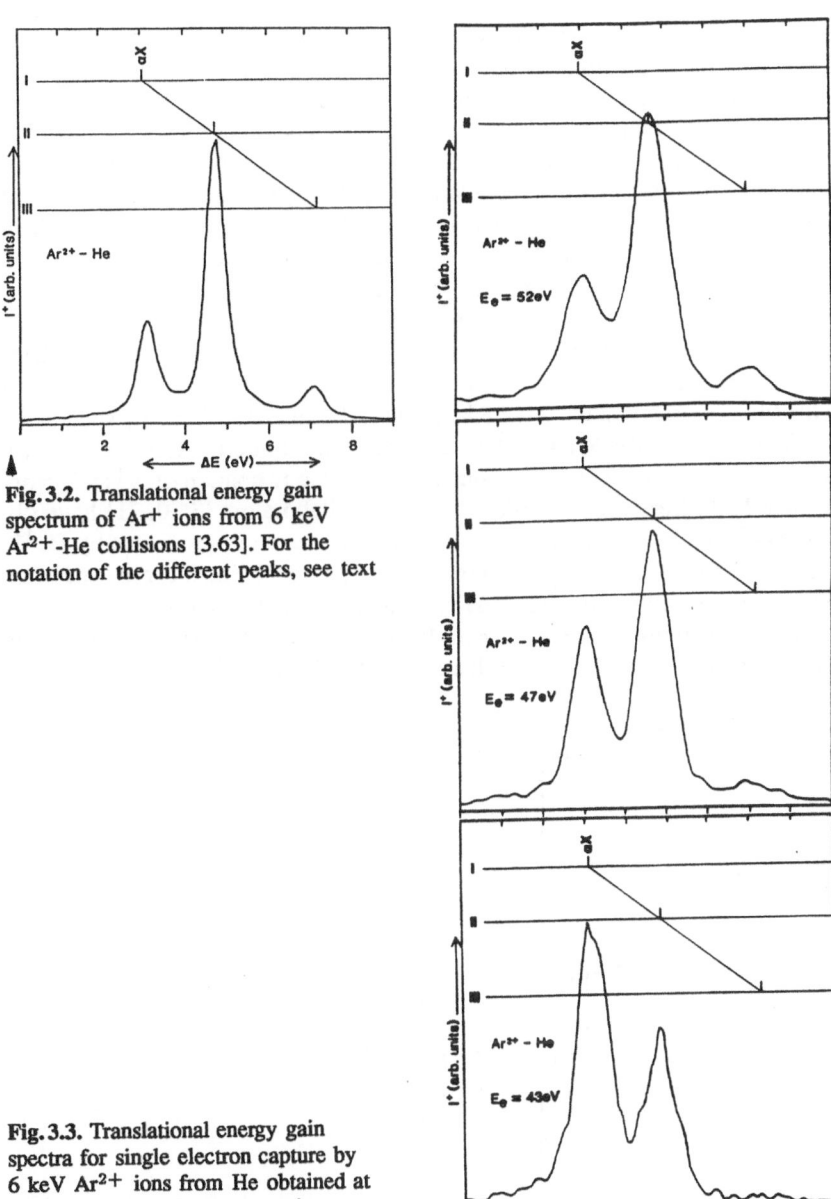

Fig. 3.2. Translational energy gain spectrum of Ar$^+$ ions from 6 keV Ar^{2+}-He collisions [3.63]. For the notation of the different peaks, see text

Fig. 3.3. Translational energy gain spectra for single electron capture by 6 keV Ar^{2+} ions from He obtained at various ionizing electron energies

of Ar$^+$. Also, there is a significant probability of forming product Ar$^+$(^{2}P) ions by reaction $I\alpha X$; to lesser extent reaction $III\alpha X$ also occurs. This is in accordance with the previous measurements [3.43, 45, 50]. The fractional populations of low-lying metastable states in Ar^{2+} ion beam have been varied by decreasing the ionizing electron energy E_e in the electron impact ion source. This results in a gradual degradation of the peaks due primarily to reaction channel $III\alpha X$ and also to $II\alpha X$. Translational energy spectra for the same collision system, taken at lower energy resolution and at several electron impact energies, are given in Fig. 3.3. Meanwhile, because the electronic transition probability depends sensitively on the internuclear separation of the collision partners, the long-lived highly excited states $3p^3$ (^4S) 3d and $3s\,3p^5$, with excitation energies between 18 and 24 eV, play an important role in Ar^{2+}-Ar collisions [3.63, 65, 66]. This observation has been clarified by reduction of E_e to isolate processes involving only the ground state ion beam.

Gilbody and his collaborators [3.67–69] have successfully demonstrated the feasibility of using translational energy spectrometry to identify the main product channels for electron capture processes by doubly charged ions from atomic hydrogen within the impact energy range 0.6–18 keV. Typical translational energy spectra for single electron capture by 2 and 6 keV N^{2+} ions in atomic hydrogen are shown in Fig. 3.4. The dominant reaction channel involves the formation of $N^+(^3D)$ from ground state N^{2+}. Very recently Winter and his collaborators [3.70, 71] have obtained translational energy spectra for single electron capture processes by He^{2+}, Ne^{2+} and Ar^{2+} ions from Li (2s). Figure 3.5 shows a translational energy spectrum for single electron capture by 10 keV He^{2+} ions on Li (2s). The dominant reaction channel is due to capture into He^+ $(n = 3)$ states. Besides the main peak, capture into $n = 2$, $n = 4$ and $n = 5$ of He^+ can be clearly identified. The peak at $\Delta E = -1.19$ eV

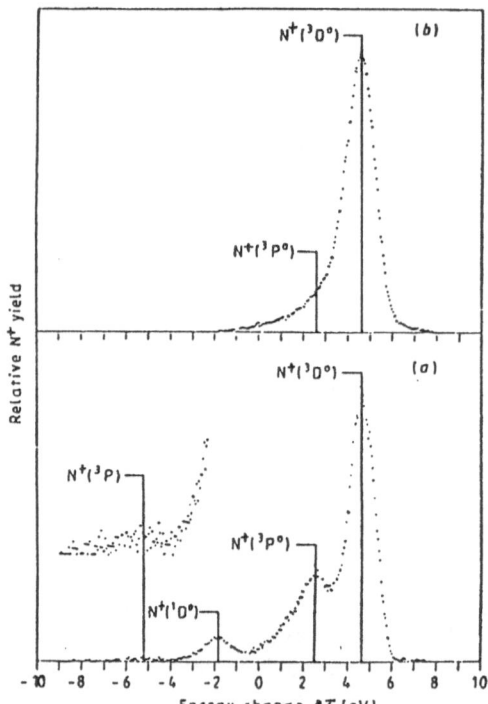

Fig. 3.4a,b. Translational energy spectra for single electron capture by (a) 6 keV and (b) 2 keV N^{2+} ions from atomic hydrogen [3.69].

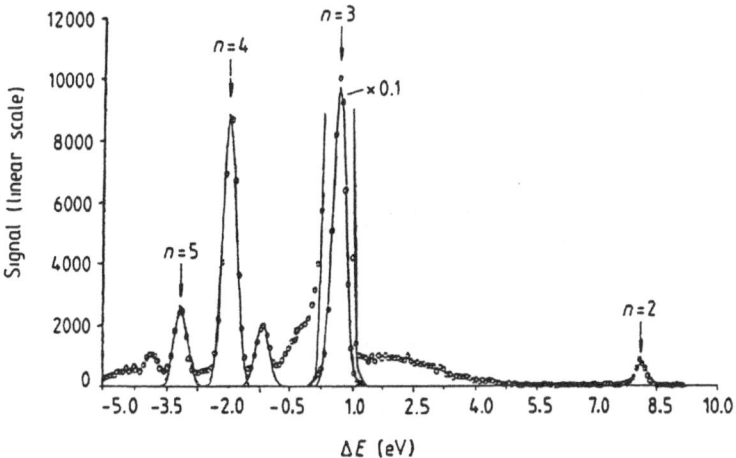

Fig. 3.5. Translational energy spectrum for single electron capture by 10 keV ^{4}He^{2+} from Li [3.71]

is due to double collisions involving Li(2s $\rightarrow$ 2p) excitation followed by capture into the He$^+$ ($n = 3$) state.

For very low collision energy ($\leq$ 200 eV), we have used a recoil ion source to investigate electron capture processes for several collision systems involving doubly charged ions [3.72, 73]. The apparatus which we have used for translational energy spectrometry is shown in Fig. 3.6 [3.73]. The ions were produced in a recoil ion source, pumped by a fast beam from the KSU tandem Van de Graaff accelerator, extracted perpendicular to the pump beam and mass analyzed by a 180° double-focusing magnet. After momentum analysis, the beam passed through a gas cell 3 mm long and with 1- and 2-mm diameter entrance and exit apertures respectively. The scattered ions that had undergone capture were energy analyzed by means of a 90° double focusing electrostatic analyzer (ESA). The ions were then detected by a one-dimensional position-sensitive channel plate detector located at the focal plane of the ESA. The scattering angle is selected by means of an aperature A1 (1 mm diameter) in front of the ESA.

Translational energy spectra for single electron capture by O^{2+} and N^{2+} ions colliding with He were carried out for impact energies in the range 60–200 eV and scattering angles between 0° and 4°. Figure 3.7 shows the translational energy-gain spectra of the product O$^+$ ions in the O^{2+}-He collision system at an impact energy of 80 eV and scattering angles of 0° and 2°. At 0° scattering angle, one peak is clearly seen; this peak arises from single electron capture into the 2p^3 ^{2}P excited state of O$^+$ from the O^{2+} (2p^2 ^{3}P) ground state incident beam. This process is exoergic by 5.21 eV, with an avoided-crossing at $R_x = 5.22$ au. For capture into the O$^+$ (2p^3 ^{4}S) ground state and O$^+$ (2p^3 ^{2}D) excited state, the avoided-crossings are located, respectively, at 2.57 and 3.75 au. The scattering angles, estimated on the basis of Coulomb potentials (neglecting the polarization) for impact parameters equal to these avoided-crossings are 4.43° and 2.95°, respectively, substantially outside the angular acceptance (1.6°) of the ESA, when the apparatus is set at 0°. Thus, one does

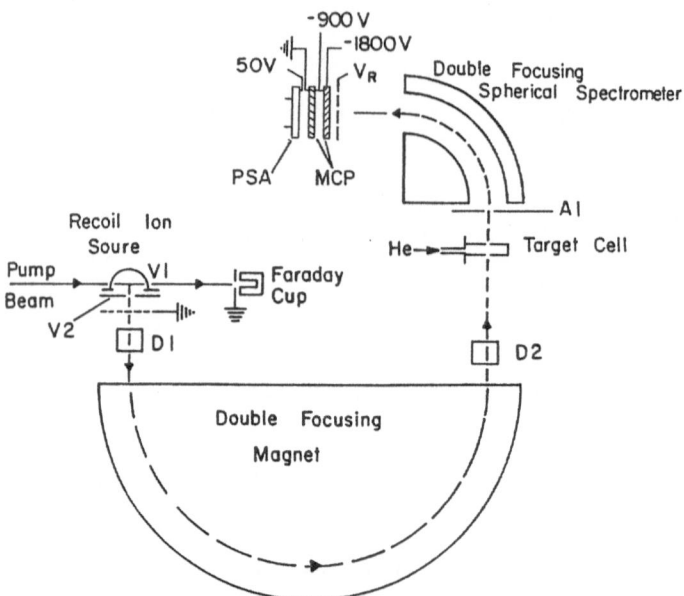

Fig. 3.6. Schematic diagram of the doubly differential energy gain spectrometer used by *Kamber* et al. [3.73] to study state-selective electron capture

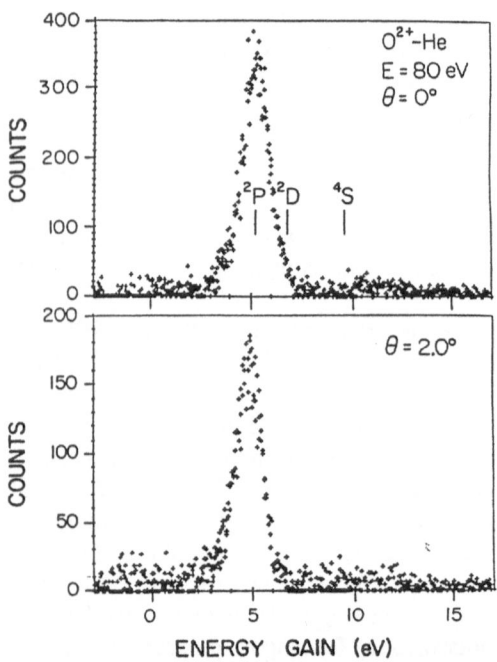

Fig. 3.7. Translational energy gain spectrum for single electron capture by 80 eV O^{2+} ions from He at scattering angles of 0° and 2° [3.73]

not expect these channels to give appreciable contributions at this impact energy. At a scattering angle of $2°$, where one might expect contributions from avoided-crossings at smaller radii to became important, the dominant reaction channel remains the same. Thus no evidence for capture from the ground state of the O^{2+} ion into the ^4S and ^2D states of O^+ was observed at any angle. However, a small shift in the energy of the ^2P is observed, which was attributed to the translational energy given to the target. For zero scattering angle, and at an impact energy of 80 eV, the translational energy given to the target $\Delta K = 0.33$ eV, while at a scattering angle of $2°$, $\Delta K = 0.69$ eV [(3.4)]. This difference of 0.36 eV was found to be enough to be observed as a shift in the energy spectrum. For this collision system, good agreement is obtained with the calculations of *Bienstock* et al. [3.74], which predicted that in our impact energy range, capture preferentially populates the O^+ $(2p^3\ ^2P)$ state.

3.4.2 State-Selective Single Electron Capture by Multiply Charged Ions

The important role played by multiply charged ions in a very high temperature medium is connected with some specific features of their structure and of collisions with atomic targets. Perhaps the main topic of interest in the study of electron capture by multiply charged ions is the distribution of captured electrons over the final state quantum numbers (n, l, m). Both the absolute values of the partial cross sections as functions of the collision energy and the shapes of the n- and l-distributions of captured electrons at a fixed energy, present a sensitive test of the theoretical calculations for the process. Electron capture into excited states is one of the most prominent features of the electron capture process by multiply charged ions from atomic targets. This feature results from the fact that the quasi-resonant energy condition is always fulfilled when the charge state of the projectile is $q \gg 1$.

Cocke and his collaborators [3.47, 75, 76] have carried out a systematic study of state-selective electron capture in slow Ne^{q+} ions on He collisions for q between 3 and 10 (where q is the projectile charge state). The results indicate that the energy gain of the capture reaction is the most important parameter in determination of the levels which are most strongly populated in the process. If curve crossings occur, for which the strength of the potential coupling between incident and final channels is optimised, those states will be selectively populated. For cases where configurations based on the projectile ground-state core have crossings near the favoured radius ($Ne^{4+,5+,7+,8+,10+}$), population of those configurations is selected. For cases where crossings with favoured configurations lie sufficiently far from the favoured radius ($Ne^{3+,6+}$), population of core excited configurations become competitive. Figure 3.8 shows the translational energy gain spectra of single electron capture for Ne^{q+} ($q = 3–6$) on He. In Ne^{3+}-He collisions, the main population is of the core-excited $(2s\ 2p^5)^1P$ state with $\Delta E = 3.62$ eV, while in Ne^{4+}-He collisions one peak is clearly seen; this peak is due to the population of the unresolved $^{2,4}P$ states formed by coupling a single 3s electron to the ^3P projectile ground-state core.

The Nagoya group have systematically studied the final state distributions for single electron capture processes in slow collisions of highly stripped C, N, O, F, S, Kr and I ions with He [3.44, 77–81]. They have found that the electron is selectively

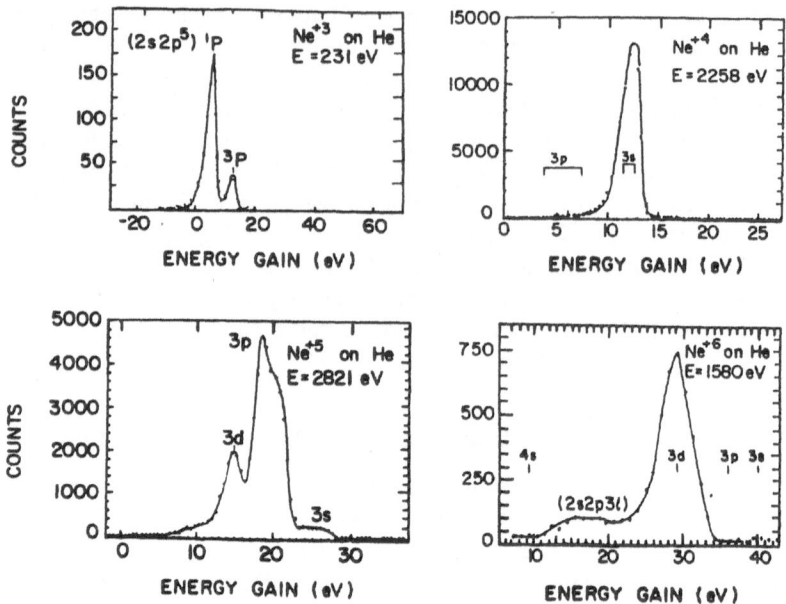

Fig. 3.8. Translational energy gain spectra for single electron capture by Ne^{q+} ions ($q = 3$–6) from He [3.76]

captured into a particular quantum number n of the product ion, and they also found that there is good similarity among the translational energy spectra for projectile ions having the same charge state. In Fig. 3.9, we show the translational energy spectra for X^{6+} + He collisions (where X = C, N, O and F). In each spectrum only a single peak is observed at the energy gain of around 30 eV, which corresponds to single electron capture into the $n = 3$ level of the product X^{5+} ions. This similarity is considered to be due to the similarity of the potential energy curves for the X^{6+}-He collision systems, since these curves do not depend on the number of core electrons of incident ions.

Kamber et al. [3.33, 63, 82–84] have measured the translational energy spectra for single electron capture processes for collisions of Ar^{q+} ions ($q = 2$–5), Kr^{q+} ions ($q = 2$–5) and Xe^{q+} ions ($q = 2$–4) with rare-gas atoms at collision energies of $3q$ keV, where q is the projectile charge state. Figure 3.10 shows the translational energy spectra for Kr^{5+}-He, Ne and Ar collisions at keV impact energy. The possible exit channels, following single electron capture, are listed in Table 3.1. In the Kr^{5+}-He collision system, the dominant reaction channel is $I\theta X$ due to capture into the $4s4p^3$ ^3S state of Kr^{4+}, with an avoided-crossing at $R_x = 6.88$ au. There are also contributions from the $I\iota X$ and $I\kappa X$ channels. The peaks at about 20 eV are due to capture into the $4p^3$ ^1D, ^3P and ^3D states via reaction channels $I\eta X$, $I\zeta X$ and $I\epsilon X$ respectively. In the Kr^{5+}-Ne collision system, capture into 4p4d states dominates the electron capture process via reaction channels $I\kappa X$, $I\lambda X$ and $I\mu X$, with significant contributions from capture into 5s ^3P and $4p^3$ ^3S states, while the Kr^{5+}-Ar collision

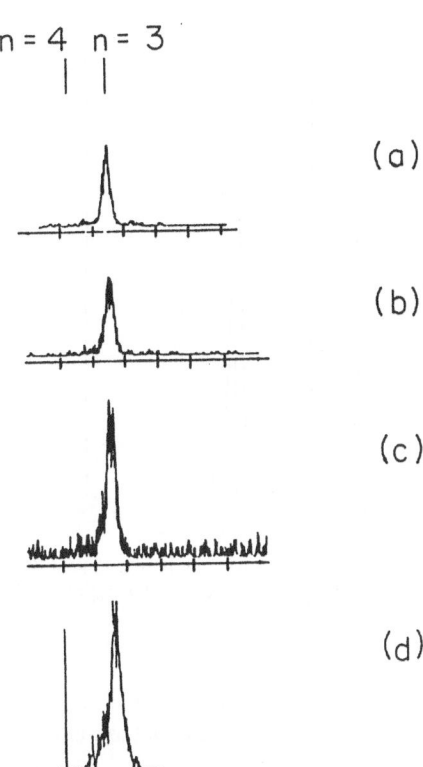

n = 4 n = 3

(a)

(b)

(c)

(d)

0 20 40 60

Energy Gain (eV)

Fig. 3.9. Translational energy spectra for single electron capture by X^{6+} ions from He (X = C (a), N (b), O (c) and F (d)) [3.77–81]

appears to be dominated by capture into the 5p ^{3}D state. In the single-electron capture for the Kr^{5+}-He, Ne and Ar collision systems, measurements show that as the outermost shell number of the target atom increases, that is, as its ionization energy decreases, capture into higher excited states of the Kr^{4+} ions becomes more important.

State-selective cross sections for single electron capture by Kr^{5+} ions from He, Ne and Ar have been calculated using a multichannel Landau-Zener (MCLZ) model. The calculated MCLZ cross sections for the observed reaction channels are compared with the measured energy gain spectra in Fig. 3.10. The value of the largest calculated cross section has been normalized to the height of the dominant peak observed in the spectrum. For the Kr^{5+}-He, Ne and Ar collision systems, the MCLZ calculations correctly predict the positions of the dominant reaction channels $I\theta X$, $I\lambda X$ and $I\rho X$ respectively. However the theoretical calculations underestimate the contributions of capture into the $4p^3$ ^{4}S, ^{3}D and ^{3}P states of Kr^{4+} in the Kr^{5+}-He and Ne collisions. On the other hand, in the case of the collision of Kr^{5+} with Ar,

107

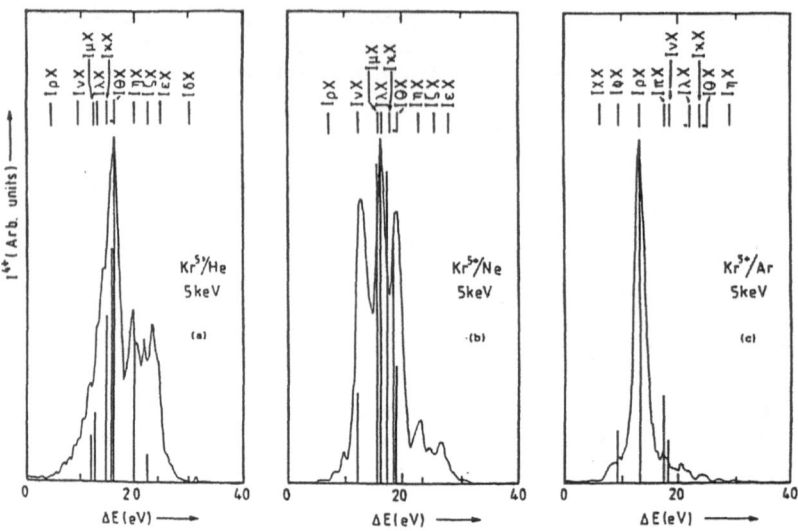

Fig. 3.10 a–c. Translational energy gain spectrum of Kr^{4+} obtained respectively from 5 keV Kr^{5+}-He (a), Ne (b) and Ar (c) collisions, together with the MCLZ calculations for each reaction channel involved [3.84]

the theory overestimates the contributions of the reaction channels $I\nu X$, $I\pi X$ and $I\phi X$. The calculated MCLZ cross sections for each reaction channel are tabulated in Table 3.1.

Giese et al. [3.31] have used translational energy-gain spectrometry to determine the final-state populations following single electron capture by Ar^{q+} ($q = 4$–8) and Ne^{q+} ($q = 4$–7) projectiles in D, D_2 and Ar at an energy of $545q$ eV (where q is the projectile charge state). The apparatus which they used is shown in Fig. 3.11. The ions are produced from a recoil ion source pumped by 20–25 MeV F^{q+}. The recoil ions are extracted with an energy of 35 to $60q$ eV. After momentum analysis, they enter a resistively heated atomic hydrogen oven which also serves as a normal gas cell for other targets. The collision energy was set by applying a negative potential V_0 to the entire oven assembly. After passing through the grid S_2, the ions were retarded by applying a voltage V_R to the entrance slit of a hemispherical double-focusing electrostatic spectrometer. Thereafter, the ions pass into the analyzer for charge and energy analysis and are then detected by a channel electron multiplier. The atomic deuterium target is obtained by heating the oven to a temperature between 2000 K and 2100 K, for which a dissociation fraction of approximately 0.8 is obtained. The translational energy scale calibration was performed by using the collision system Ar^{q+}-Ar as a standard. In Fig. 3.12 is shown the translational energy gain spectra for the Ar^{6+}-Ar, D_2 and D single electron capture collision systems at 3270 eV impact energy. In the Ar^{6+}-Ar collision, the two largest peaks are due to capture into 5s and 4f states of Ar^{5+}, with contributions coming from capture into 4d and 5p states. In the Ar^{6+}-D_2 collision, capture into 4d, 4f and 5s states dominates the electron capture process. In the Ar^{6+}-Ar and D_2 collision systems, the broad peaks,

Table 3.1. The single electron capture reaction channels for Kr^{5+}/He, Ne and Ar collisions, together with MCLZ calculations of cross sections

Reactants and initial states	Products and final states	ΔE [eV]	Designation of reaction process	Calculated σ(MCLZ) [cm²] × 10⁻¹⁶	Calculated $f_{n,l}$
$Kr^{5+}+(4p\,^2p)$ $+He(1s^2\,^1S)$	$Kr^{4+}(4p^2\,^3p)+He^+(1s\,^2S)$	38.82	$I\alpha X$	negligible	0.671
	$Kr^{4+}(4p^2\,^1D)+He^+$	36.1	$I\beta X$	negligible	0.671
	$Kr^{4+}(4p^2\,^1S)+He^+$	33.18	$I\gamma X$	negligible	0.671
	$Kr^{4+}(4s\,4p^3(^4S)5s)+He^+$	29.82	$I\delta X$	negligible	0.671
	$Kr^{4+}(4s\,4p^3(^3D))+He^+$	24.7	$I\varepsilon X$	0.134	0.671
	$Kr^{4+}(4s\,4p^3\,^3p)+He^+$	22.25	$I\zeta X$	1.38	0.671
	$Kr^{4+}(4s\,4p^3\,^1D)+He^+$	19.56	$I\eta X$	7.97	0.671
	$Kr^{4+}(4s\,4p^3\,^3S)+He^+$	15.81	$I\theta X$	15.98	0.671
	$Kr^{4+}(4p\,4d\,^3F)+He^+$	15.47	$I\iota X$	10.74	-0.5
	$Kr^{4+}(4p\,4d\,^3D_1)+He^+$	14.52	$I\kappa X$	7.7	-0.5
	$Kr^{4+}(4p\,4d\,^3P)+He^+$	12.84	$I\lambda X$	3.17	-0.5
	$Kr^{4+}(4p\,4d\,^3D_3)+He^+$	12.15	$I\mu X$	2.06	-0.5
	$Kr^{4+}(4p\,5s\,^3P_0)+He^+$	9.34	$I\nu X$	0.06	0.447
	$Kr^{4+}(4p\,5p\,^3D)+He^+$	4.06	$I\rho X$	negligible	-0.632
$Kr^{5}+(4p\,^2p)$ $+Ne(2p^6\,^1S)$	$Kr^{4+}(4s\,4p^3\,^3D)+Ne^+(2p^5\,^2p)$	27.72	$I\varepsilon X$	negligible	0.6708
	$Kr^{4+}(4s\,4p^3\,^3P)+Ne^+$	25.27	$I\zeta X$	0.004	0.6708
	$Kr^{4+}(4s\,4p^3\,^1D)+Ne^+$	22.58	$I\eta X$	0.16	0.6708
	$Kr^{4+}(4s\,4p^3\,^3S)+Ne^+$	18.83	$I\theta X$	5.68	0.6708
	$Kr^{4+}(4p\,4d\,^3F)+Ne^+$	18.5	$I\iota X$	12.06	-0.5
	$Kr^{4+}(4p\,4d\,^3D_2)+Ne^+$	17.54	$I\kappa X$	15.14	-0.5
	$Kr^{4+}(4p\,4d\,^3P)+Ne^+$	15.86	$I\lambda X$	16.68	-0.5
	$Kr^{4+}(4p\,4d\,^3D_3)+Ne^+$	15.16	$I\mu X$	15.444	-0.5
	$Kr^{4+}(4p\,5s\,^3P_0)+Ne^+$	12.36	$I\nu X$	4.4	-0.5
	$Kr^{4+}(4p\,5p\,^3D)+Ne^+$	7.08	$I\rho X$	0.005	-0.632
$Kr^{5}+(4p\,^2p)$ $+Ar(3p^6\,^1S)$	$Kr^{4+}(4s\,4p^3\,^1D)+Ar^+(3p^5\,^2P)$	28.35	$I\eta X$	negligible	0.671
	$Kr^{4+}(4s\,4p^3\,^3S)+Ar^+$	24.6	$I\theta X$	negligible	0.671
	$Kr^{4+}(4p\,4d\,^3D)+Ar^+$	23.34	$I\kappa X$	0.003	-0.5
	$Kr^{4+}(4p\,4d\,^3P)+Ar^+$	21.66	$I\lambda X$	0.035	-0.5
	$Kr^{4+}(4p\,5s\,^3P_0)+Ar^+$	18.24	$I\nu X$	3.66	0.447
	$Kr^{4+}(4p\,5s\,^3P_2)+Ar^+$	17.33	$I\pi X$	7.04	0.447
	$Kr^{4+}(4p\,5p\,^3D)+Ar^+$	12.88	$I\rho X$	27.83	-0.632
	$Kr^{4+}(4p\,5p\,^1S)+Ar^+$	9.27	$I\phi X$	4.33	-0.632
	$Kr^{4+}(4p\,4f\,^3G)+Ar^+$	6.16	$I\chi X$	0.002	0.224

centred around 30 eV, are due to transfer ionization processes, presumably due to double electron capture into doubly-excited states which autoionize back to the Ar^{5+} channel before detection. In Ar^{6+}-D collisions, the observed spectrum exhibits a number of incompletely resolved peaks and illustrates that capture into the 5p state dominates, while capture into the 4f, 5s and 5d states show less contributions. A full theoretical description of this spectrum has been given recently by *Hansen* and *Taulbjerg* [3.85].

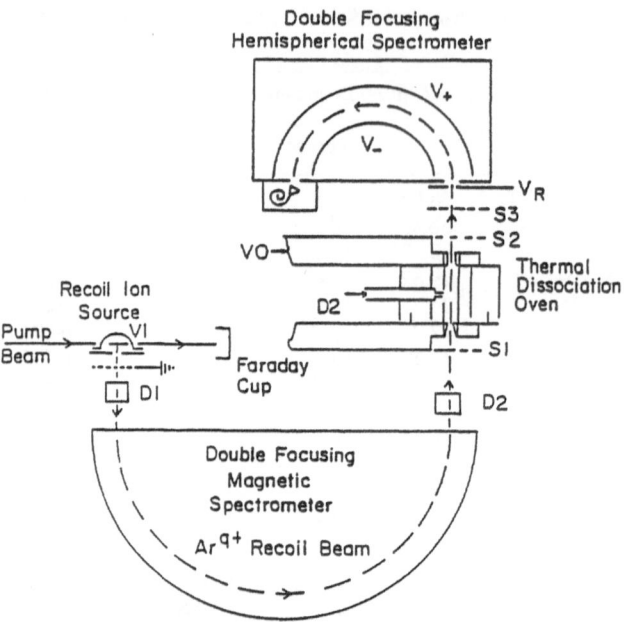

Fig. 3.11. Experimental set-up used by *Giese* et al. [3.31] for translational energy spectrometry of inelastic Ar^{q+} and Ne^{q+}-D and D_2 collisions

Several experimental measurements on single electron capture by multiply charged ions have been recently reported. The Belfast group has used the technique of translational energy spectrometry to study state-selective electron capture processes by multiply charged C, N, O and Ar ions in He, H_2 and H [3.68, 69, 86]. The Aarhus group has studied electron capture processes by Ar^{q+} ions ($q = 6$–10), C^{4+} and Xe^{q+} ions ($q = 10$–20) from rare-gas atoms [3.27, 29, 46, 87]. Very recently *Andersson* et al. [3.88] have measured the translational energy spectra for single electron capture by A^{6+} ions (A = Ne, Ar, Kr and Xe) from He at impact energies between 200 and 2000 eV. The observed spectra were compared with the corresponding theoretical spectra constructed using a simple semiclassical curve-crossing model based on a classical deflection function and Landau-Zener transition probabilities. Figure 3.13 shows the experimental spectra together with the theoretical calculations for the product Ar^{5+} ions in the Ar^{6+}-He collision system. At an impact energy of 2000 eV, they found that capture occurs mainly into the 4p and 4s states of Ar^{5+}. However, at energies below 500 eV, they observed a structure on the lower energy side of the 4p peak which could not attributed to the population of the 3d state. This structure was reproduced in the theoretical spectra where the 4s and 4p channels were taken into account, and was interpreted as due to effects of angular scattering.

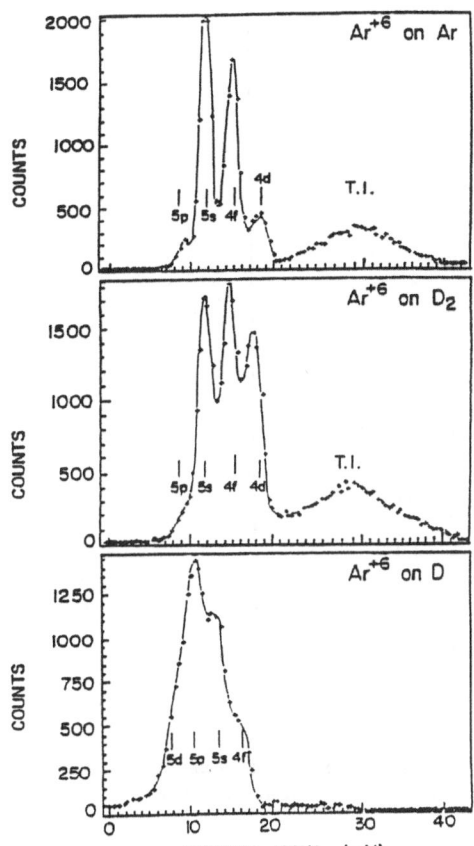

Fig. 3.12. Translational energy gain spectra for Ar^{6+} on Ar, D_2 and D targets at 3270 eV [3.31]

3.4.3 Multiple-Electron Capture Processes by Multiply Charged Ions

At low energies multiple-electron capture processes are a characteristic feature of collisions of multiply charged ions with atomic targets. This feature stems from the multitude of relaxation paths of the excited states of the multiply charged ion-atom complex. The theoretical description of these processes is extremely difficult and the existing theoretical predictions are very limited; they are either only qualitative or are based on very crude models. The existence of several active electrons in the collision of the multiply charged ions with rare-gas atoms allows a great variety of possible multiple-electron capture processes, namely:

a) True multiple-electron capture (TMC)

$$A^{q+} + B \rightarrow A^{(q-k)+*} + B^{k+}$$

b) Autoionizing multiple-electron capture (AMC)

$$A^{q+} + B \rightarrow A^{(q-n)+**} + B^{n+} \rightarrow A^{(q-k)+*} + B^{n+} + (n-k)e \,,$$

111

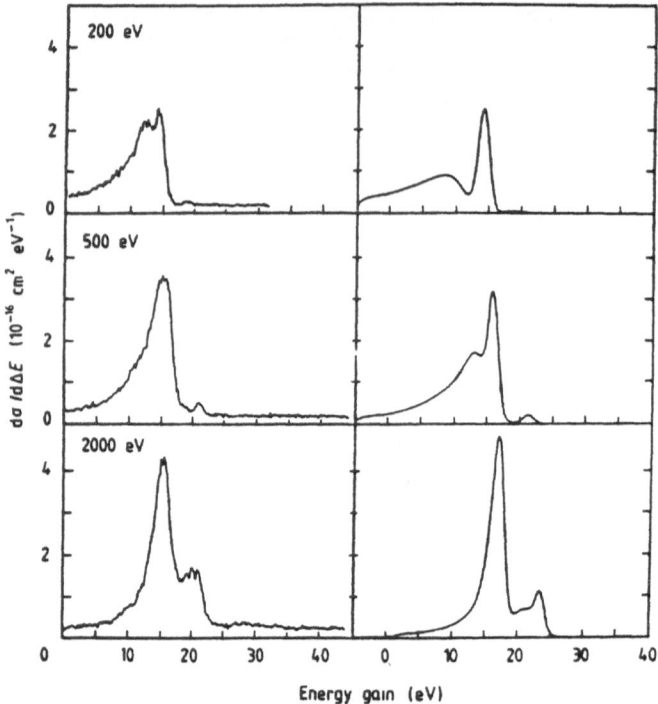

Fig. 3.13. Translational energy gain spectra for single electron capture by Ar^{6+} from He. The left-hand side shows experimental spectra at projectile energies of 200, 500 and 2000 eV. The right-hand side shows the corresponding theoretical spectra at the same projectile energies [3.88]

where k is the number of electrons captured by the projectile and n is the number of electrons removed from the target.

Although a large number of studies have been made on state-selective single electron capture by multiply charged ions from atomic targets, relatively little work has been reported on state-selective multiple electron capture processes. *Kamber* et al. [3.33, 84, 89] measured state-selective double electron capture by Ar^{q+} ions ($q = 3–5$), Kr^{q+} ions ($q = 3–5$) and Xe^{q+} ions ($q = 3, 4$) from rare-gas atoms. Figure 3.14 shows the translational energy spectra observed for the formation of Kr^{+} ions from the reaction of Kr^{3+} with Ar and Kr at an impact energy of 9 keV [3.84]. For the Kr^{3+}-Ar collision system, the dominant double-electron capture channel is $I\alpha X$, due to capture from the ground state of Kr^{3+} into the ground state of Kr^{+}; this process is exoergic by 18.08 eV with an avoided crossing at $R_x = 3.01$ au (see Table 3.2). There are probably contributions from unresolved processes $I\alpha A$ and $I\alpha B$ arising from the excitation of the target product. The broad peak at $\Delta E = 4.6$ eV is due to capture into first and higher excited states of Kr^{+}. In Kr^{3+}-Kr collisions, the reaction channels due to capture into the first ($I\beta X$) and higher excited states are found to be most important; the ground-state exit channel $I\alpha X$, and capture accompanied by target excitation ($I\alpha A$ and $I\alpha B$), make only a small contribution.

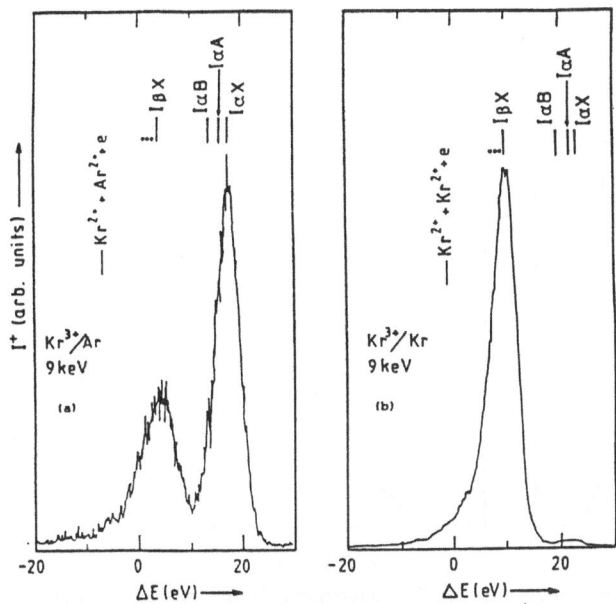

Fig. 3.14 a,b. Translational energy gain spectrum of Kr^+ ions obtained respectively from 9 keV Kr^{3+}-Ar (a) and Kr (b) collisions [3.84]

Table 3.2. The double electron capture reaction channels for Kr^{3+}/Ar and Kr collisions

Reactants and initial states	Products and final states	ΔE [eV]	Designation of reaction process
$Kr^{3+}(4p^3\ ^4S)+Ar(3p^6\ ^1S)$	$Kr^+(4p^5\ ^2P)+Ar^{2+}(3p^4\ ^3P)$	18.08	$I\alpha X$
	$Kr^+(4p^6\ ^2S)+Ar^{2+}$	4.58	$I\beta X$
	$Kr^+(4p^5\ ^2P)+Ar^{2+}(3p^4\ ^1D)$	16.34	$I\alpha A$
	$Kr^++Ar^{2+}(3p^4\ ^1S)$	13.96	$I\alpha B$
$Kr^{3+}(4p^3\ ^4S)+Kr(4p^4\ ^1S)$	$Kr^+(4p^5\ ^2P)+Kr^{2+}(4p^4\ ^3P)$	22.9	$I\alpha X$
	$Kr^+(4p^6\ ^2S)+Kr^{2+}$	9.4	$I\beta X$
	$Kr^+(4p^5\ ^2P)+Kr^{2+}(4p^4\ ^1D)$	21.09	$I\alpha A$
	$Kr^++Kr^{2+}(4p^4\ ^1S)$	18.8	$I\alpha B$

In Fig. 3.15, are shown the translational energy spectra for the formation of Kr^+ ions in the collision of Kr^{4+} ions with Ar and Kr. In Kr^{4+}-Ar collisions, two peaks are clearly seen, the strongest at $\Delta E = 7.18$ eV due to the reaction channel $I\beta X$. The second peak is due to three-electron capture into the ground state ($4p^5\ ^2P$) of Kr^+ with contributions from the metastable states present in the incident beam (see Table 3.3). In Kr^{4+}-Kr collisions, the energy spectrum shows a peak centred near the eight channels in the series corresponding to the formation of $4p^4(^3P)4d$ and $4p^4(^3P)5p$ Kr^+ ions within the range ΔE between 10 and 14 eV.

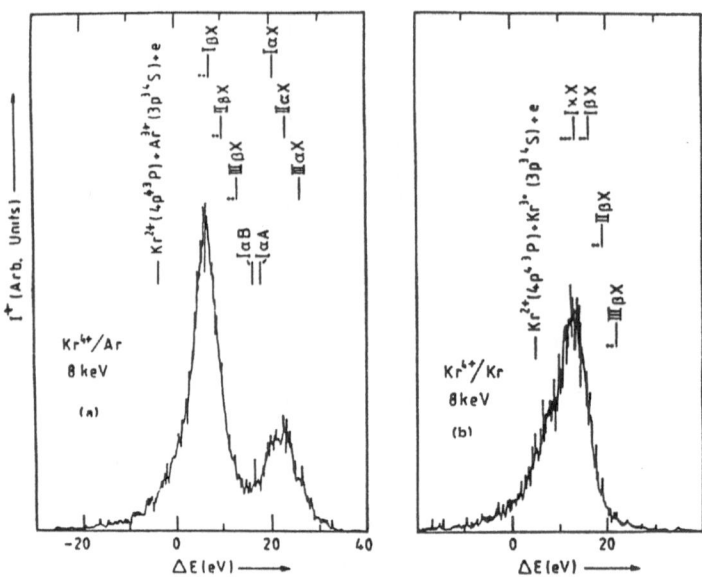

Fig. 3.15 a,b. Translational energy gain spectrum of Kr^+ ions obtained from 8 keV Kr^{4+}-Ar (a) and Kr (b) collisions [3.84]

Table 3.3. The three electron capture reaction channels for Kr^{4+}/Ar and Kr collisions

Reactants and initial states	Products and final states	ΔE [eV]	Designation of reaction process
$Kr^{4+}(4p^2\ ^3P)+Ar(3p^6\ ^1S)$	$Kr^+(4p^5\ ^2P)+Ar^{3+}(3p^3\ ^4S)$	20.68	$I\alpha X$
	$Kr^+(4p^6\ ^2S)+Ar^{3+}$	7.18	$I\beta X$
	$Kr^+(4p^5\ ^2P)+Ar^{3+}(3p^3\ ^2D)$	18.07	$I\alpha A$
	$Kr^++Ar^{3+}(3p^3\ ^2P)$	16.36	$I\alpha B$
$Kr^{4+}(4p^2\ ^1D)+Ar$	$Kr^++Ar^{3+}(3p^3\ ^4S)$	23.38	$II\alpha X$
	$Kr^+(4p^6\ ^2S)+Ar^{3+}$	9.88	$II\beta X$
$Kr^{4+}(4p^2\ ^1S)+Ar$	$Kr^+(4p^6\ ^2S)+Ar^{3+}$	26.34	$III\beta X$
	$Kr^+(4p^6\ ^2S)+Ar^{3+}$	12.84	$III\beta X$
$Kr^{4+}(4p^2\ ^3P)+Kr(4p^6\ ^1S)$	$Kr^+(4p^6\ ^2S)+Kr^{3+}(3p^3\ ^4S)$	16	$I\beta X$
	$Kr^+(4d\ ^2D)+Kr^{3+}$	12.95	$I\kappa X$
$Kr^{4+}(4p^2\ ^1D)+Kr$	$Kr^+(4p^6\ ^2S)+Kr^{3+}$	18.7	$II\beta X$
$Kr^{4+}(4p^2\ ^1S)+Kr$	Kr^++Kr^{3+}	21.66	$III\beta X$

The strong repulsion in the exit channels of double- and triple-electron capture processes leads to multiple crossings between the entrance channels, as well as crossings between the series of exit channels of the single electron capture process. Multiple capture can be produced by a multi-step mechanism through successive single electron capture processes during a single collision, or by a single step correlated double- or triple-electron capture. In this latter case, two-electron correlation effects

114

may play a dominant role in the dynamics of multiple-electron capture processes. It is noteworthy that multiple-electron capture processes are a particularly efficient tool to create metastable and doubly excited states of multiply charged ions and atoms, unlike excitations by electron or by photon impact where multiply excited states are produced only by the weak electron-electron correlation.

Roncin et al. [3.90] measured the differential cross sections for single- and double-electron capture by highly charged ions at low keV energies by using a coincident energy gain spectrometry technique. They found that, for electrons captured simultaneously, the scattering occurs in forward directions, whereas for electrons captured successively the scattering occurs at finite angles. *Barat* et al. [3.91] applied a multicoincidence technique to study the relative importance of single- and autoionizing double-electron capture processes. Figure 3.16 shows the translational energy gain spectra for single- and double-electron capture for collisions of O^{6+} ions with He, Ne and Ar at an impact energy of 9 keV. With a He target, single electron capture (SEC) populates O^{5+} ($n = 3$) states, while the autoionizing double-electron

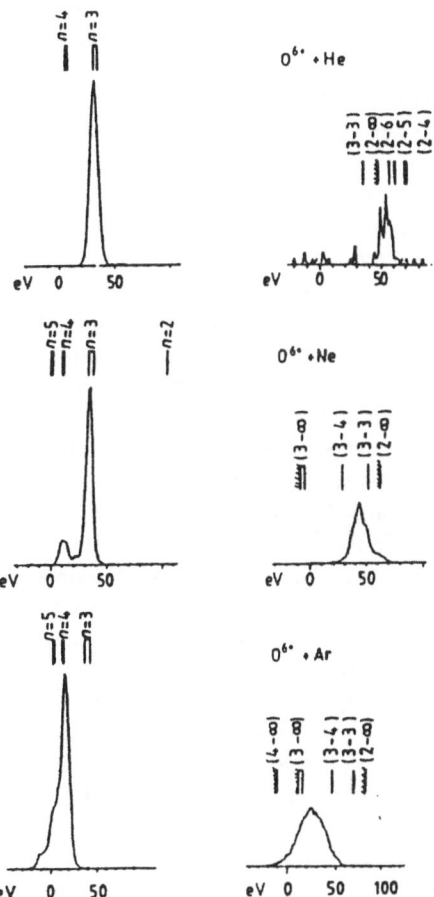

Fig. 3.16. Translational energy gain spectra for collisions of O^{6+} ions with He, Ne and Ar at 9 keV [3.91]. For each spectra the left-hand spectrum corresponds to SEC in coincidence with He^+ recoil ions and the right-hand side corresponds to ADC processes in coincidence with He^{2+}

capture (ADC) spectrum clearly shows that O^{4+} ($2l,n'l'$) states with $n' \geq 6$ are populated. With a Ne target, SEC also populates the O^{5+} ($n = 3$) state, however significant population of the $n = 4$ state is also found. For O^{6+}-Ar collision system, O^{5+} ($n = 4$) is preferentially populated with a significant contribution of $n = 5$, while the states populated in ADC move from O^{4+} ($n = 2$, $n' \geq 6$) to O^{4+} ($n = 3$, $n' = 4$–8).

3.5 Discussion of State-Selective Differential Electron Capture Cross Sections

Measurements of the differential cross section for state-selective electron capture processes generally give more precise information about the nature of the process than measurements of the total cross sections. One of the principle advantages to be gained from such measurements is the fact that it may be possible to relate the scattering angle to the classical impact parameter. The possibility of this connection depends upon the knowledge of the potential energy curves of the incoming and outgoing reaction channels that participate in the collision and on the avoided crossings at which transitions take place.

The first experiments, showing oscillations in differential cross section for singly-charged ions, were carried out without energy analysis by Everhart and coworkers [3.92–94]. First differential cross section studies for single electron capture by doubly charged ions were carried out by the Hasted group. *Hasted* et al. [3.95] measured the differential cross sections for single electron capture in collisions of C^{2+}, N^{2+} and O^{2+} ions with He, Ne and Ar over the impact energy range 1–3 keV and for scattering angles between 0.25° and 2°, but without energy analysis. Oscillatory structure in inelastic scattering was clearly observed after deconvolution of the data. *Makhdis* et al. [3.41] measured the differential single electron capture functions for the collision systems O^{2+}-He and C^{2+}-He and Ne. The exit channels were defined by the energy defect analysis using parallel plate momentum analyzers before and after the collisions. *Kamber* et al. [3.64, 96] measured the differential elastic and inelastic scattering functions and the electron capture probability $P(b)$ for the collision systems Ar^{2+}-He and Ne and Ar^{3+}-He at 540 and 1250 eV impact energies respectively. In the translational ion energy gain spectrometer (Fig. 3.17), the magnetically separated ions (MA) are monochromatized by means of an electrostatic parallel plate analyzer (ES) and steered through a gas collision chamber (CC). The scattering angle is selected by an electrical deflection system (D1, D2) after which the angle selected ions pass into another fixed-position parallel plate electrostatic analyzer (EA) before collection. Scanning of this analyzer enables translational energy spectra to be recorded. It is possible to maintain the analyzed energy constant so that a single channel is collected whilst the electrical deflection system is scanned, thus providing a state-selective differential cross section over a range of angle 0°–1.2°, as shown in Fig. 3.18.

We present here some typical results obtained with this apparatus. The inset of Fig. 3.18, shows the zero-scattering angle translational energy gain spectrum for 540

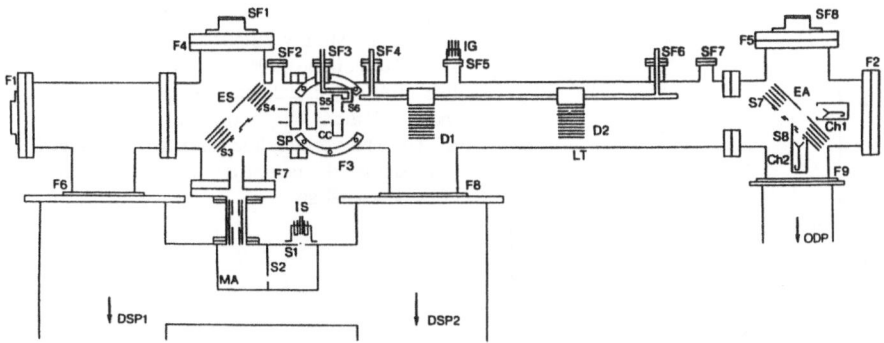

Fig. 3.17. Experimental apparatus used by Hasted and his collaborators [3.41, 64] for double differential energy gain measurements

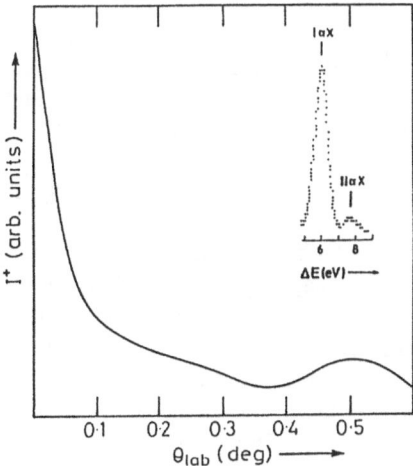

Fig. 3.18. The inelastic differential cross section for the reaction channel $I\alpha X$. Inset shows the translational energy gain for single electron capture by 540 eV Ar^{2+} from Ne [3.64]

eV Ar^{2+}-Ne collisions. Two electron capture channels are clearly identified. These channels are due to the following processes:

$$Ar^{2+}(3p^4\ {}^3P) + Ne \rightarrow Ar^+(3p^5\ {}^3P) + Ne^+ + 6.06\ eV \qquad I\alpha X$$

$$Ar^{2+}(3p^4\ {}^1D) + Ne \rightarrow Ar^+(3p^5\ {}^3P) + Ne^+ + 7.80\ eV \qquad II\alpha X$$

The $I\alpha X$ channel, due to capture from ground state incident ions Ar^{2+} (^3P) into the ground state Ar^+ (^3P), is overwhelmingly dominant. This process is exoergic by 6.06 eV, with an avoided crossing at 4.49 au. The observed $II\alpha X$ channel indicates the presence of the long lived excited state Ar^{2+} (^1D) in the incident beam. State-selective differential single electron capture cross section for the reaction channel $I\alpha X$ is a relatively smooth function with local maxima and minima, which are due either to multiple crossings or interference between scattering amplitudes developed along different paths leading to the same final state. The probability of electron cap-

117

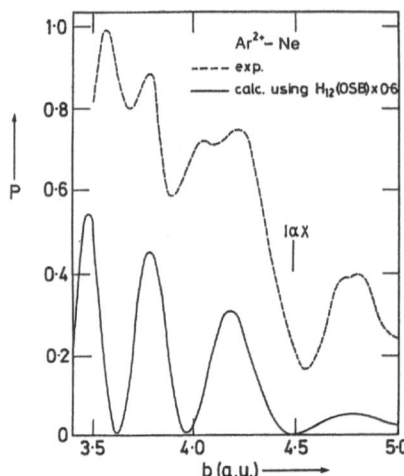

Fig. 3.19. Measured and calculated electron capture probability functions $P(b)$ for 540 eV Ar^{2+}-Ne collisions. Broken curve, experimental data; full curve, calculated data using interaction matrix element H_{12}^{OSB} multiplied by a factor of 0.6. The vertical line marked $I\alpha X$ indicates the calculated position for the avoided crossing $R_x = b = 4.49$ au

ture as a function of scattering angle, which can be classically related to impact parameter b, can be derived from the capture differential cross sections $I^+(\theta)$ and the elastic differential cross sections $I^{2+}(\theta)$, using the following relation [3.64]

$$P = \frac{I^+}{I^+ + I^{2+}} . \tag{3.24}$$

In a semi-classical approximation, at impact parameter b larger than the nuclear separation R_x at which the avoided crossing occurs, little capture occurs at low energies. For smaller b, capture can occur, but its probability $P(b)$ is an oscillating function of the impact parameter b, owing to quantum interference effects. Comparison of a $P(b)$ function derived from experiment with those predicted on the basis of the semi-classical impact parameter method of *Bates* et al. [3.97], on the assumption of a single avoided crossing, is made in Fig. 3.19. The value of the interaction matrix element H_{12} was calculated using the empirical expression of *Olson* et al. [3.98], and adjusted to give the best fit to the experimental data. Reasonably good agreement between the two has been achieved using a value for $H_{12} = 0.6 H_{12}^{OSB}$.

Cocke and his colleagues [3.76] have constructed an apparatus for measuring angular distributions of slow multiply charged ions following capture. A position sensitive channel plate chevron followed by a resistive anode was used to determine the scattering angle and a simple retarding grid system was used to select the final charge states. Multiply charged projectiles were obtained from recoil ion sources at KSU and the Oak Ridge ECR source. The systems studied include partially stripped first-row elements, Ar^{q+} ($q = 6-8$), and O^{2+} on He [3.73, 76, 99]. The general features of the distributions were interpreted in terms of classical deflection functions for the heavy-particle motion. As discussed by *Ford* and *Wheeler* [3.100], the motion of the heavy particles may be considered classical if the product of the scattering angle (θ) of the projectile and angular momentum divided by $\hbar$ is greater than unity, if the angular momentum is large, and if a stationary phase approximation is valid for the evaluation of the scattering amplitude.

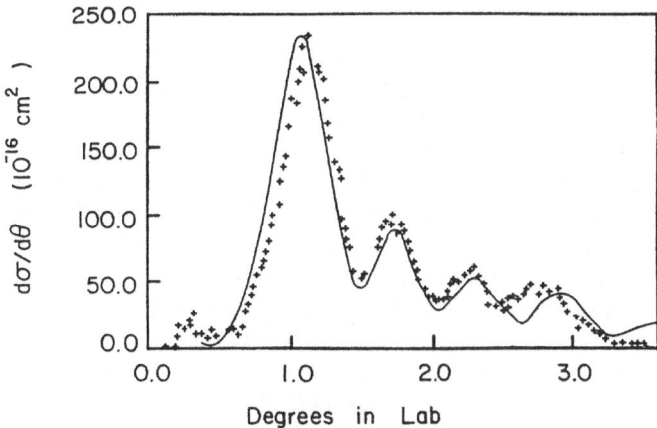

Fig. 3.20. Differential cross sections for double electron capture by 1520 eV C^{4+} ions from He. The data points are from [3.101], normalized to the theoretical peak, and the theory (solid line), folded with experimental resolution, is from [3.102]

Figure 3.20 displays the differential cross section for double electron capture by C^{4+} from He, which shows an oscillatory structure. In this collision system, the dominant reaction channel is double electron capture into the $(1s^2 2s^2\ ^1S)$ state of C^{2+} from the $C^{4+}(1s^2\ ^1S)$ incident ion, which essentially means that there are only two states that are active during the collision. These oscillations were well reproduced and described as Stuckelburg oscillations by *Barany* et al. [3.101] using a two-state curve-crossing model, and recently by *Tan* et al. [3.102], using a quantal two-channel molecular orbital close coupling expansion method.

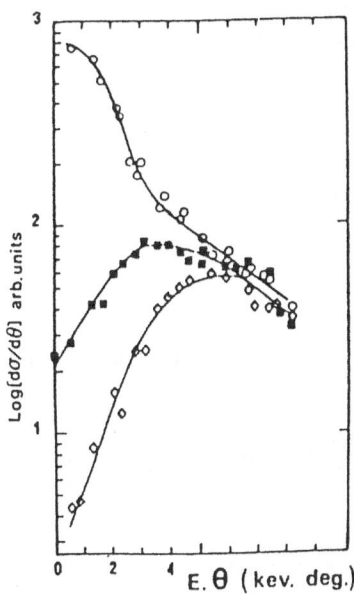

Fig. 3.21. Differential cross sections for 8 keV O^{8+}-He collisions; ○ SEC ($n = 4$); ■ ADC ($n = 3$, $n' = 4$); ◇ ADC ($n = 3$, $n' = 3$) [3.90]

The Orsay group [3.48, 90, 91] have measured the differential cross sections for single- and multiple-electron capture processes by multiply charged ions from rare-gas atoms using a coincident energy gain spectrometry technique. In Fig. 3.21 is shown the differential cross sections for O^{8+}-He collisions at impact energy of 8 keV. They found that the single electron capture takes place almost exclusively into $n = 4$ states of O^{7+}, while autoionizaing double-electron capture (ADC) takes place into the $(n - 3, n' = 4)$ and $(n = 3, n' = 3)$ states. Figure 3.21 illustrates that the differential cross sections for single electron capture are strongly peaked in the forward direction at a reduced angle $\tau = 0$ keV deg, whereas differential cross sections for the double-electron capture channels exhibit maxima at respectively $\tau = 3.5$ and $\tau = 6$ keV deg. From their measurements and by using the potential energy curves for the collision systems, they conclude that double-electron capture occurs mainly via a two-step mechanism, that is, two successive single electron capture during a single collision.

Acknowledgements. One of us (CLC) acknowledges support from the Division of Chemical Sciences, Office of Basic Energy Sciences, Office of Energy Research, U.S. Department of Energy.

References

3.1   J. B. Hasted: Proc. Roy. Soc. A **205**, 421 (1951)
3.2   J. B. Hasted: Proc. Roy. Soc. A **212**, 235 (1952)
3.3   J. B. Hasted, J. B. H. Stedeford: Proc. Roy. Soc. A **227**, 466 (1955)
3.4   J. B. Hasted, R. A. Smith: Proc. Roy. Soc. A **235**, 354 (1956)
3.5   J. B. Hasted, A. Y. J. Chang: Proc. Phys. Soc. **80**, 893 (1962)
3.6   R. G. Cooks (ed.): *Collision Spectroscopy* (Plenum, New York 1978) pp. 252
3.7   R. E. Olson, M. Kimura: J. Phys. B **15**, 4231 (1982)
3.8   R. K. Janev, L. P. Presnyakov, V. P. Shevelko (eds.): *Physics of Highly Charged Ions* (Springer, Berlin, Heidelberg 1985)
3.9   R. K. Janev, L. P. Presnyakov: Phys. Rep. **70**, 1 (1981)
3.10   P. T. Greenland: Phys. Rep. **81**, 131 (1982)
3.11   B. H. Bransden, R. K. Janev: Adv. At. Mol. Phys. **19**, 1 (1983)
3.12   Dz. Belkic, R. Gayet, A. Salin: Phys. Rep. **56**, 279 (1979)
3.13   L. D. Landau: Phys. Z. Sowetunion **2**, 46 (1932)
3.14   C. Zener: Proc. Roy. Soc. London A **137**, 696 (1932)
3.15   E. C. G. Stueckelberg: Helv. Phys. Acta **5**, 369 (1932)
3.16   D. R. Bates: Proc. Roy. Soc. London A **257**, 22 (1960)
3.17   S. S. Gershtein: Zh. Eksp. Teor. Fiz. **43**, 706 (1962) [English transl.: Sov. Phys. JETP **16**, 501 (1962)]
3.18   R. E. Olson, A. Salop: Phys. Rev. A **14**, 579 (1976)
3.19   D. R. Bates, B. L. Moiseiwitsch: Proc. Phys. Soc. A **67**, 805 (1954)
3.20   M. Kimura, T. Iwai, Y. Kaneko, N. Kobayashi, A. Matsumoto, S. Ohtani, S. Takagi, H. Tawara, S. Tsurubuchi: J. Phys. Soc. Japan **53**, 2224 (1984)
3.21   K. Taulbjerg: J. Phys. B **19**, L369 (1986)
3.22   A. Salop, R. E. Olson: Phys. Rev. A **13**, 1312 (1976)
3.23   R. K. Janev, D. S. Belic, B. H. Bransden: Phys. Rev. A **28**, 1293 (1983)

3.24 H. Ryufuku, K. Sasaki, T. Watanabe: Phys. Rev. A **21**, 745 (1980)
3.25 H. F. Beyer, K. H. Schartner, F. Folkmann: J. Phys. B **13**, 2459 (2459)
3.26 N. Bohr, J. Lindhard: K. Dan. Vidensk. Selsk. Mat. Fys. Medd. **28**, 1 (1954)
3.27 P. Hvelplund, L. H. Andersen, A. Barany, H. Cederquist, J. Heinemeier, H. Knudsen, K. B. MacAdam, E. H. Nielsen, J. Sorensen: Nucl. Instrum. Methods B **9**, 421 (1985)
3.28 H. Winter: Phys. Scr. **T3**, 159 (1983)
3.29 H. Cederquist, L. H. Andersen, A. Barany, P. Hvelplund, H. Knudsen, E. H. Nielsen, J. O. K. Pedersen, J. Sorensen: J. Phys. B **18**, 3951 (1985)
3.30 F. W. Meyer, A. M. Howald, C. C. Havener, R. A. Phaneuf: Phys. Rev. Lett. **54**, 2663 (1985)
3.31 J. P. Giese, C. L. Cocke, W. Waggoner, L. N. Tunnell, S. L. Varghese: Phys. Rev. A **34**, 3770 (1986)
3.32 R. W. McCullough, S. M. Wilson, H. B. Gilbody: J. Phys. B **20**, 2031 (1987)
3.33 E. Y. Kamber: J. Phys. B **21**, 4185 (1988)
3.34 A. Niehaus: J. Phys. B **19**, 2925 (1986)
3.35 E. W. P. Bloemen, D. Dijkkamp, F. J. de Heer: J. Phys. B **15**, 1391 (1982)
3.36 D. Dijkkamp, A. Brazuk, A. G. Drentje, F. J. de Heer, H. Winter: J. Phys. B **17**, 4327 (1984)
3.37 P. H. Woerlee, T. M. El Sherbini, F. J. de Heer, F. W. Saris: J. Phys. B **14**, L235 (1979)
3.38 R. Morgenstern, A. Niehaus, G. Zimmermann: J. Phys. B **13**, 4811 (1980)
3.39 A. Bordenave, P. Benoit Cattin, A. Gleizes, S. Dousson, D. Hitz: J. Phys. B **18**, L195 (1985)
3.40 Y. H. Chen, R. E. Johnson, R. R. Humphris, M. W. Siegel, J. W. Boring: J.Phys. B **8**, 1527 (1975)
3.41 Y. Y. Makhdis, K. Birkinshaw, J. B. Hasted: J. Phys. B **9**, 111 (1976)
3.42 Y. Sato, J. H. Moore: Phys. Rev. A **19**, 495 (1979)
3.43 B. A. Huber, H. J. Kahlert: J. Phys. B **16**, 4655 (1983)
3.44 K. Okuno, H. Tawara, T. Iwai, Y. Kaneko, M. Kimura, N. Kobayashi, A. Matsumoto, S. Ohtani, S. Takagi, S. Tsurubuchi: Phys. Rev. A **28**, 127 (1983)
3.45 M. Lennon, R. W. McCullough, H. B. Gilbody: J. Phys. B **16**, 2191 (1983)
3.46 E. H. Nielsen, L. H. Andersen, A. Barany, H. Cederquist, J. Heinemeier, P. Hvelplund, H. Knudsen, K. B. MacAdam, J. Sorensen: J. Phys. B **18**, 1789 (1985)
3.47 C. Schmeissner, C. L. Cocke, R. Mann, W. Meyerhof: Phys. Rev. A **30**, 1661 (1984)
3.48 P. Roncin, M. Barat, H. Laurent, J. Pommier, S. Dousson, D. Hitz: J. Phys. B **17**, L521 (1984)
3.49 V. V. Afrosimov, A. A. Basalaev, M. N. Panov, A. V. Samoilov: Zh. Eksp. Teor. Fiz. **91**, 465 (1986) [English transl.: Sov. Phys. -JETP **64**, 273 (1986)]
3.50 U. Jellen-Wutte, J. Schweinzer, W. Vanek, H. Winter: J. Phys. B **18**, L779 (1985)
3.51 H. S. W. Massey, H. B. Gilbody: *Electronic and Ionic Impact Phenomena*, Vol. 4 (Clarendon Press, Oxford 1974)
3.52 J. B. Hasted: *Physics of Atomic Collisions*, 2nd ed. (Butterworths, London 1972)
3.53 R. K. Janev, H. Winter: Phys. Rep. **117**, 265 (1985)
3.54 Phys. Scr. **24** (1981) and **T3** (1983)
3.55 J. H. Moore, C. C. Davis, M. A. Coplan (eds.): *Building Scientific Apparatus* (Addison-Wesley, London 1983)
3.56 P. Varga and H. Winter: Phys. Rev. A **18**, 2453 (1978)
3.57 A. Matsumoto, S. Ohtani, T. Iwai: J. Phys. B **15**, 1871 (1982)
3.58 A. Brazuk, H. Winter: J. Phys. B **15**, 2233 (1982)
3.59 B. A. Huber, H.-J. Kahlert: J. Phys. B **16**, 4655 (1983)
3.60 T. Nakamura, N. Kobayashi, Y. Kaneko: J. Phys. Soc. Jpn **54**, 2774 (1985)
3.61 N. Kobayashi, T. Nakamura, Y. Kaneko: J. Phys. Soc. Jpn **52**, 2684 (1983)
3.62 E. Y. Kamber, A. G. Brenton, S. Hughes: J. Phys. B (To be published)
3.63 E. Y. Kamber, P. Jonathan, A. G. Brenton, J. H. Beynon: J. Phys. B **20**, 4129 (1987)
3.64 E. Y. Kamber, D. Mathur, J. B. Hasted: J. Phys. B **15**, 263 (1982)
3.65 J. Stevens, R. S. Peterson, E. Pollack: Phys. Rev. A **27**, 2396 (1983)
3.66 J. Puerta, B. A. Huber: J. Phys. B **18**, 4445 (1985)
3.67 R. W. McCullough, M. Lennon, F. G. Wilkie, H. B. Gilbody: J. Phys. B **16**, L173 (1983)
3.68 R. W. McCullough, F. G. Wilkie, H. B. Gilbody: J. Phys. B **17**, 1373 (1984)

3.69 F. G. Wilkie, F. B. Yousif, R. W. McCullough, J. Geddes, H. B. Gilbody: J. Phys. B **18**, 479 (1985).

3.70 J. Schweinzer, U. Jellen-Wutte, W. Vanek, H. Winter, J. E. Hansen: J. Phys. B **21**, 315 (1988)

3.71 F. Aumayr, J. Schweinzer, H. Winter: J. Phys. B **22**, 1027 (1989)

3.72 E. Y. Kamber, C. L. Cocke, J. P. Giese, J. O. K. Pedersen, W. Waggoner, S. L. Varghese: Nucl. Instrum. Methods B **24/25**, 288 (1987)

3.73 E. Y. Kamber, C. L. Cocke, J. P. Giese, J. O. K. Pedersen, W. Waggoner: Phys. Rev. A **36**, 5575 (1987)

3.74 S. Beinstock, T. G. Heil, A. Dalgarno: Phys. Rev. A **29**, 503 (1984)

3.75 R. Mann, C. L. Cocke, A. S. Schlachter, M. Prior, R. Marrus: Phys. Rev. Lett. **49**, 1329 (1982).

3.76 L. N. Tunnell, C. L. Cocke, J. P. Giese, E. Y. Kamber, S. L. Varghese, W. Waggoner: Phys. Rev. A **35**, 3299 (1987)

3.77 S. Ohtani, Y. Kaneko, M. Kimura, N. Kobayashi, T. Iwai, A. Matsumoto, K. Okuno, S. Takagi, H. Tawara, S. Tsurubuchi: J. Phys. B **15**, L533 (1982)

3.78 S. Tsurubuchi, T. Iwai, Y. Kaneko, M. Kimura, N. Kobayashi, A. Matsumoto, S. Ohtani, K. Okuno, S. Takagi, H. Tawara: J. Phys. B **15**, L733 (1982)

3.79 M. Kimura, T. Iwai, Y. Kaneko, N. Kobayashi, A. Matsumoto, S. Ohtani, K. Okuno, S. Takagi, S. Tsurubuchi: J. Phys. B **15**, L851 (1982)

3.80 T. Iwai, Y. Kaneko, M. Kimura, N. Kobayashi, S. Ohtani, K. Okuno, S. Takagi, S. Tsurubuchi: Phys. Rev. A **26**, 105 (1982)

3.81 H. Tawara, T. Iwai, Y. Kaneko, M. Kimura, N. Kobayashi, A. Matsumoto, S. Ohtani, O. Okuno, S. Takagi, S. Tsurubuchi: Phys. Rev. A **29**, 1529 (1984)

3.82 E. Y. Kamber, W. G. Hormis, A. G. Brenton, J. B. Hasted, J. H. Beynon: J. Phys. B **20**, 105 (1987)

3.83 W. G. Hormis, E. Y. Kamber, A. G. Brenton, J. B. Hasted, J. H. Beynon: Int. J. Mass Spectrom. Ion Phys. **76**, 263 (1987)

3.84 E. Y. Kamber, W. G. Hormis, J. B. Hasted, A. G. Brenton, J. H. Beynon: J. Phys. B **21**, 3423 (1988)

3.85 J. P. Hansen, K. Taulbjerg: J. Phys. B **21**, 2459 (1988)

3.86 R. W. McCullough, S. M. Wilson, H. B. Gilbody: J. Phys. B **20**, 2031 (1987).

3.87 P. Hvelplund, A. Barany, H. Cederquist, J. O. K. Pedersen: J. Phys. B **20**, 2515 (1987)

3.88 L. R. Andersson, J. O. K. Pedersen, A. Barany, J. P. Bangsgaard, P. Hvelplund: J. Phys. B **22**, 1603 (1989)

3.89 E. Y. Kamber, W. G. Hormis, A. G. Brenton, J. B. Hasted, J. H. Beynon: J. Phys. B **18**, 117 (1985)

3.90 P. Roncin, M. Barat, H. Laurent: Europhys. Lett. **2**, 371 (1986)

3.91 M. Barat, M. N. Gaboriaud, L. Guillemot, P. Roncin, H. Laurent, S. Andriamonje: J. Phys. B **20**, 5771 (1986)

3.92 F. P. Ziemba, E. Everhart: Phys. Rev. Lett. **2**, 299 (1959)

3.93 G. J. Lockwood, E. Everhart: Phys. Rev. **124**, 567 (1962)

3.94 G. H. Morgan, E. Everhart: Phys. Rev. **128**, 667 (1962)

3.95 J. B. Hasted, S. M. Iqbal, M. M. Yousif: J. Phys. B **4**, 343 (1971)

3.96 E. Y. Kamber, J. B. Hasted: J. Phys. B **16**, 3025 (1983)

3.97 D. R. Bates, H. C. Johnston, I. Stewart: Proc. Phys. Soc. **84**, 517 (1964)

3.98 R. E. Olson, F. T. Smith, E. Bauer: Appl. Opt. **10**, 1848 (1971)

3.99 W. Waggoner, C. L. Cocke, L. N. Tunnell, C. C. Havener, F. W. Meyer, R. A. Phaneuf: Phys. Rev. A **37**, 2386 (1988)

3.100 K. W. Ford, J. A. Wheeler: Ann. Phys. **7**, 259 (1959)

3.101 A. Barany, H. Danared, H. Cederquist, P. Hvelplund, H. Knudsen, J. O. K. Pedersen, C. L. Cocke, L. N. Tunnell, W. Waggoner, J. P. Giese: J. Phys. B **19**, L427 (1986)

3.102 J. Tan, C. D. Lin, M. Kimura: J. Phys. B **20**, L91 (1987)

4. Energy Spectrometry of Fine-Structure Transitions in Ion-Atom Collisions

N. Kobayashi

With 15 Figures

The ground electronic states of singly charged ions of rare gas atoms (except He) have p^5, ^2P_J ($J = 3/2$ and $1/2$) configuration. Energy splittings of the fine-structure states (different J states) are 0.098, 0.186, 0.660 and 1.30 eV for Ne^+, Ar^+, Kr^+ and Xe^+ [4.1], respectively. In the case of doubly charged ions of the rare gases, the ground states are ^3P_J ($J = 2$, 1 and 0) where the $^3P_2$ is the lowest level. These ions also have low-lying long-lived metastable states, $^1D_2$ and $^1S_0$. The energies of the $^3P_1$, $^3P_0$, $^1D_2$ and $^1S_0$ states of Ne^{2+}, Ar^{2+}, Kr^{2+} and Xe^{2+} ions are listed in Table 4.1 [4.1].

The existence of fine-structure states and low-lying metastable states are not limited to the rare gas ions but are found, in general, for almost all ions except a few cases such as H-like and He-like ions. Recently it has been revealed that the fine-structure states play an important role in inelastic processes in ion-atom and ion-molecule collisions. For example, cross sections for charge transfer reactions depend strongly on the fine-structure states of the incident ions [4.2–5]. Very recently, it has been found that excitation processes in ion-molecule collisions also depend strongly on the fine-structure states of the ions [4.6–9].

When ions are produced by conventional electron impact and proton impact techniques a number of fine-structure states and metastable states may also be formed.

Table 4.1. Energy levels in p^4 configurations of doubly charged ions of rare gases [4.1]

Level	Energy [eV]			
	Ne^{2+}	Ar^{2+}	Kr^{2+}	Xe^{2+}
$^3P_2$	0	0	0	0
$^3P_1$	0.08	0.14	0.56	1.21
$^3P_0$	0.11	0.19	0.66	1.01
$^1D_2$	3.20	1.74	1.81	2.12
$^1S_0$	6.91	4.12	4.10	4.64

Springer Series in Chemical Physics, Vol. 54
Deepak Mathur (ed.): Physics of Ion Impact Phenomena
© Springer-Verlag Berlin Heidelberg 1991

Therefore, it is necessary to know the populations of these states in the ion beam. Usually, the populations are measured by the ion attenuation method [4.2, 3, 10]. However, the accuracy of the resulting measured values is not very good. Furthermore, in cases where more than two states are contained in the ion beam, determination of the relative populations of all the states is not easy. *Adams* et al. [4.2] have measured the fractional populations of ^3P, $^1D_2$ and $^1S_0$ in Xe^{2+} by applying an elegant technique in which two different reactant gases are mixed in a buffer gas in a selected ion flow tube and ion-neutral reaction rate coefficients are measured which are dependent upon the fine-structure states of the ion. However, the different J states of ^3P could not be distinguished in this experiment.

A study of collision induced transitions among these states seems important in order to obtain a better understanding of ion collision processes. As mentioned above, in the case of doubly charged ions of the rare gases, the ground states are ^3P and the low-lying metastable states are ^1D and ^1S. The transitions ^3P-^1D and ^3P-^1S are spin non-conservative and the transition ^1D-^1S is forbidden in electric dipole interaction, but is allowed in electric quadrupole interaction. Transition probablities for radiative transitions among the ^3P, ^1D and ^1S states of Ne^{2+}, Ar^{2+} and Kr^{2+} are listed in Table 4.2 [4.11–13].

Fine-structure transitions of ions were first studied by *Johnson* [4.13, 14] theoretically for Ar^++Ar, Kr^++Kr and Xe^++Xe collisions in the 1970s. However, no further experimental study has been performed for a long time, because there has been no useful technique to observe the fine-structure transitions. In 1981, *Itoh* et al. [4.15] reported the first successful direct measurement of the fine-structure transitions in Ar^++Ar by means of high resolution ion translational energy spectrome-

Table 4.2. Transition probabilities for p^4 configurations of Ne^{2+}, Ar^{2+} and Kr^{2+} in units of s^{-1}

Transition	Type [a]	Ne^{2+} [b]	Ar^{2+} [c]	Kr^{2+} [d]
$^3P_2$-$^3P_1$	q	2.6×10^{-9}	3.6×10^{-7}	7.6×10^{-4}
	m	6.0×10^{-3}	3.1×10^{-2}	2.0
$^3P_2$-$^3P_0$	q	2.0×10^{-8}	2.7×10^{-6}	2.5×10^{-3}
$^3P_1$-$^3P_0$	m	1.1×10^{-3}	5.2×10^{-3}	2.3×10^{-2}
$^3P_2$-$^1D_2$	q	3.0×10^{-4}	1.4×10^{-3}	4.3×10^{-2}
	m	1.7×10^{-1}	3.2×10^{-2}	4.7
$^3P_1$-$^1D_2$	q	3.8×10^{-5}	1.3×10^{-4}	1.0×10^{-3}
	m	5.2×10^{-2}	8.3×10^{-2}	5.3×10^{-1}
$^3P_0$-$^1D_2$	q	1.2×10^{-5}	2.9×10^{-5}	3.7×10^{-4}
$^3P_2$-$^1S_0$	q	5.1×10^{-3}	4.3×10^{-2}	6.9×10^{-1}
$^3P_1$-$^1S_0$	m	2.2	4.0	53
$^3P_0$-$^1S_0$				
$^1D_2$-$^1S_0$	q	2.8	3.1	4.5

[a] q: electric quadrupole transition, m: magnetic dipole transition
[b] Ref. [4.11]
[c] Ref. [4.12]
[d] Ref. [4.13]

try. They observed excitation and de-excitation processes between $^2P_{3/2}$ (the lower state) and $^2P_{1/2}$ of Ar^+ $(3p^5, ^2P_J)$ and proposed a method to evaluate the fractional populations of the fine-structure states contained in their ion beam.

After that, further systematic studies have been carried out on the excitation and de-excitation processes between $^2P_{3/2}$ and $^2P_{1/2}$ states of the singly charged ions of rare gases and among ^3P_J, $^1D_2$ and $^1S_0$ of the doubly charged ions of rare gases in collisions with rare gas atoms, which are $^1S_0$, with a high resolution ion translational energy spectrometer. In addition to that, we have made a perfect state-to-state study of one-electron capture processes involving rare gas ions.

In this chapter the study of inelastic processes involving the fine-structure states of rare gas ions and the present status of the high resolution translational energy spectrometer at Tokyo Metropolitan University will be described.

4.1 High-Resolution Ion Translational Energy Spectrometry

4.1.1 Apparatus

The translational energy spectrometer used at Tokyo consists of an ion source, a momentum selector, an energy selector, a collision cell, an energy analyzer and a detector. Ions are produced within a conventional electron impact ion source and are mass selected with the momentum selector made of 90° ferrite magnets whose mean radius is 30 mm. The ion source and the momentum selector precede the entrance of the energy selector. The momentum selector is enclosed within a μ-metal shield to prevent stray magnetic fields from disturbing the focusing condition of the energy selector. The resolution of the momentum selector $(\Delta M/M)$ is about 1/10.

The energy selector and analyzer are electrostatic hemispherical condensers with real entrance and exit apertures. The mean radii of each of them is 75 mm, and the entrance and exit apertures are 0.5 mm in diameter. The distance between the inner and the outer electrodes is 30 mm. Aluminium alloy is used for both selector and analyzer in order to reduce the weight of the rather large condensers.

The acceleration voltage applied to the ions is limited to produce kinetic energies in the range from about 100 eV to 2.5 keV. Since the collision energy is rather high, almost all the scattered ions are concentrated in the forward direction. The post-collision energy analyzer is located at 0°. Electric power to all spectrometer components is supplied through noise filters.

The entire system is housed in an all-metal bell jar of 600 mm inner diameter and pumped with a 1200 $1s^{-1}$ oil diffusion pump with a liquid N_2 cooled trap. The ultimate pressure is 5×10^{-10} Torr. This ultrahigh vacuum provides stable operation and a long lifetime of a few years. Ultimate energy resolution of 10 meV (FWHM) has been achieved with Li^+ ions produced with a thermionic emission type source [4.16]. Full details of the spectrometer design have been described elsewhere [4.17–19].

4.1.2 Broadening of Line Shape

Translational energies of scattered ions are broadened because of recoil of targets in collisions between particles having almost equal mass. Therefore, high resolution in translational energy spectrometry can only be achieved at very low energies or very small scattering angles.

Detailed consideration of the broadening of line shape has been presented by *Lorents* and *Conklin* [4.20]. The observed spectral line width, δE, may arise from three sources: the energy spread of the primary beam, δE_p, thermal motion of target particles, δE_{th}, and the angular spread of scattered ions detected, δE_{an}. Assuming that these three sources are independent, the overall width of the line shape is given by

$$\delta E = \left[(\delta E_p)^2 + (\delta E_{th})^2 + (\delta E_{an})^2 \right]^{1/2} . \tag{4.1}$$

In studies of collisional excitation of ions, the excitation energy, Q, coincides with the energy change of the projectiles before and after the collision, ΔW, if the energy analyzer is set at $0°$ and the incident energy is much higher than the excitation energy, and the target can be assumed to be at rest. Furthermore, in the case when δE_{an} is 0, $\delta E \simeq \delta E_p$ if the energy analyzer is located at $0°$.

When the target mass is much lighter than that of the projectile, and hence the effect of momentum transfer can not be neglected, or when the excitation energy is rather large, the measured change of translational energy does not coincide with the change of the internal energy, even though the energy analyzer is set at $0°$. Moreover, the width of the line shape may also be broadened by the thermal motion of the target atoms. The full width at half-maximum (FWHM) of the measured spectrum is given by

$$\delta E(\text{FWHM}) = \left[4 \ln 2 \frac{M_1 kT}{M_0 E_1} \Delta W^2 + \delta E_p^2(\text{FWHM}) \right]^{1/2} , \tag{4.2}$$

where T is the temperature of the target gas and k is the Boltzmann constant [4.21]. For a 1 keV Ne^+ ion incident on H_2 and for $\Delta W = 5$ eV, the energy spread due to the thermal motion of target molecules becomes 200 meV. Therefore, it becomes necessary to cool the target gases in order to achieve high resolution in collisions of heavy ions with light targets.

4.2 Fine-Structure Transitions in Ne^+, Ar^+ and Kr^+

4.2.1 Translational Energy Spectra

Itoh et al. [4.15] have observed excitation and de-excitation between fine-structure states of the ground electronic state of Ar^+ in collisions with Ar. Subsequently, in 1986, *Kobayashi* et al. [4.22] studied the same process in collisions of Kr^+ with He, Ne and Kr. At the time when these studies were made, the energy resolution of the

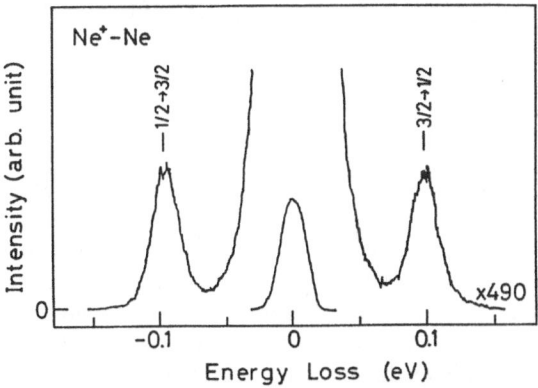

Fig. 4.1. A translational energy spectrum obtained for collisions of Ne$^+$ with Ne at 700 eV. The peak located at zero energy loss corresponds to the primary beam. The width of this peak is about 24 meV (FWHM). Peaks appearing in energy loss (right hand) and energy gain (left hand) sides are due to the excitation $^2P_{3/2} \rightarrow {}^2P_{1/2}$ and de-excitation $^2P_{3/2} \leftarrow {}^2P_{1/2}$, respectively

translational energy spectrometer used was about 40 meV (FWHM). Therefore, the fine-structure transitions in Ne$^+$ could not be resolved from the tail of the primary beam. Very recently, *Fukuroda* et al. [4.23] have succeeded in improving the energy resolution of the spectrometer to 20 meV, and in observing the fine-structure transitions in Ne$^+$.

A typical translational energy spectrum obtained for 700 eV Ne$^+$ ions incident on Ne is shown in Fig. 4.1. The peak located at an energy loss of zero corresponds to the primary beam, and peaks in the energy loss and gain sides are due to the excitation $^2P_{3/2} \rightarrow {}^2P_{1/2}$ and de-excitation $^2P_{3/2} \leftarrow {}^2P_{1/2}$, respectively. An important characteristic feature is seen in the spectrum: the intensities of the excitation and de-excitation peaks seem almost the same. Moreover, the intensity ratios of the two peaks are also found to be independent of the collision energy used. Almost the same type of spectra have also been obtained for Ar$^+$ and Kr$^+$ projective ions.

4.2.2 Fractional Populations of $^2P_{3/2}$ and $^2P_{1/2}$ States

Itoh et al. [4.15] measured the peak height ratios of the two peaks, $I(3/2 \rightarrow 1/2)/I(3/2 \leftarrow 1/2)$, obtained in collisions of Ar$^+$, which is produced by electron impact with the energy of 150 eV, with Ar over a wide collision energy range from 100 to 1500 eV. They observed that the ratios are just unity within an accuracy of $\pm 2\%$ and are independent of the collision energy. The ratios were also measured as a function of the electron impact energy, keeping the collision energy fixed. They again observed that the measured ratios are just unity, and independent of the electron impact energy in the ion source above 25 eV.

On the basis of these observations, they pointed out that the energy dependences of the cross sections for excitation and de-excitation are the same and detailed balance between the two processes should hold. If we can assume the detailed balance between the excitation and de-excitation processes, it becomes possible to deduce the fractional population of the fine-structure states in the primary ion beam.

In a single collision condition, the intensity ratio of the peaks corresponding to the excitation and de-excitation is given by

$$\frac{I(3/2 \to 1/2)}{I(3/2 \leftarrow 1/2)} = \frac{\sigma_{ex}}{\sigma_{de-ex}} \frac{N(3/2)}{N(1/2)} , \tag{4.3}$$

where σ_{ex} and σ_{de-ex} are the cross sections for excitation and de-excitation, respectively, and $N(J)$ is the fraction of the J state contained in the primary beam. The ratio of the cross sections for excitation and de-excitation from state 1 to state 2, $\sigma_{ex}/\sigma_{de-ex}$ is given by the ratio of statistical weights of pseudo-molecule states, f_2/f_1, from the principle of detailed balance. In the present case, the statistical weight is given by $f_i = (2J_i + 1)$, since the target is $^1S_0$. Therefore, the ratio $\sigma_{ex}/\sigma_{de-ex}$ for a fine-structure transition must be 1/2. If so, the fractional ratio of the two fine-structure states contained in the primary ion beam can be determined from the peak intensity ratio by using the relation

$$\frac{N(1/2)}{N(3/2)} = \frac{1}{2} \frac{I(3/2 \leftarrow 1/2)}{I(3/2 \to 1/2)} \tag{4.4}$$

and

$$N(3/2) + N(1/2) = 1 . \tag{4.5}$$

As mentioned above, *Itoh* et al. [4.15] obtained the value 1.00 ± 0.02 for the intensity ratio of the excitation and de-excitation peaks. From this fact, the following result is obtained: the fractional ratio of $J = 1/2$ to $J = 3/2$ state of Ar^+ ions produced by electron impact with electron energy above 25 eV must be 1/2 in order to coincide with the statistical ratio. In other words, $N(3/2) = 0.666$ and $N(1/2) = 0.333$.

In the case of Ar^+ ions, the energy separation between $J = 3/2$ and $J = 1/2$ states is very small (0.18 eV). Therefore, the fractional population coincides well with the statistical ratio. However, in the case of Kr^+, the energy separation (0.66 eV) is much larger. A large number of Rydberg states of the neutral atom converging to the $^2P_{1/2}$ state of the ion exist above the $^2P_{3/2}$ state, and the atoms excited to such Rydberg states rapidly autoionize. Therefore, an appreciable deviation of the experimental fractional ratios from the statistical one would be expected in such a case.

Kobayashi et al. [4.22] have measured the fractional ratios of the two fine-structure states of Kr^+ as a function of the electron impact energy in the ion source in the range from 18 to 150 eV. The measured fractional rations of $J = 1/2$ to $J = 3/2$ states is shown in Fig. 4.2. The measurements have been made with different target gases, Ne, Ar and Kr. In the low electron energy region, the ratio increases sharply as the electron energy increases. Above about 40 eV, the ratio becomes almost flat and a little lower than the statistical one, 0.5. Averaging all data obtained above the electron energy of 40 eV, the value of 0.479 ± 0.002 has been deduced for the ratio.

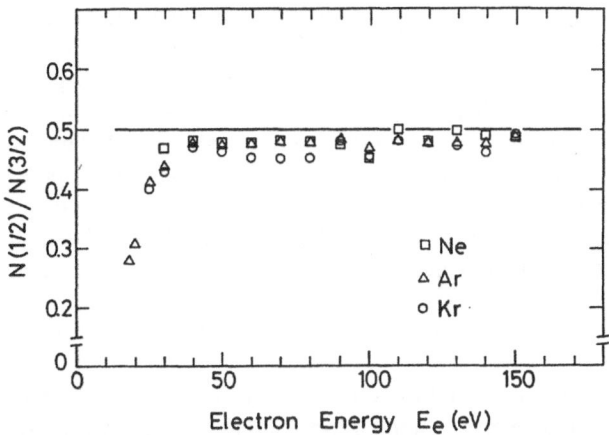

Fig. 4.2. Fractional ratios of $J = 1/2$ to $J = 3/2$ states of $Kr^+ (4p^5, {}^2P_J)$ ions produced by electron bombardment. The measurements were made with different target gases. The value expected from the statistical model, 0.5, is indicated by a solid line

In the case of Ne^+, we have not precisely measured the fractional ratio. However, the intensities of the excitation and de-excitation peaks were almost the same as shown in Fig. 4.1. Therefore, the population of the two fine-structure states in the Ne^+ beam is expected to coincide well with the statistical one for electron energies above several tens of eV.

For Xe^+ ions, the energy separation of the fine-structure states is very large (1.3 eV). It is expected that a large discrepancy will exist between the fractional ratio obtained by experiment and that expected on the basis of the statistical model. Therefore, we have tried to measure the translational energy spectra using Xe^+ ions. However, we could not succeed in observing the fine-structure transition, since the cross section was too small to be detected at a collision energy up to 2 keV, which is the highest energy possible in our spectrometer.

4.2.3 Cross Sections for Fine-Structure Transitions in Ne^+, Ar^+ and Kr^+

The translational energy spectrum shown in Fig. 4.1 was obtained by adjusting the energy analyazer to have an acceptance angle of $\pm 0.45°$ at $0°$. If the inelastically scattered ions are concentrated within a very narrow angle, most of the scattered ions will then be collected by the detector. In such a case, we can obtain absolute integral cross sections from the intensities of the primary and inelastic peaks. The cross section for the transition from J_1 to J_2 is given by

$$\sigma(J_1 \rightarrow J_2) = \frac{I(J_1 \rightarrow J_2)}{I_0} \frac{1}{nl} \left[1 + \frac{N(J_2)}{N(J_1)} \right] , \tag{4.6}$$

where I_0 and $I(J_1 \rightarrow J_2)$ are the intensities of the primary and inelastic peaks, respectively, n is the density of the target gas and l is the collision length.

For Ar$^+$-Ar collisions, *Itoh* et al. [4.15] compared the angular profiles of the primary ions and the inelastically scattered ions. The angular distribution of the inelastically scattered ions was found to be somewhat broader than that of the primary ion beam. Therefore, they were not able to obtain the total cross sections. However, they measured the partial cross sections for forward scattering within $\pm 0.45°$ by setting the energy analyzer at $0°$, in the energy range from 50 to 1500 eV, and compared the result with the theoretical calculation of *Johnson* [4.14]. The magnitude of the measured cross sections was found to be a little smaller than the calculated values. However, the energy dependance of both cross sections resembled each other. From this fact, we suppose that the inelastically scattered ions are concentrated within a very narrow angle because the transition energy is very small compared to the collision energy.

Kobayashi et al. [4.22] have measured the cross section for the fine-structure transition in Kr$^+$+He collisions at energies ranging from 500 to 1300 eV. They have compared the measured cross section with the theoretical results of *Johnson* [4.13] for the symmetric system Kr$^+$+Kr. Although the magnitude of the former was found to be much smaller than the latter, the energy dependences resembled each other.

Very recently, *Fukuroda* et al. [4.23] have performed a systematic study of fine-structure transitions. They have measured the excitation and de-excitation cross sections in collisions of Ne$^+$ and Ar$^+$ with rare gases Ne, Ar, Kr and Xe. Measured cross sections, which are partial cross section for scattering within $\pm 0.45°$ for the de-excitation, $\sigma(3/2 \leftarrow 1/2)$, in Ar$^+$ and Ne$^+$ are shown in Figs. 4.3 and 4.4, respectively, as a function of the relative velocity v. The excitation cross sections, $\sigma(3/2 \rightarrow 1/2)$, shown in the figures are half the values expected on the basis of the

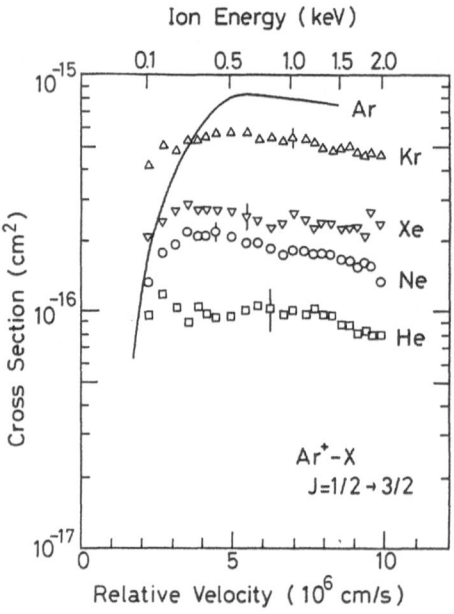

Fig. 4.3. Measured cross sections for the fine-structure transition $\sigma(3/2 \leftarrow 1/2)$ in Ar$^+$ collisions with rare gases. The excitation cross sections $\sigma(3/2 \rightarrow 1/2)$ are half of the $\sigma(3/2 \leftarrow 1/2)$ values. The cross section for the symmetric system Ar$^+$+Ar previously measured is shown by a solid curve [4.15]

130

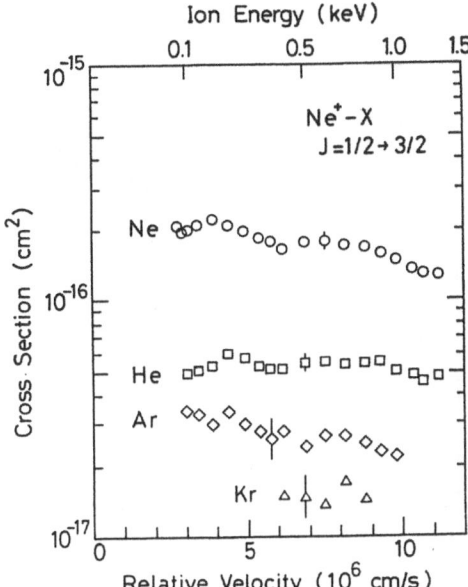

Fig. 4.4. Measured cross sections for the fine-structure transition $\sigma(3/2 \leftarrow 1/2)$ in Ne$^+$

principle of detailed balance. In Fig. 4.3 the cross section for the symmetric system Ar$^+$+Ar, measured by *Itoh* et al. [4.15], is shown by a solid line. *Fukuroda* et al. [4.23] have re-measured these cross sections and have obtained good agreement with the previous experimental result, thus confirming the above noted discrepancy.

In the case of the fine-structure transition in Ar$^+$ ions, the velocity dependences of the measured cross sections resemble each other regardless of the target atoms, except for the case of Ar target. In the low velocity region, the cross sections increase rapidly as the velocity increases, and have maxima at almost the same velocity of about 4×10^6 cms^{-1} ($E_{lab} = 350$ eV). Above the maximum velocity, they gradually decrease with velocity. In the case of the symmetric system Ar$^+$+Ar, the cross section has a maximum at a rather higher velocity than those for the other target atoms. However, the velocity dependence of the cross sections above maximum velocity is very similar to those for the asymmetric systems.

Magnitudes of the cross sections above the maximum velocity depend strongly on the nature of the target species. The symmetric system, Ar$^+$+Ar has the largest cross section. On the other hand, the cross section for the asymmetric systems are much smaller than that for the symmetric system. When the difference of the ionization energies of the projectile and target atoms is the largest, the cross section is the smallest. At a glance, the magnitude of the cross sections seem to be inversely proportional to the difference between the absolute value of the ionization energies of the projectile, I_p, and the target I_t: $|\Delta E| = |I_t - I_p|$.

In the case of Ne$^+$ ions, the magnitude of the measured cross sections are very small compared with those for Ar$^+$ ions, although the energy separation between the fine-structure states is very small. For Xe target, the cross section remains too small to be measured; it has been estimated to have an upper limit of 1×10^{-17} cm^2.

131

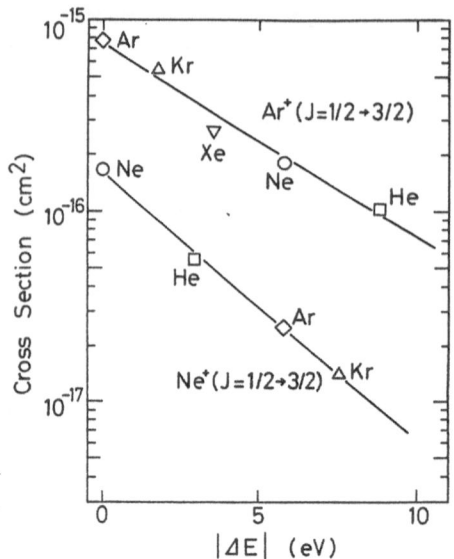

Fig. 4.5. The dependence of the cross sections for fine-structure transitions on the energy difference between the ionization energies of the projectile and target atoms, $|\Delta E| = |I_p - I_t|$. The data are values at 1 keV and 0.8 keV collision energy for Ar^+ and Ne^+, respectively. The solid lines are drawn by the method of least squares

The velocity dependence of the measured cross sections are similar to those of Ar^+ above the maximum velocity, except for the $Ne^+ + He$ collision for which the cross section is almost flat. The magnitudes of the cross section are again inversely proportional to ΔE. The cross section for the symmetric system is the largest, while that for the system $Ne^+ + Xe$, which has the largest $|\Delta E|$, is the smallest.

In order to examine the dependence of the cross section on $|\Delta E|$, the values at $v = 7 \times 10^{-6}$ cms^{-1} (1 keV) for Ar^+ and $v = 8.8 \times 10^{16}$ cms^{-1} (800 eV) for Ne^+ are plotted as a function of $|\Delta E|$ in Fig. 4.5. The cross sections for the fine-structure transition of the ions seem to be well represented by the relationship:

$$\sigma = \sigma_0 \exp\left[-\frac{|\Delta E|}{\alpha}\right] , \tag{4.7}$$

where σ_0 is the cross section for the fine-structure transition of a symmetric system and α is a constant which depends on the ion species. This supports the conjecture that the interaction responsible for the fine-structure transition in ion-atom collisions depends strongly on the absolute value of the difference between ionization energies of projectile and target species. The parameters obtained by a least squares method are $\sigma_0 = 7.5 \times 10^{-16}$ cm^2 [$\sigma_0(\exp) = 7.7 \times 10^{-16}$ cm^2], $\alpha = 4.3$ eV for Ar^+, and $\sigma_0 = 1.6 \times 10^{-16}$ cm^2 [$\sigma_0(\exp) = 1.7 \times 10^{-16}$ cm^2], $\alpha = 3.1$ eV for Ne^+.

So far, theoretical calculations of the cross sections for fine-structure transitions have been performed only by Johnson for the symmetric systems $Ar^+ + Ar$, $Kr^+ + Kr$ and $Xe^+ + Xe$. No calculation has yet been carried out for the asymmetric systems described in this chapter. At present we have precise information only on the potential energies for these collision systems. Theoretical studies for these asymmetric systems seem necessary.

On the other hand, from the experimental viewpoint, it must be noted that the cross sections obtained are not total ones but partial ones scattered into the forward direction within $\pm 0.45°$. Therefore, it is possible that the apparent dependence of the measured cross sections on $|\Delta E|$ is affected by possible differences in angular distribution functions; consequently, we must be careful in concluding that cross sections for the fine-structure transitions are fully represented by the relation (4.7). More detailed studies, such as the measurement of differential cross sections, seem an urgent necessity.

4.3 Excitation and De-excitation Processes in Doubly Charged Rare Gas Ions

4.3.1 Translational Energy Spectra

The ground state of doubly charged ions of the rare gases is ^3P_J, with the state $J = 2$ being the lowest in energy. Metastable states $^1D_2$ and $^1S_0$ exist near the ground state. Nearly twenty years ago, *Moore* [4.24, 25] first observed the spin non-conservative transition $^3P \rightarrow {}^1D$ in N^+ collisions with the rare gases by means of translational energy spectrometry. The energy resolution of his apparatus was somewhat inferior to that of our spectrometer. Therefore, he was not able to distinguish the fine-structure states of ^3P.

Kobayashi et al. [4.19] first succeeded in resolving the different spin-orbit states of ^3P in collisions of Kr^{2+} with He and Ne. A typical translational energy spectrum for 2016 eV Kr^{2+} ions incident on Ne is shown in Fig. 4.6. The ions were produced by electron impact at an electron energy of about 150 eV. The peaks located at en-

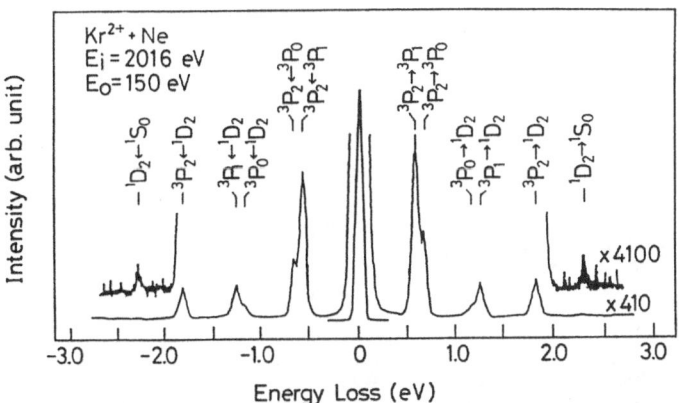

Fig. 4.6. A translational energy spectrum of scattered Kr^{2+} from Ne. The peak located at zero energy loss is mainly due to the primary beam. The peaks located in energy loss and energy gain sides correspond, respectively, to excitation and de-excitation among $^3P_2$, $^3P_1$, $^3P_0$, $^1D_2$ and $^1S_0$ states of Kr^{2+}

ergy loss and gain (negative loss) sides correspond to excitation and de-excitation, respectively. All transitions among $^3P_2$, $^3P_1$, $^3P_0$, $^1D_2$ and $^1S_0$ can be distinguished.

4.3.2 Fractional Populations of ^3P_J, $^1D_2$ and $^1S_0$

Peak heights for excitation and de-excitation between the two states are appreciably different to those seen in the case of fine-structure transitions in singly charged ions. This fact suggests that the populations of the fine-structure states in the doubly charged ion beam are different from those expected from statistical considerations.

Assuming that the principle of detailed balance holds for excitation and de-excitation between two states i and k, the ratio of the populations of i and k in the primary beam is easily obtained by

$$\frac{N(i)}{N(k)} = \frac{(2J_i + 1)}{(2J_k + 1)} \frac{I(i \rightarrow k)}{I(i \leftarrow k)} . \tag{4.8}$$

In general, it is possible that highly metastable states may also be present in the ion beam. However, it is known that the population of such long-lived metastable states in Kr^{2+} ions produced by electron bombardment is only a few percent at most [4.26]. Therefore, we can neglect the existence of such highly excited long-lived metastable states and assume that

$$N(^3P_2) + N(^3P_1) + N(^3P_0) + N(^1D_2) + N(^1S_0) = 1 . \tag{4.9}$$

If the fractional ratios, $N(^3P_1)/N(^3P_2)$, $N(^3P_0)/N(^3P_2)$, $N(^1S_0)/N(^3P_2)$ are obtained, we can evaluate all of the fractional populations.

No peak corresponding to the transition $^3P_2 \rightarrow {}^1S_0$ is seen in Fig. 4.6. The measured fractional populations of $^3P_2$, $^3P_1$, $^3P_0$, $^1D_2$ and $^1S_0$ states in the ion beam produced by electron impact are shown in Fig. 4.7 as a function of the electron en-

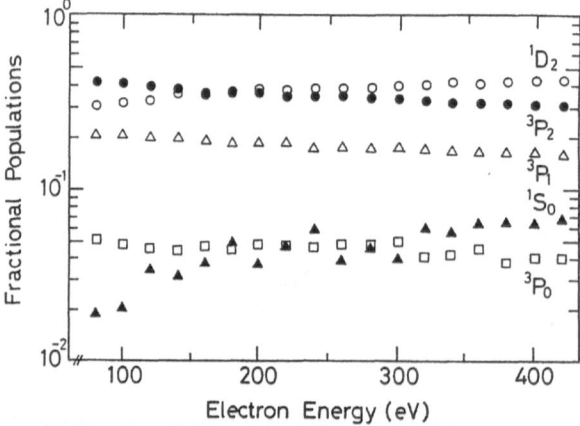

Fig. 4.7. Measured fractional populations of $^3P_2$, $^3P_1$, $^3P_0$, $^1D_2$ and $^1S_0$ states in the Kr^{2+} ion beam generated by electron bombardment. The populations expected from the statistical model are 0.333 for $^3P_2$ and $^1D_2$, 0.200 for $^3P_1$ and 0.066 for $^3P_0$ and $^1S_0$

ergy. If the production of these states obeys the simple statistical rule, the fractional populations of these states must be 0.333, 0.200, 0.066, 0.333 and 0.066, respectively, and should be independent of the electron energy.

The measured fractional populations do not coincide with the values expected from the statistical model, and, moreover, they appear to depend on the electron energy. The values of $^3P_2$, $^3P_1$ and $^3P_0$ decrease monotonically while those of $^1D_2$ and $^1S_0$ increase as the electron energy increases. The fractional population of $^1S_0$ is particularly strongly dependent on the electron energy. Kobayashi et al. [4.19] have discussed in detail the origin of the dependence of the measured fractional populations on the electron energy. Subsequently, Nakamura et al. [4.27] determined the fractional populations of $^3P_2$, $^3P_1$, $^1D_2$ and $^1S_0$ of Ar^{2+} states by the same technique. They reported that, in contrast to the situation in Kr^{2+}, the measured fractional populations are very close to the statistical ones above an electron impact energy of 100 eV.

Very recently, Fukuroda [4.28] has tried to measure the fractional populations of these states in Ne^{2+}. However, he has not succeeded in obtaining the translational energy spectra showing fine-structure state excitation and de-excitation. This will be discussed below.

4.3.3 Cross Sections for Excitation and De-excitation Among $^3P_2$, $^3P_1$, $^3P_0$, $^1D_2$ and $^1S_0$ States

Study of the excitation and de-excitation processes in doubly charged ions of rare gases is very important; not only the transitions among fine-structure states of ^3P but also forbidden transitions, ^3P_J-$^1D_2$, ^3P_J-$^1S_0$ (spin non-conservative) and $^1D_2$-$^1S_0$ (electric quadrupole allowed) are directly observable by means of translational energy spectrometry. These transitions cannot be observed by conventional photoemission spectroscopy.

Since the fractional populations of all the low-lying states of Kr^{2+} have been determined, Kobayashi et al. [4.19] have evaluated the cross sections for transitions among these states. They have measured the cross section for excitation and de-excitation in Kr^{2+}+He and Kr^{2+}+Ne. The measured cross sections for the excitation processes $^3P_2$-$^3P_1$, $^3P_2$-$^3P_0$, $^3P_{0,1}$-$^1D_2$, $^3P_2$-$^1D_2$ in collisions of Kr^{2+} with He and Ne are shown in Figs. 4.8 and 4.9, respectively. Since detailed balance between excitation and de-excitation is assumed, the cross section for de-excitation from the state k to i is given by multiplying the factor $(2J_i + 1)/(2J_k + 1)$ with the cross section shown in the figure.

An interesting feature is seen in the figure in that the spin non-conservative transitions $^3P_2$-$^1D_2$ and $^3P_{0,1}$-$^1D_2$ have rather large cross sections. The magnitudes of the cross sections for these transitions are almost the same as those for the spin conservative transitions. Almost the same feature has also been observed in Kr^{2+}+Ne collisions. In the case of Ne target, the cross sections for the transition $^1D_2$-$^1S_0$ has been measured, although the magnitude was found to be very small.

Very recently, Fukuroda [4.28] has measured the cross sections for the transitions $^3P_2$-$^3P_1$, $^3P_2$-$^1D_2$ and $^1D_2$-$^1S_0$ in collisions of Ar^{2+} with He and Ne in the energy

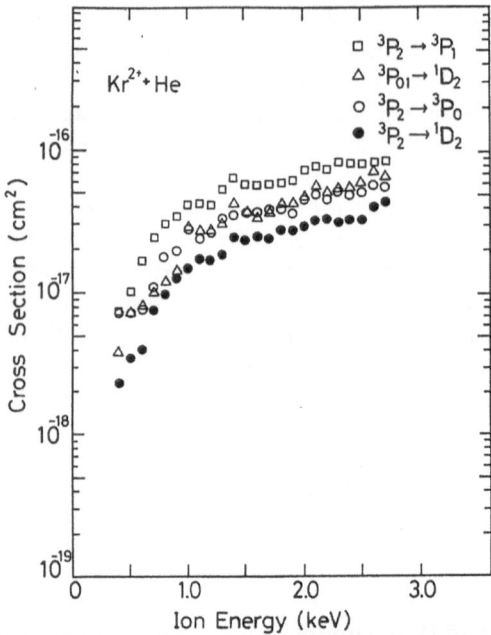

Fig. 4.8. Measured cross sections for the excitation in collisions of Kr^{2+} with He. Cross sections for de-excitation from state k to i are given by multiplying the factor $(2J_i + 1)/(2J_k + 1)$ to the excitation cross sections from the state i to k

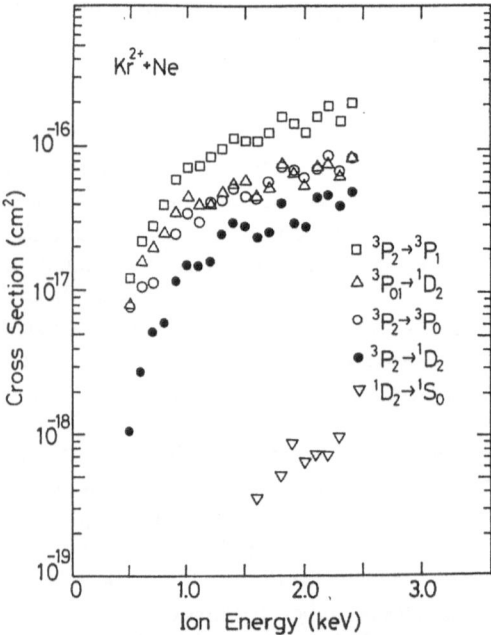

Fig. 4.9. Measured cross sections for excitation processes in collisions of Kr^{2+} with Ne. Cross sections for de-excitation are given by the same procedures as described in the caption of Fig. 4.8

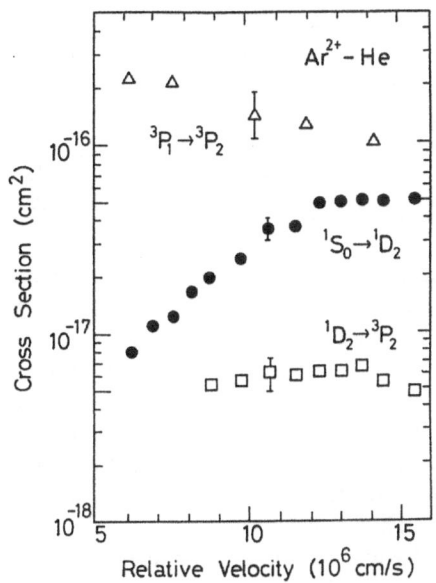

Fig. 4.10. Measured cross sections for excitation processes in collisions of Ar^{2+} with He. Cross sections for de-excitation are given by the same procedures as described in the caption of Fig. 4.8

range from 0.8 to 5 keV. The measured cross sections for these excitation processes are shown in Fig. 4.10. Quite different features are seen in this figure compared to the case of Kr^{2+} collisions. The cross section for the fine structure transition $^3P_2$-$^3P_1$ has a magnitude of the order of 10^{-16} cm^2, which is almost the same as that in the Kr^{2+} collisions at the highest energy studied. It has an increasing trend even at the lowest energies studied, in contrast to the case of Kr^{2+}.

The magnitude of the cross section for the spin non-conservative transition $^3P_2$-$^1D_2$ is very small, about 1 % of that for the fine-structure transition. However, it increases as the collision energy decreases. The cross section for the electric quadrupole transition $^1D_2$-$^1S_0$ has a quite different energy dependence compared with those for the other two processes. It strongly depends on the collision energy and increases sharply as the energy increases. Almost the same results have also been obtained in Ar^{2+} + Ne collisions.

The characteristic features observed in Ar^{2+} are quite different from those in Kr^{2+}. In the case of Kr^{2+} collisions the magnitudes of the cross sections for the spin non-conservative transitions are comparable with those for the fine-structure transitions, and the energy dependences are almost the same. Furthermore, the electric quadrupole transition $^1D_2$-$^1S_0$ has not been observed in Kr^{2+} + He although it has been weakly observed in Kr^{2+}+Ne.

The transition energy of the processes $^3P_2$-$^1D_2$ and $^1D_2$-$^1S_0$ are 3.62 and 2.37 eV, respectively, for Kr^{2+}, and 3.98 and 2.38 eV, respectively, for Ar^{2+}. The values are not very much different for the two ion species. Therefore, the differences observed in cross sections for Kr^{2+} and Ar^{2+} cannot be simply attributed to the differences between the transition energies.

Fukuroda [4.28] has also studied the fine structure transition in Ne^{2+} collisions with the rare gases. However, it has so far not been possible to resolve fine-structure transitions because the transition energies are very close. In addition to that, a somewhat peculiar feature has been seen. The transitions among ^3P, ^1D and ^1S states have not been observed, not only for rare gas targets but also for diatomic molecules such as N_2, O_2 and H_2. It is concluded that the cross sections for the spin non-conservative and electric quadrupole transitions are smaller than 10^{-18} cm^2 in the case of Ne^{2+} collisions.

From these observations, it is suggested that the cross sections for spin non-conservative transitions are very small in light ions and they become large in heavy ions. *Moore* [4.24, 25] has studied the spin non-conservative transitions in collisions of N^+ ions with the rare gases. He reported that such transitions occur with large probability in the case of scattering from heavy atoms, like Kr and Xe, in contrast to scattering with light atoms such as Ne and Ar. He has explained these observations by analogy to optical spectroscopy where LS coupling dominates in light atoms while jj coupling becomes more important in heavy atoms. The results of *Kobayashi* et al. [4.19] and *Fukuroda* [4.28] seem to support his explanation.

The electric quadrupole transition $^1D_2$-$^1S_0$ has been unambiguously observed only in Ar^{2+} collisions. Although the same transition was also observed in the Kr^{2+}+Ne collision system, the measured cross section was found to be extremely small. At present, this peculiar feature is not well explained. Accumulation of further data for such transitions seem necessary.

4.4 The Role of Fine-Structure States in Electron Capture Reactions

4.4.1 Relative Cross Sections for Reactions in $Kr^{2+}(^1D_2)$ + Ne

Electron capture processes play an important role in collisions of multiply charged ions with atoms. Many state-resolved studies have been performed (see Chap. 3). High-resolution translational energy spectrometry enables us to distinguish both the initial and final states of such reactions.

Nakamura et al. [4.29] have succeeded in partly resolving the initial and final states of the one-electron capture reaction in Kr^{2+}+He and Kr^{2+}+Ne collisions. They reported that only the $^1S_0$ state in the primary ion beam captures an electron in Kr^{2+}+He collisions, and the product ions, Kr^+ and He^+, are both in the ground electronic state. They also mentioned that, in the case of Kr^{2+}+Ne collisions, all the low-lying Kr^{2+} states can capture an electron from Ne, and product ions are also in the ground electronic states. For each initial state of Kr^{2+}, the following four reaction channels are possible:

$$Kr^{2+}(X) + Ne\ (^1S_0)$$
$$\rightarrow Kr^+(^2P_{1/2}) + Ne^+(^2P_{1/2}) + 4.057\,eV \ldots (a)$$
$$\rightarrow Kr^+(^2P_{1/2}) + Ne^+(^2P_{3/2}) + 4.154\,eV \ldots (b)$$

$$\rightarrow Kr^+(^2P_{3/2}) + Ne^+(^2P_{1/2}) + 4.723\,eV \dots \textbf{(c)}$$
$$\rightarrow Kr^+(^2P_{3/2}) + Ne^+(^2P_{3/2}) + 4.820\,eV \dots \textbf{(d)}.$$

Nakamura et al. [4.29] have measured the cross sections for the various re-action channels and found that the process starting from $X = {}^1D_2$ has the largest cross section. In these experiments, they were also successful in resolving the fine-structure states of $Kr^+(^2P_{3/2,1/2})$. However, they could not distinguish the fine-structure states of $Ne^+(^2P_{3/2,1/2})$, and they observed the peaks corresponding to the reaction channels (a) + (b) and (c) + (d) from the initial state $Kr^{2+}(^1D_2)$. Suc-cess in resolving the fine-structure states of Ne^+ would have enabled a 'perfect' state-to-state study to be performed.

Recently, *Fukuroda* et al. [4.30] have tried to improve the energy resolution of the spectrometer and have succeeded in obtaining an energy resolution of about 50 meV (FWHM) for doubly charged ions (25 meV for singly charged ions). All of the four reaction channels from the initial state $Kr^{2+}(^1D_2)$+Ne $(^1S_0)$ have been distinguished. A typical translational energy spectrum is shown in Fig. 4.11. Here, only spectra corresponding to the reactions starting from the $^1D_2$ state are shown and those from the ^3P_J and $^1S_0$ states are not shown. The peaks are not clearly sep-arated, but they are easily deconvoluted using Gaussian distributions with a width of 80 meV and assigned to reactions (a) – (d). Although the energy resolution of the spectrometer is 50 meV, the spread of the measured peaks are 80–110 meV (FWHM). This is explained by the thermal motions of the target atoms in the collision cell, as discussed in Sect. 4.1.2.

Relative cross sections for the reactions (a) – (d) have been measured for collision energies in the range 400 to 2000 eV (77 to 385 eV in the centre-of-mass system). The relative cross sections have been normalized at each collision energy by the signal

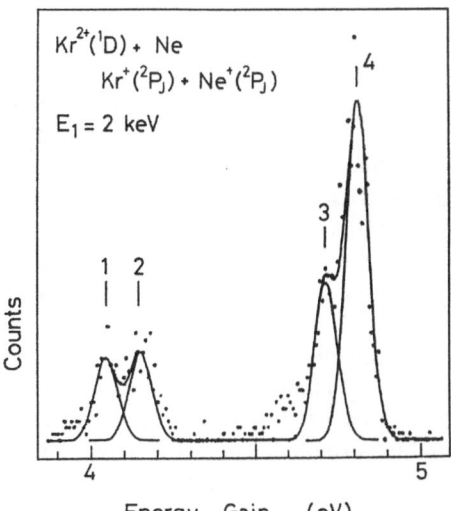

Fig. 4.11. Translational energy spectrum of Kr^+ produced by one-electron capture reaction in collisions of Kr^{2+} with Ne at a collision energy of 2 keV. Curves show the calculated spectrum composed of Gaussian distributions having FWHM of 80 meV. The numbers labeling the peaks refer to the reactions (a) – (d) (see text)

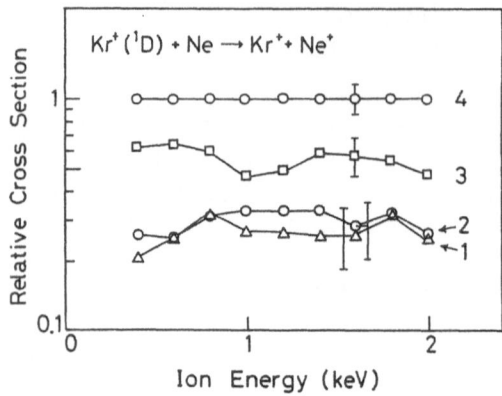

Fig. 4.12. Measured relative cross sections. Numbers shown in the figure correspond to the reactions (a) – (d). The data are normalized by the cross section value for reaction (d) at each energy (see text)

counts for reaction (d). The results are shown in Fig. 4.12. Some interesting features may be noticed in the figure. The relative cross sections measured are almost flat over the energy range studied. The magnitudes of the cross sections for reactions (a) and (b) are almost the same and about one third of that for reaction (d). The relative value of the cross section for reaction (c) is about one half of that for (d).

Since the energy separation of the fine-structure states in Ne^+ is very small, it may be expected that these states are mixed during the collision. If so, the cross section should be proportional to the statistical weight of the $Ne^+(^2P_J)$ states, and it is expected that the cross section for reaction (a) should be one half of that for (b), and the cross section for (c) should also be half of that for (d). The measured cross section for (c) is, indeed, roughly one half of that for (d), in accordance with the statistical ratio. However, the cross sections for (a) and (b) are almost the same, in discord with the statistical prediction. This suggests that the adjacent reaction channels are not mixed during the collision and are independent of each other.

4.4.2 Diabatic Potential Energy Curves

Since the collision velocity ($v_i < 10^{-7}$ cms^{-1}) in this experiment is much smaller than the classical orbital electron velocity, it can be reasonably assumed that the reaction occurs through crossing points of diabatic potential energy curves of the quasimolecule formed during the collision. The Landau-Zener model, which has been widely used and has been successful in explaining the mechanism of electron capture reactions in collisions of multiply charged ions with atoms and molecules, is used here to discuss the role of the fine-structure states in one-electron capture reactions.

Diabatic potential energy curves generated from the initial states $Kr^{2+}(^3P_J)+Ne$ and $Kr^{2+}(^1D_2)+Ne$, and the product states $Kr^+(^2P_J)$ and $Ne^+(^2P_J)$, are shown in Fig. 4.13. Here, we assume that the interactions between the initial and final systems are only due to the polarization force and the Coulomb force, respectively. The crossing radius of the potential curves for the reaction with exoergicity, ΔE (in eV), is calculated by the equation

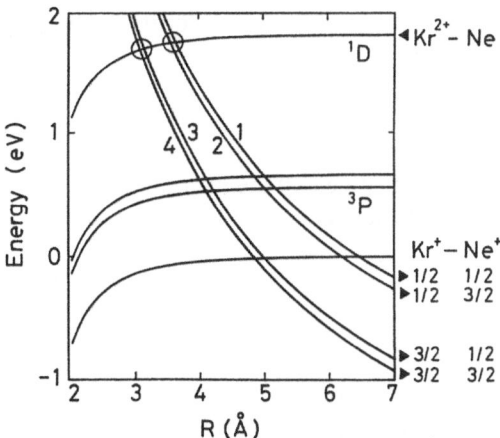

$$\Delta E = 14.4 \left[\frac{1}{R_c} + \frac{2\alpha}{R_c^4} \right] , \tag{4.10}$$

where α (in Å^3) is the polarizability of Ne and R_c (in Å) is the internuclear distance at which crossing occurs. Since the relative cross sections measured are different from those expected by the statistical model, we also assume that the adjacent potential curves and the crossing points are independent.

4.4.3 Landau-Zener Model for Single Crossing

First, we consider the case where two states generated from the initial state 1 and the final state 2 cross at the point, a, at an internuclear distance R_c. In the Landau-Zener approximation, the probablity of the diabatic transition is given by

$$p_a = \exp \left(\frac{-2\pi H_{12}^2}{V_b \Delta F} \right) \tag{4.11}$$

and that of the adiabatic one is given by

$$\bar{p}_a = 1 - p_a , \tag{4.12}$$

where v_b is the radial velocity, H_{12} is the value of the interaction matrix element between the two potential curves at R_c and ΔF is the difference in the slopes of the two curves. For the interaction matrix element H_{12}, the empirical formula given by *Olson* et al. [4.31] for an ion charge $q = 2$

$$H_{12} = 56.5 R_c \exp(-2.11 R_c) \text{eV} , \tag{4.13}$$

is used throughout the following calculation.

Fukuroda et al. [4.21] have proposed a matrix representation for the transition probablity $P = P^a P^a$ where

$$P^a = \begin{vmatrix} p_a & \bar{p}_a \\ \bar{p}_a & p_a \end{vmatrix} . \tag{4.14}$$

The transition probability from state i to j is given by the matrix element P_{ij}; then the transition probablity of one-electron capture is

$$P_{12} = 2p_a\bar{p}_a \tag{4.15}$$

and the cross section is given by

$$\sigma = 2\pi \int_0^{R_c} P_{12} b\, db . \tag{4.16}$$

The cross section for one-electron capture calculated using (4.16) is shown in Fig. 4.14 as a function of R_c. Similar curves have been reported in many other studies. The cross section has an appreciable magnitude in a certain range of R_c. This range is often called the 'reaction window'. *Nakamura* et al. [4.29] have reported the relative cross sections for the different reaction channels of the one-electron capture process in $Kr^{2+} + Ne$ in which the electronic states of the primary Kr^{2+} ions and the fine-structure states of the product Kr^+ ions have been distinguished, but the fine-structure states of Ne^+ have not been resolved. Their results are shown in Fig. 4.14. The calculated reaction window appears to agree well with their experimental results. The results obtained by *Fukuroda* et al. [4.30], in which the initial and the final states have been fully resolved, are also plotted in the same figure. These appear to deviate from the calculated ones. This fact suggests that a more careful calculation is necessary.

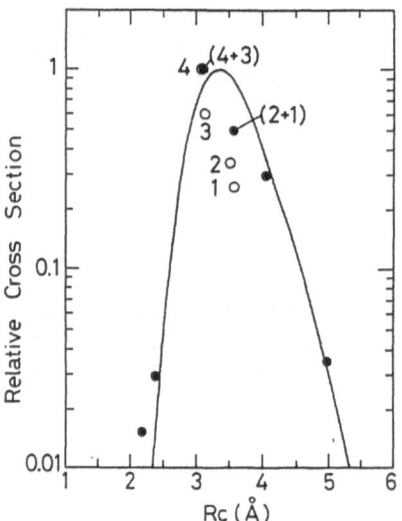

Fig. 4.14. The R_c dependence of measured relative cross sections normalized to the maximum value of the measured cross sections. Open circles represent the present results at 1.4 keV and closed circles represent the previous results obtained by *Nakamura* et al. [4.29] at 1334 eV. The full curve represents the calculated relative cross section using the two-state Landau-Zener model. The numbers 1, 2, 3 and 4 shown in the figure correspond to reactions (a), (b), (c) and (d), respectively, described in the text

4.4.4 Multichannel Landau-Zener Model

As seen in Fig. 4.13, the potential curve corresponding to the incident $Kr^{2+}+Ne$ channel crosses four potential curves corresponding to the $Kr^{+}+Ne^{+}$ product channels. Here, we label the crossing points by a, b, c and d from the outermost one, and the initial and final states by numbers 1 to 5. The probablity matrix at the crossing, a, is given by

$$P^{a} = \begin{vmatrix} p_a & \bar{p}_a & 0 & 0 & 0 \\ \bar{p}_a & p_a & 0 & 0 & 0 \\ 0 & 0 & 1 & 0 & 0 \\ 0 & 0 & 0 & 1 & 0 \\ 0 & 0 & 0 & 0 & 1 \end{vmatrix} \tag{4.17}$$

The transition probablity from state i to j is given by the element P_{ij} of the matrix

$$P = P_a P_b P_c P_d P_d P_c P_b P_a , \tag{4.18}$$

where the matrix P is a 5×5 symmetric one. The transition cross section is obtained from (4.16.). The upper limit of the integration is taken as the R_c value of the outermost crossing.

The calculated cross sections for reaction (a) – (d) are shown in Fig. 4.15 by broken curves. The cross sections are normalized to the value obtained for reaction (d). Two pairs of cross sections having almost the same value have been obtained, in contrast to the experimental result. Given that we have utilised the interaction energy given in (4.13), which depends only on R_c, it is quite natural that we have obtained two pairs of similar cross sections, since the crossing for reaction (a) and (b) are very close to those for reactions (c) and (d).

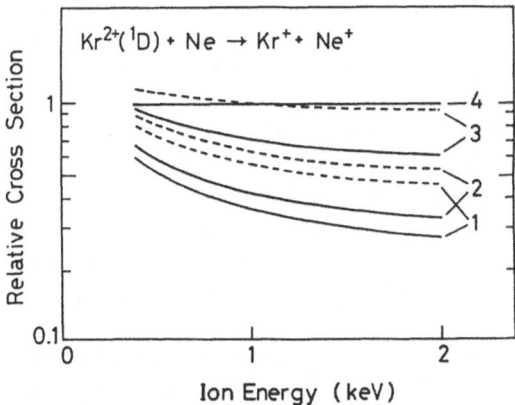

Fig. 4.15. Calculated relative cross sections using the five-state Landau-Zener model. The broken curves represent the values calculated without weight factors. The full curves represent the values calculated with weight factors. The numbers 1–4 labeling the lines correspond to reaction (a) – (d), respectively (see text). The data are normalized to the value for reaction (d) at each energy

4.4.5 Landau-Zener Calculation with Weighted Transition Probability

In the above calculation we adopted the empirical interaction energy obtained by *Olson* et al. [4.31]. It should be noted that this interaction energy was empirically obtained by comparing measured cross sections with the Landau-Zener model with only a single crossing. Therefore, their interaction energy depends only on R_c. However, in the present case, we discuss the partial cross sections which are distinguished for particular sublevels. It therefore seems necessary to modify their empirical formula in order to explain the experimental results within the Landau-Zener formalism.

The interaction energy H_{12} in (4.11) is the coupling strength between two molecular states. Therefore, we try to introduce a weighted transition probablity which takes into account the correlation between the quasimolecular states. Then, the diabatic transition probability of (4.11) is corrected as

$$P_k = \exp\left(-\frac{2\pi H_{12}^2 w_k}{v_b \Delta F}\right) \tag{4.19}$$

for the multicrossing model, where w_k is the weight at the crossing k.

Fukuroda et al. [4.21] have studied the vibrational distribution of H_2^+ produced by the one-electron capture reaction in $Ne^{2+}+H_2$ collisions. They found that the Franck-Condon factor is a good weighting factor to reproduce the experimental results. Similar attempts have been made by *Taulbjerg* [4.32] in explaining the reaction window obtained by the experimental study of one-electron capture processes in $Ar^{8+}+Ne$ collisions. He has corrected the interaction energy of *Olson* and *Salop* [4.33] by multiplying by a function of the quantum numbers n and l, the principal and angular momentum quantum numbers of the captured electron.

For the present, we assume that the transition between two diabatic potentials at a crossing occurs when both molecular states are of the same symmetry and the same spin multiplicity. We think this assumption is quite natural. If so, the number of molecular states common to the initial and final states should be related to the transition probablity. The molecular states resulting from the separated atoms $Kr^{2+}(^1D_2)+Ne\,(^1S_0)$ are in singlet states $^1\Sigma^+$, $^1\Pi$ and $^1\Delta$. On the other hand, the molecular states resulting from $Kr^+(^2P)+Ne^+(^2P)$ are $^1\Sigma^+$, $^1\Sigma^+$, $^1\Sigma^-$, $^1\Pi$, $^1\Pi$, $^1\Delta$, $^3\Sigma^+$, $^3\Sigma^+$, $^3\Sigma^-$, $^3\Pi_i$, $^3\Pi_i$ and $^3\Delta$. In order to correlate these 12 molecular states with the four final states for the reactions (a) – (d), we consider that correlation from (J_{Kr}, J_{Ne}) to (Λ, S) couples with the molecular states resulting from $^2P + ^2P$ of the separated atoms. In this correlation, the electronic angular momentum about the internuclear axis is conserved. The intermediate molecular states through which the transition from the initial $^1D + ^1S$ to the final $^2P + ^2P$ can occur are only five states $^1\Sigma^+$, $^1\Sigma^+$, $^1\Pi$, $^1\Pi$ and $^1\Delta$; the angular momentum quantum numbers of the intermediate molecular states for reactions (a) – (d) are 1, 1, 1 and 2, respectively. Thus the weight factors for the transition probablity are given by $w_1 = w_2 = w_3 = 0.2$ and $w_4 = 0.4$.

The relative cross sections have been calculated by the Landau-Zener model again with these values of the weight factors. The calculated relative cross sections are shown in Fig. 4.15 as the full curves. The results agree well with the experimental ones, except at low energy. From the view point of the weighted transition probability, it is quite understandable that the present experimental cross sections deviate from the reaction window calculated for a single crosssing. The total sum of the cross sections for sublevels is seen to agree with the calculated window.

In the above calculation, we have taken into account only the incoming potential for $Kr^{2+}(^1D)+Ne$. The outgoing potentials cross the potentials resulting from the $Kr^{2+}(^1S)+Ne$ and $Kr^{2+}(^3P_J)+Ne$ incoming channels. The molecular states resulting from the states $^1S+^1S$ and $^3P+^1S$ are singlet and triplet states, respectively. In this model, since electron spin conservation is assumed, there is no interaction between the states $^1D+^1S$ and $^3P+^1S$, even if the potentials for these states cross that for the one-electron capture state $^2P+^2P$. Therefore, we need to consider only the effect of the state $Kr^{2+}(^1S)$. The effect has been estimated by the multichannel Landau-Zener model having two incoming and four outgoing potentials. However, the effect is found to be very small. The difference between the results obtained using the six-state model and those from the five-state model are within 2 % of each other at a collision energy of 2 keV. This is consistent with the experimental result of *Nakamura* et al. [4.29], who observed only a small cross section for electron capture from $Kr^{2+}(^1S)$ to $Kr^+(^2P)$. Therefore, we may expect that the transition occurring through the state $Kr^{2+}(^1S)$ has little effect on the one-electron capture cross section.

Acknowledgements. The support of the Grant in Aid from the Ministry of Education, Science and Culture, Grant No. 63606003, is gratefully acknowledged.

References

4.1   C. E. Moore: *Atomic Energy Levels*, (National Bureau of Standards, Washington, 1971)
4.2   N. G. Adam, D. Smith, D. Griff: J. Phys. B **12**, 791 (1979)
4.3   N. G. Adam, D. Smith, E. Alge: J. Phys. B **13**, 3235 (1980)
4.4   T. Tanaka, J. Durup, T. Kato, I. Koyano: J. Chem. Phys. **74**, 5561 (1981)
4.5   I. Koyano, K. Tanaka, T. Kato: *Electronic and Atomic Collisions*, eds. D.C. Lorents, M. Meyerhof, J.R. Peterson, (North-Holland, Amsterdam, 1986), p. 529
4.6   M. R. Spalburg, E. A. Gislason: Chem. Phys. **94**, 339 (1985)
4.7   G. Parlant, E. A. Gislason: *Electronic and Atomic Collisions*, eds. H. B. Gilbody, W. R. Newell, F. H. Read, A. C. H. Smith, (Elsevier, Amsterdam, 1988), p. 357
4.8   T. Nakamura, N. Kobayashi, Y. Kaneko: J. Phys. Soc. Japan. **55**, 3831 (1986)
4.9   N. Kobayashi: *Electronic and Atomic Collisions*, eds. H. B. Gilbody, W. R. Newell, F. H. Read, A. C. H. Smith, (Elsevier, Amsterdam, 1988), p. 343
4.10   A. Matsumoto, S. Othani, T. Iwai: J. Phys. B **15**, 4655 (1983)
4.11   W. L. Wiese, M. W. Smith, B. M. Glennon: *Atomic Transition Probablities* Vol. 1, (National Bureau of Standards, Washington, 1966)
4.12   W. L. Wiese, M. W. Smith, B. M. Miles: *Atomic Transition Probablities*, Vol. 2 (National Bureau of Standards, Washington, 1966)
4.13   R. E. Johnson: J. Phys. B **3**, 539 (1970)
4.14   R. E. Johnson: J. Phys. Soc. Japan. **32**, 1612 (1972)
4.15   Y. Itoh, N. Kobayashi, Y. Kaneko: J. Phys. Soc. Japan. **50**, 3541 (1981)
4.16   Y. Itoh, N. Kobayashi, Y. Kaneko: J. Phys. B **14**, 679 (1981)

4.17 N. Kobayashi, Y. Itoh, Y. Kaneko: J. Phys. Soc. Japan **45**, 617 (1978)

4.18 N. Kobayashi: *Electronic and Atomic Collisions*, ed. S.Datz, (North-Holland, Amsterdam, 1982) p. 355

4.19 N. Kobayashi, T. Nakamura, Y. Kaneko: J. Phys. Soc. Japan **52**, 2684 (1983)

4.20 D. C. Lorents, G. M. Conklin: J. Phys. B **5**, 950 (1972)

4.21 A. Fukuroda, N. Kobayashi, Y. Kaneko: J. Phys. B **22**, 3457 (1989)

4.22 N. Kobayashi, A. Nakamura, Y. Kaneko: J. Phys. Soc. Japan **52**, 1581 (1983)

4.23 A. Fukuroda, N. Kobayashi, Y. Kaneko: J. Phys. Soc. Japan **59**, 898 (1990)

4.24 J. H. Moore: Phys. Rev. A **8**, 2359 (1973)

4.25 J. H. Moore: Phys. Rev. A **10**, 724 (1974)

4.26 P. Varga, H. Winter: Phys. Rev. **18**, 2453 (1978)

4.27 T. Nakamura, N. Kobayashi, Y. Kaneko: J. Phys. Soc. Japan **54**, 2774 (1985)

4.28 A. Fukoroda: private communication

4.29 T. Nakamura, N. Kobayashi, Y. Kaneko: J. Phys. Soc. Japan **54**, 1743 (1985)

4.30 A. Fukuroda, N. Kobayashi, Y. Kaneko: J. Phys. B **22**, 3471 (1989)

4.31 R. E. Olson, F. T. Smith, E. Bauer: Appl. Opt. **10**, 1848 (1971)

4.32 K. Taulbjerg: J. Phys. B **19**, L367 (1986)

4.33 R. E. Olson, A. Salop: Phys. Rev. A **14**, 579 (1976)

5. Probing Interaction Potentials: Small Angle Differential Scattering of H^+ and H with He

L. K. Johnson and R. F. Stebbings

With 9 Figures

Absolute measurements of heavy-particle angular differential scattering cross sections at keV energies and small angles permit determination of interaction potential energy curves and provide data required to model physical processes in plasmas and planetary atmospheres. Furthermore, these cross sections are typically so strongly forward-peaked that absolute small angle measurements contain the major part of the total cross section. The structure observable at small angles (due primarily to diffraction effects, classical rainbow scattering, and quantal interference oscillations) is also worthy of study in its own right. Despite the substantial body of data on angular differential scattering, there exist very few *absolute* measurements. A programme of measurement of absolute differential cross sections for scattering of atoms and ions was begun at Rice University in 1983. In this chapter, absolute differential cross sections determined by this group [5.1–3] for direct scattering of keV-energy H atoms from neutral helium and direct and charge transfer scattering of keV-energy H^+ ions from neutral helium are reviewed.

The small number of electrons and the large energy differences between quantum states make the $H^{(0,+)}$-He system relatively amenable to calculations of differential cross sections. When one quantum state of the collision complex dominates the interaction between particles, simple and rapid modeling techniques (in particular, potential scattering calculations) may be used to describe the structure observed [5.4, 5]. When more than one quantum state is important to a collision process (for example, in charge transfer), characteristic oscillatory structure is frequently observed in scattering cross sections. Semiclassical models of these processes assume that a transition between two states takes place in a well-localized region as the two particles approach or recede from each other [5.6]. Interferences between the scattering amplitudes for these two cases is responsible for oscillatory structure in the cross sections. M. Kimura and N. F. Lane in *Johnson* et al. [5.3] have used fully quantum mechanical, close-coupling techniques to investigate H^+-He charge transfer and direct scattering, while potential scattering techniques were used by *Gao* et al. [5.1] and *Johnson* et al. [5.3] to assess the H^+-He ground state potential energy curve of *Helbig* et al. [5.7] and to optimize the parameters for analytic forms of the H-He ground state potential energy curve.

Springer Series in Chemical Physics, Vol. 54
Deepak Mathur (ed.): Physics of Ion Impact Phenomena
© Springer-Verlag Berlin Heidelberg 1991

5.1 Experimental Method

Figure 5.1 shows a schematic of the apparatus [5.4, 5, 8, 9] used by *Gao* et al. [5.1], *Newman* et al. [5.2] and *Johnson* et al. [5.3]. Protons emerging from the electron-impact ion source were accelerated to the desired energy and focused by an electostatic lens. The resulting beam was momentum analyzed by a pair of confocal 60° sector magnets and passed through a charge transfer cell (CTC). For experiments requiring an H atom beam, krypton gas in the charge transfer cell converted about 10 % of the protons into neutrals via charge transfer. The near-resonant nature of the H^+-Kr charge transfer and the electric field applied between deflection plates DP1 (quenching H(2s) atoms by Stark-mixing) together ensured that the fast hydrogen atoms were predominantly in the ground state. The deflection plates DP1 also removed ions in the beam energing from the CTC. The projectile beam passed through a target cell (TC) before striking an axial position-sensitive detector (PSD), which was available to monitor both the primary beam and the fast collision products. Deflection plates DP2 were available to prevent ions from striking the detector.

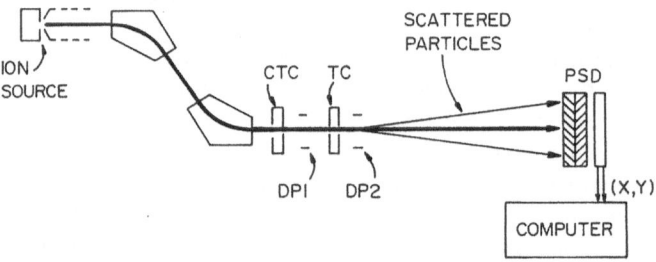

Fig. 5.1. Schematic of the apparatus used by *Gao* et al. [5.1], and *Johnson* et al. [5.3]

Two different configurations of the apparatus were employed; one [5.1, 3] to collect data from very small-angle scattering (0°–1°) and another [5.2] from larger-angle scattering (0.5°–4.5°). In the very-small-angle configuration, the collimating aperture and the entrance aperture of the TC were 20 μm and 30 μm in diameter, respectively, and were separated by 49 cm, thereby collimating the ion beam to less than 0.003° divergence. It is perhaps useful to note that these apertures are so small as to be virtually invisible to the naked eye, and that this degree of collimation is comparable to that provided by two 1 cm apertures separated by a distance of 200 m. Such stringent constraints necessarily require high stability in the beam generation and transport system and result in rather tenuous projectile beams. The TC had a length of 0.4 cm and an exit aperture 300 μm in diameter. For very-small-angle H^+-He measurements, a PSD with an active area 4.0 cm in diameter was axially located 109 cm beyond the TC, geometrically limiting the maximum observable scattering angle to about 1.3°. For very-small-angle H-He scattering, a 2.5 cm diameter PSD was used, limiting the maximum angle to about 0.7°. In the larger-angle configuration, the CTC and TC apertures were both 25 μm in diameter and were separated by 10 cm, limiting the beam divergence to 0.03°, and the TC was shortened to 0.2 cm.

A 2.5-cm diameter PSD located 15 cm beyond the TC limited the maximum observable scattering angle to approximately 5°.

An LSI 11/2 microcomputer monitored the output of the PSD electronics, sorting the arrival coordinates of each detected particle into bins in a 90×90 array. The physical area corresponding to the bin size was variable, so that the conflicting requirements of high resolution and large angular coverage could be met. The minimum physical bin size for the experiments was $109 \times 109 \ \mu m^2$ (for the 4.0 cm detector) or $68 \times 68 \ \mu m^2$ (for the 2.5 cm detector). The bin size was measured by observing the shadow of a nickel grid of known dimensions placed directly in front of the detector as an ion beam was swept over the detector suface. This technique was also used to determine the position-finding accuracy of the PSD. For the charge transfer experiment, the primary ion beam flux was measured intermittently during the neutral particle accumulation by removing the electric field established between plates DP2.

The establishment of an absolute scale for the differential cross sections requires only knowledge of the relative detection efficiency for primary and scattered particles, rather than knowledge of the absolute detection efficiency for each type of particle. Concerning the direct scattering measurements, it was observed that the PSD efficiency is, in general, somewhat dependent on the local count rate. It was, however, possible to equalize the detection efficiencies for primary and scattered particles to within a few percent by careful selection of the PSD operating voltage. This involved measurement of the relative detection efficiency as a function of PSD operating voltage for both diffuse and localized ion beams of appropriate intensity, a procedure carried out each time cross sections were measured. The issue of possibly differing detection efficiencies for neutral and charged species was examined previously by *Gao* et al. [5.9] using the cancellation of interference terms in the He^+-He charge transfer and direct scattering (a technique established by *Nagy* et al. [5.10]) with the conclusion that the two efficiencies were equal at 5.0 keV. However, this is not to say that ion neutral efficiencies were equal at other impact energies. In fact, a series of independent measurements of the H^+-Kr charge transfer cross section at Rice [5.11] using four different PSDs resulted in integrated cross sections (0°–1°) varying by $\pm 5 \%$ at 1.5 keV and $\pm 10 \%$ at 0.5 keV. No systematic effects were observed, although a variety of configurations (microchannel plate size, age, history, and manufacturing quality) were used. The apparent explanation for this result is that at these energies, the ratio of the ion to neutral efficiencies varied slightly from one PSD to another.

Under the thin target conditions used in this experiment, the differential cross section was determined from the measured quantities by the expression

$$\frac{d\sigma(\theta)}{d\Omega} = \frac{\Delta S(\theta)}{S\tau\Delta\Omega} \,, \tag{5.1}$$

where S is the primary ion beam flux in particles per second, $\Delta S(\theta)$ is the neutral flux scattered at angle θ into a solid angle $\Delta\Omega$ steradians, and τ is the target thickness. For the present geometry, τ was determined [5.12, 13] to be accurately given by the product nL, where n is the number density obtained from a measurement of gas pressure in the TC, and L is the physical length of the cell. At a typical target cell pressure of 5 mTorr, residual vacuum chamber pressure was maintained below 2×10^{-7} Torr. Under these conditions, only 5 % of the beam was scattered by the target gas, making multiple collision effects negligible.

For charge transfer and direct scattering, two 90×90 data files, one with gas in the target cell and one without, were taken. The scattered flux, $\Delta S'(\theta)$, was obtained by organizing the 90×90 data arrays into concentric rings and subtracting the gas-out data from the gas-in data. This procedure permitted discrimination between counts due to scattering from the target gas and counts arising from other sources, such as PSD dark counts or scattering from the background gas or from edges of apertures. Measurements at several values of target cell pressure yielded essentially identical differential cross sections; therefore, the presence of target gas did not have a significant effect on background count rates. For ion-neutral direct scattering, additional files were accumulated to properly account for counts on the detector due to neutral collision products.

The experimental uncertainty in the number of counts at a particular angle was primarily statistical, ranging from 1 % near 0° to 10 % at the largest angles. The angular uncertainty arose from the finite width of the primary ion beam, the discrete width of the analysis rings, and electronic errors in the detector's position encoding circuits, amounting to about 0.02° at the smallest scattering angles. In general, PSD spatial resolution is related to the size of the electron pulse impinging on the anode. This phenomenon was studied by *Nitz* et al. [5.4] with a single channel analyzer (SCA), who recorded the contributions to the electronic image of the primary beam from different portions of the pulse height spectrum. Large pulses provided good signal-to-noise ratio for the position recording electronics and, therefore, resulted in accurate positions, but smaller pulses (amounting to a few percent of the total counts) were registered as much as 1000 μm outside the geometrically-limited impact region. The details of the distribution depended on operating conditions: the problem was accentuated by high local count rates and by low PSD operating bias, both of which increased the relative number of small output pulses. This effect primarily interfered with measurement of the scattered signal at the smallest angles ($\theta < 0.05°$), where the spurious primary beam counts increased the apparent diameter of the primary beam. It was possible to eliminate these counts (enhancing the angular resolution) by using the SCA to reject small pulses, and cross sections were so obtained by *Gao* et al. [5.1] for 0.5 keV H-He scattering. However, cross sections obtained in this way were no longer absolute and, therefore, were normalized to cross sections obtained at larger angles. In principle, the angular resolution was also influenced by the length of the target cell, the beam divergence, and thermal motion of the (room temperature) target, but these factors were not significant relative to the beam size, ring size, and position-encoding errors The effect of the finite angular resolution of the apparatus was estimated [5.4, 9] by calculating the convolution of theoretical cross sections with an apparatus function which accounts for the above-mentioned effects. The convolution raised the minima in the strongly oscillating He$^+$-He cross sections only at very small angles ($\sim 0.05°$). Convolution would not have appreciably changed the $H^{(0,+)}$-He calculated cross sections in [5.1, 3], since their oscillations were less pronounced.

5.2 Theoretical Considerations

The classical approach to scattering describes the forces between particles as a function of time as the particles approach and recede from each other. However, fully quantum mechanical descriptions of atomic scattering use a time-independent approach, such as that used by Kimura and Lane in [5.3]. In their close coupling treatment they solved the total system Schrödinger equation by expanding the total wavefunction in terms of Born-Oppenheimer electronic state wavefunctions and solving the resultant set of coupled time-independent equations for the S-matrix which yields the differential scattering cross sections. The close coupling method is particularly appropriate for processes in which more than one quantum state contributes significantly to the scattering event.

The Born-Oppenheimer electronic wavefunctions, which are required to exhibit the appropriate symmetries, are represented by a superposition of configurations, each a Slater determinant of molecular orbitals (MO). It is convenient to write the molecular orbitals as linear combinations of atomic orbitals (LCAO) on each of the nuclear centers; an equivalent modified valence bond method was used by *Kimura* and *Lane* [5.3], although it generally takes a relatively large number of atomic orbitals (50 at 1 keV for the H^+-He calculation of Kimura and Lane) to accurately form the molecular orbital. A more detailed discussion is presented in Chap. 8.

A formally correct alternative to the molecular orbital expansion approach to the collision problem is to use atomic orbitals (AO) to expand the scattering wavefunctions. The AO technique has greater applicability at high collision energies, where the collision time is too short to allow the electrons to form well-defined molecular states. Naturally the molecular orbital approach is more attractive as the collision energy is reduced, since the molecular nature of the collision system becomes more apparent. In practice, it is frequently a question of how many states are necessary to obtain a reasonable scattering wavefunction expansion. At low energies, the molecular orbital method used by Kimura and Lane provides reasonable results with a small number (5 for their 1 keV H^+-He calculation) of scattering wavefunction expansion states, while the number required in an atomic orbital calculation would be significantly greater; the reverse is true at high energies (> 100 keV). The fully quantum mechanical molecular orbital approach undertaken by Kimura and Lane for H^+-He scattering is discussed further below.

Simpler methods may sometimes be used when more than one quantum state is important to a collision process; for example, potential scattering methods may often be used to describe exact resonance symmetric collisions. Inelastic scattering processes, including charge transfer, have been studied by using the models of Landau, Zener, and Stueckelberg (LZS), and Demkov, for potential energy curve crossing and pseudo-crossings, respectively, involving two states [5.6]. In these models, it is assumed that transitions between states takes place in a well-localized range of internuclear separation as the two heavy particles approach or recede from each other. These models have the advantage that no lengthy ab initio calculations are required to discuss multiple-state processes. However they provide only qualitative information about the collision processes of interest. A very select group of collision processes exists in which the scattering can be thought of as arising from the interaction of two structureless particles through a single interaction potential energy curve. In this single-channel case, the initial and final quantum state of the system are identical, and no electronic transitions occur in the course of the collision. H-He and H^+-He direct scattering events such as those reviewed here are good candidates for the time-independent, single-channel theory, since the initial and final states are identical. If other quantum states are widely separated in energy from the initial state, then non-adiabatic coupling between states during the collision is small, and a satisfactory description of the scattering need only include the single interaction potential, with the advantage that the calculations can be performed relatively quickly on an inexpensive computer, such as an IBM PC.

5.2.1 Molecular Orbital Expansion Method

Molecular orbital expansion studies [5.14] have revealed that in H^+-He collisions below 20 keV, charge transfer to the H (1s) state dominates all other charge transfer and excitation processes, having a cross section that is more than an order of magnitude larger than that for H (2p) charge transfer or ionization. As the energy is increased above 60 keV, ionization becomes the most significant process. Thus a two-state molecular orbital close-coupling calculation should provide satisfactory cross sections for the dominant channels in the energy range $E \leq 5$ keV.

The total time-independent scattering wavefunction of the system is described by Kimura and Lane in an adiabatic representation [5.15] as as expansion in electronic Born-Oppenheimer MO wavefunctions

$$\Psi(R, r) = \sum_i F_i(R, r)\chi_i^a(R) , \tag{5.2}$$

where $F_i(R, r)$ represents the Born-Oppenheimer electronic wavefunction for fixed internuclear coordinates R and all electronic coordinates r, and includes electron translation factors (ETFs); the expansion coefficents $\chi_i^a(R)$ represent the nuclear (scattering) wavefunctions that correspond to the respective electronic states. Configuration interaction calculations were perfomed to obtain eigenvalues and eigenfunctions of the electronic Hamiltonian, and Slater-type orbitals (STO) were employed as basis functions. Values of the orbital exponents for the STOs have been given by *Kimura* [5.14]. Substitution of (5.2) into the stationary Schrödinger equation for the collision system yields coupled, second-order differential equations for the $\chi_i^a(R)$. Molecular orbital electron translation factors (MO-ETFs) were introduced to correctly localize the active electrons with one or the other of the nuclei at large separations, and were optimized by minimizing the coupling to high-lying states. It is computationally convenient to solve the coupled equations in a diabatic representation. The unitary transformation matrix $C(R)$, satisfying $\chi_i^a(R) = C(R)\chi^d(R)$, where $\chi^{a,d}(R)$ are column vectors, is introduced to transform to the diabatic representation, in which the radial derivative coupling terms are zero [5.15]. The resulting coupled equations are, in matrix form,

$$\left[\frac{1}{2\mu}\nabla_R^2 I - V^d(R) + EI\right]\chi^d(R) = 0 , \tag{5.3}$$

where μ is the reduced mass of the system, I is the identity matrix, and V^d, the diabatic potential matrix, is related to its adiabatic counterpart (V^a) by $V^d = C^{-1}V^aC$. Since only two states are considered here, (5.3) consists of two coupled scalar equations, which are solved to obtain the scattering S matrix for each partial wave (for the present spherically symmetric case) using the log-derivative method [5.16]. The differential cross section for charge transfer is then obtained from the formula (in the centre of mass system):

$$\frac{d\sigma(\Theta)}{d\Omega} = \frac{1}{2k^2}\left[\sum_l (2l + 1)S_{12}^l P_l(\cos\Theta)\right]^2 , \tag{5.4}$$

where S_{12}^l is the scattering S matrix element for partial wave l, Θ is the scattering angle in center-of-mass coordinates, and k is the momentum of the projectile. A similar equation with $(1-S_{11}^l)$ in place of S_{12}^l describes the direct elastic scattering cross sections. The number of partial waves needed to obtain reasonable convergence of cross sections was found to be 1800 for 1.5 keV and 3200 for 5.0 keV (laboratory frame energies). The calculated adiabatic potential curves for the initial H^+-He (1^1S) and final H(1s)-He$^+$ (1s) states are plotted in Fig. 5.2, along with the non-adiabatic

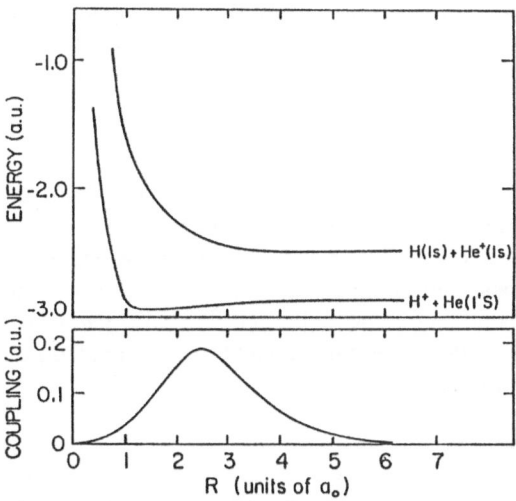

Fig. 5.2. Adiabatic potentials and corresponding radial coupling between the two states calculated by Kimura and Lane in [5.3] using the molecular orbital technique

Table 5.1. Values derived by *Kimura* and *Lane* in [5.3] for the initial H$^+$-He($1s^2$) and final H($1s$)-He$^+$($1s$) state potential energy curves, obtained from the full configuration interaction method. Internuclear distance is given in a_0, while energies are stated in a.u.

Internuclear r[a_0]	Coupling	Potential energy V [a.u.]	
		H$^+$-He($1s^2$)	H($1s$)-He$^+$($1s$)
0.10		12.87	15.05
0.20		3.156	5.246
0.30	0.0	0.1395	2.121
0.40	0.002	−1.219	0.6579
0.50	0.006	−1.934	−0.1652
0.60	0.010	−2.344	−0.6812
0.70	0.016	−2.590	−1.027
0.80	0.021	−2.742	−1.273
0.90	0.029	−2.836	−1.457
1.00	0.036	−2.895	−1.599
1.50	0.089	−2.967	−2.028
2.00	0.150	−2.947	−2.256
2.50	0.185	−2.926	−2.382
3.00	0.150	−2.913	−2.446
3.50	0.100	−2.907	−2.477
4.00	0.060	−2.903	−2.490
4.50	0.035	−2.902	−2.496
5.00	0.021	−2.902	−2.497
6.00	0.008	−2.902	−2.498
7.50	0.002	−2.902	−2.499
10.0	0.0	−2.902	−2.499

radial coupling matrix element (where the optimized MO-ETF was used [5.14] between these states). Representative values of the potentials have been tabulated in Table 5.1 as an aid to other investigators.

5.2.2 Potential Scattering

The complexity of the potential scattering problem is reduced if the interaction potential is spherically symmetric; in this case, the method of partial waves [5.17] was particularly useful to *Johnson* et al. [5.3] and *Gao* et al. [5.1]. The wavefunction Ψ describes the motion of a structureless particle under the influence of a potential, where Ψ satisfies the time-independent Schrödinger equation. The radial part of the equation may be solved beyond the range of the potential and $u_l(r)$, the radial part of the wavefunction, expressed asymptotically as

$$u_l(r)_{r \to \infty} \to N_l \sin\left(kr - \frac{l\pi}{2} + \delta_l\right), \tag{5.5}$$

introducing the phase shift δ_l, and where N_l is a normalization constant. The phase shift is interpreted as the difference in phase between the free particle solution and the solution in the presence of the potential. The normalization constant N_l is determined by the required asymptotic behavior of Ψ.

Far from the scattering centre, the potential is negligible, and Ψ is appropriately represented by an incident plane wave and an outgoing scattered wave. The asymptotic form of Ψ is given by

$$\Psi(r)_{r \to \infty} \to e^{ikz} + \frac{1}{r} e^{ikr} f(\Theta) \, , \tag{5.6}$$

where $f(\Theta)$ is the complex scattering amplitude. With the normalization constant determined, the differential cross section can be written as

$$\frac{d\sigma(\Theta)}{d\Omega} = |f(\Theta)|^2 = \frac{1}{2k^2} \left| \sum_l (2l+1)[e^{2i\delta_l} - 1]P_l(\cos\Theta) \right|^2 , \tag{5.7}$$

where Θ is the scattering angle in the center of mass frame, k is the wave number, δ_l is the phase shift of the lth partial wave, and $P_l(\cos\Theta)$ is the lth Legendre polynomial. The above expression is known as the Rayleigh-Faxén-Holtsmaark equation [5.18]. Phase shifts were obtained using the semi-classical JWKB approximation in the form [5.19]

$$\delta_l^{JWKB} = kb \left[\frac{\pi}{2} - \frac{r_0}{b} + I_1 \right] , \tag{5.8}$$

where $b = (l + \frac{1}{2})/k$ is the classical impact parameter, r_0 is the associated turning point, and I_1 is the integral

$$I_1 = \frac{1}{b} \int_{r_0}^{\infty} \left[\left[1 - \frac{V(r)}{T} - \frac{b^2}{r^2} \right]^{\frac{1}{2}} - 1 \right] dr \tag{5.9}$$

in which T represents the centre-of-mass frame collision energy. In the limit of large l the phase shifts become small and were determined using the Jeffreys-Born approximation [5.20]

$$\delta_l^{JB} = \frac{-k}{2T} \int_b^{\infty} V(r) \left[1 - \frac{b^2}{r^2} \right]^{-\frac{1}{2}} dr \, . \tag{5.10}$$

Typically several thousand partial waves were used, and the JWKB and Jeffreys-Born method yielded very similar results.

The potential energy curves used for H^+-He direct scattering [5.3] are those found in [5.7]:

$$V(r, A, B, C) = \frac{Z_1 Z_2}{r} e^{-r/A} \left[1 + \frac{r}{A} + r^2 \left(\frac{1}{2A^2} - \frac{U}{BZ_1 Z_2} \right) \right]$$
$$- U \left[1 + \frac{r}{B} + \frac{r^2}{C^2} + \frac{2Ur^4}{\alpha} \right]^{-1} \tag{5.11}$$

where $U = 4.37311$ hartree and $\alpha = 1.3835$ bohr^3; these are derived from analytical fits to the theoretical results of *Michels* [5.21] ($A = 0.423$, $B = 0.483$, $C = 0.441$)

and *Wolniewicz* [5.22] ($A = 0.442$, $B = 0.505$, $C = 0.451$). For H-He collisions [5.1], the measured differential cross sections were used to obtain empirical interaction potentials. Two potential forms

$$V_1(r) = A_1 \frac{e^{-\alpha_1 r}}{r} + A_2 e^{-\alpha_2 r} \tag{5.12}$$

and

$$V_2(r) = e^{-\beta r} \sum_{n=-1}^{4} B_n r^n \tag{5.13}$$

were used to calculate cross sections, using the potential scattering method. A non-linear least-squares fit of calculated cross sections to the data determined the best values for the parameters in (5.12) and (5.13), with the cross sections at all three energies fitted simultaneously using the same parameters for a given form of the interaction potential energy. Phase shifts calculated from the potentials were used to determine the classical deflection function

$$\theta(l) = 2 \frac{d\delta_l}{dl} . \tag{5.14}$$

The deflection function was used in turn to identify the impact parameters b_{min} and b_{max} corresponding to θ_{max} at highest energy and θ_{min} at lowest energy. Then b_{min} and b_{max} were used to calculate the classical distance of closest approach, leading to a range within which the derived potentials are considered valid. The uncertainty in the potential derived from these data arose principally from the finite angular resolution of the apparatus, and was estimated as $\pm 20\%$.

5.3 Results and Discussion

5.3.1 H-He Direct Scattering

The best fit potential $V_2(r)$ determined by *Gao* et al. [5.1] is shown in Fig. 5.3 along with other recent experimental [5.23] determinations and with the potential calculated by *Theodorakopoulos* et al. [5.24]. Measured differential cross sections by *Gao* et al. [5.1] and *Newman* et al. [5.2] at 0.5, 1.5, and 5.0 keV lab energy are shown in Fig. 5.4. Cross sections calculated using the best fit $V_2(r)$ and from the potential of *Theodorakopoulos* et al. [5.24] are given in Fig. 5.4 as solid and dashed lines, respectively. The cross sections generally decrease smoothly with increasing scattering angle except for an undulation seen in several cases in the vicinity of $0.1°$. Similar structure was observed at Rice in earlier He-rare gas measurements and potential scattering calculations by *Nitz* et al. [5.4] and *Gao* et al. [5.5], and *Beier* [5.25] discussed this structure by analogy to optical diffraction from a disc.

Turning to the potential scattering calculations, the final parameters for both analytic forms of potential energy curve derived by *Gao* et al. [5.1], listed in Table

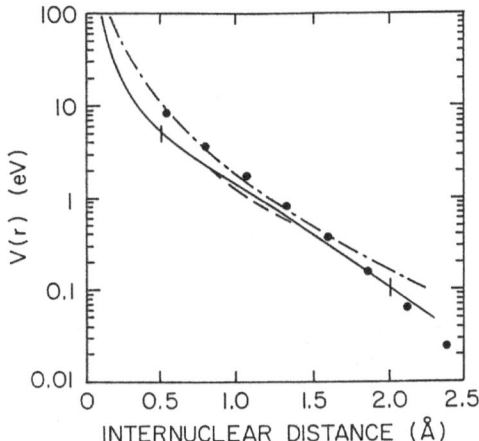

Fig. 5.3. H-He ground state interaction potential ($X^2 \Sigma^+$). Solid line: *Gao* et al. [5.1] (vertical bars represent $\pm 20\%$ uncertainty); closed circles: theoretical results of *Theodorakopoulos* et al. [5.24]; dashed-dotted line: universal potential of *O'Connor* and *Biersack* [5.23]; dashes: fit by *Belyaev* and *Leonas* [5.23] to their experimental results

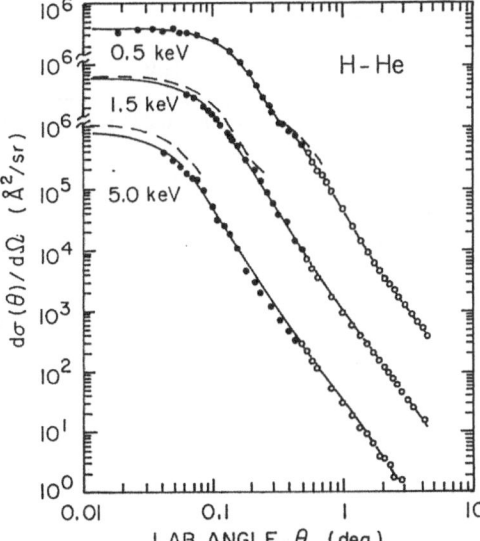

Fig. 5.4. Differential cross sections for H-He scattering at projectile energies of 0.5, 1.5, and 5.0 keV. Experimental data were obtained by *Gao* et al. [5.1] (closed circles) and by *Newman* et al. [5.2] (open circles). Solid and dashed lines are differential cross sections calculated from the empirical H-He ground state potential $V_2(r)$ of *Gao* et al. [5.1] and from the potential calculated by *Theodorakopoulos* et al. [5.24], respectively. Notice the shift of the y-axes.

5.2, give satisfactory and nearly identical results, suggesting that these empirical potentials provide an adequate representation of the ground-state interaction and that inelastic processes are not important in H-He scattering below 5 keV. Some data on inelastic processes occurring in this system, complied by *Gao* et al. [5.1], appear in Table 5.3, supporting this contention. Table 5.4 lists the values of the integral cross sections for H-He scattering [5.1] which were obtained by integrating the differential cross sections over the experimental angular range. The uncertainties in Table 5.4 were obtained by combining systematic effects, such as uncertainties in target thickness, primary beam flux and detector calibration.

Table 5.2. Parameters derived by *Gao* et al. [5.1] for the empirical H-He potential energy curve. V and r are in eV and Å, respectively. Effective range of the potentials is $0.05\,\text{Å} \leq r \leq 2.3\,\text{Å}$

$V_1(r) = A_1 \frac{e^{-\alpha_1 r}}{r} + A_2 e^{-\alpha_2 r}$	
$A_1\ [\text{eV·Å}] = 19.37$	$\alpha_1\ [\text{Å}^{-1}] = 8.578$
$A_2\ [\text{eV}] = 16.19$	$\alpha_2\ [\text{Å}^{-1}] = 2.477$

$$V_2(r) = e^{-\beta r} \sum_{n=-1}^{4} B_n r^n$$

$\beta\ [\text{Å}^{-1}] = 4.082$	$B_{-1}\ [\text{eV·Å}] = 16.36$	$B_2\ [\text{eV·Å}^{-2}] \doteq 43.70$
	$B_0\ [\text{eV}] = -35.07$	$B_3\ [\text{eV·Å}^{-3}] = -30.89$
	$B_1\ [\text{eV·Å}^{-1}] = 67.24$	$B_4\ [\text{eV·Å}^{-4}] = 21.18$

Table 5.3. Cross sections for H-He scattering: experimental results obtained by *Gao* et al. [5.1] and *Newman* et al. [5.2] (integrated from 0.02° to 4.5° in the lab frame), projectile excitation to H(2p), H(2s), and H($n = 3$) and target ionization. All cross sections are in Å^2

	0.5 keV	1.5 keV	5.0 keV
Experimental results	11. ($0.02° \leq \theta \leq 4.5°$)	4.4 ($0.05° \leq \theta \leq 4.5°$)	1.5 ($0.05° \leq \theta \leq 4.5°$)
H(2p)	0.6 [a]	0.5 [a]	0.3 [b]
H(2s)	–	0.1 [b]	0.08 [b]
H($n = 3$)	0.09 [c]	0.07 [c]	–
He ionization	< 0.0001 [d]	~ 0.0001 [d]	~ 0.001 [d]
Contribution from inelastic processes to total cross section	< 7 %	< 16 %	< 25 %

[a] [5.30]
[b] [5.31]
[c] [5.32]
[d] [5.33]

Table 5.4. Integral cross sections obtained by *Gao* et al. [5.1] and *Newman* et al. [5.2] for H-He direct scattering

Process	Cross section [Å^2]	Angular range	Other work
H(0.5 keV)-He	11. ± 0.6	0.02°–4.5°	
H(1.5 keV)-He	4.4 ± 0.2	0.05°–4.5°	5.2 [a], 3.2 [b] (0.12°–180°)
H(5.0 keV)-He	1.5 ± 0.08	0.05°–4.5°	

[a] [5.34]
[b] [5.23]

5.3.2 H+-He Charge Transfer and Direct Scattering

Measured and calculated cross sections from *Johnson* et al. [5.3] are shown in Figs. 5.5–5.7 along with the 5.0 keV experimental results of *Fitzwilson* and *Thomas* [5.26]. Charge transfer cross sections were not experimentally obtained at energies lower than 5.0 keV since the total cross section rapidly decreases with decreasing

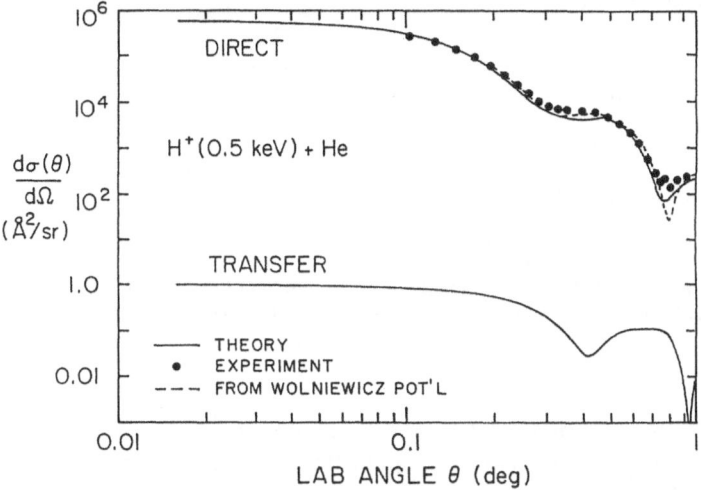

Fig. 5.5. Experimental data for H+ (0.5 keV) – He direct scattering and theoretical predictions by Kimura and Lane for differential cross sections of H+ (0.5 keV) – He direct and charge transfer scattering, from [5.3]

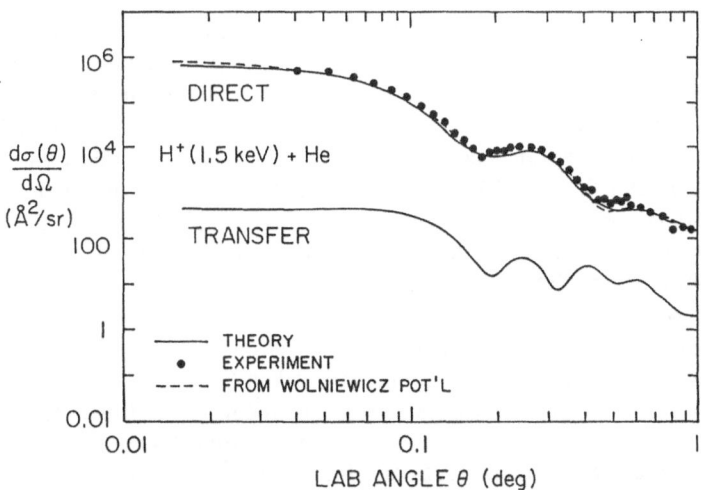

Fig. 5.6. Experimental data for H+ (1.5 keV)-He direct scattering and theoretical predictions by Kimura and Lane for differential cross sections of H+ (1.5 keV)-He direct and charge transfer scattering, from [5.3]

159

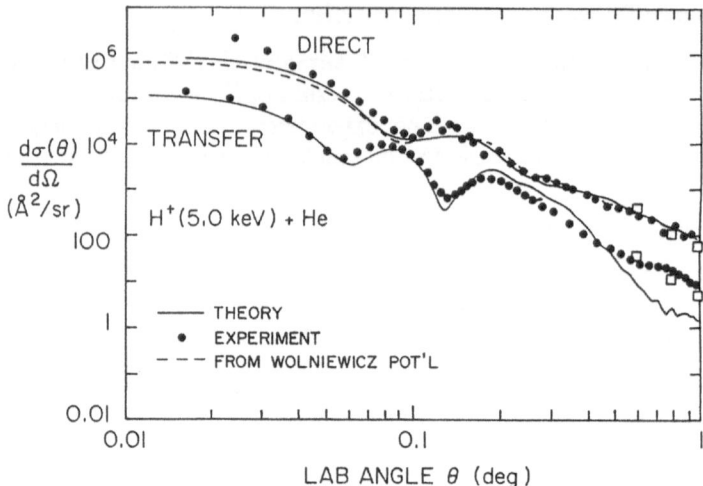

Fig. 5.7. Experimental data and theoretical predictions by Kimura and Lane for differential cross sections of H^+ (5.0 keV)-He direct and charge transfer scattering, from [5.3], including data from [5.26] (open squares)

impact energy. Direct scattering results are shown with the result of Kimura and Lane's two-state, molecular orbital close-coupling theory and with the results of the JWKB calculation from the potential of Wolniewicz. The result from the potential of Michels lies very close to that of Wolniewicz, which is not surprising since these two potentials differ primarily in well depth, and there only by about 0.2 eV. Experimental charge transfer results are shown together with calculated cross sections from the MO theory.

The cross section structure in *Johnson* et al. [5.3] may be interpreted as a consequence of several effects. Classical one-potential, trajectory-dependent effects (such as rainbow scattering from the attractive part of the H^+-He ground state potential) are responsible for some of the structure in the data. *Johnson* et al. [5.3] calculated the semiclassical deflection function, and thereby located the classical rainbow maxima at $\theta = 0.32°$ (0.5 keV), $0.11°$ (1.5 keV) and $0.03°$ (5.0 keV). Gentle undulations are also present in the cross sections due to quantum diffraction [5.25].

Interference due to coupling between the direct and charge transfer channels is responsible for 'Demkov oscillations' in the cross sections. In the Demkov model [5.6], a pseudo-crossing radius R_x is defined as the internuclear separation where the nonadiabatic coupling reaches a maximum; in the present case R_x is about $2.5a_0$ (1.3 Å), as seen in Fig. 5.2. If the classical turning point is less than R_x, the charge transfer channel becomes important, as shown by the S-matrix elements [5.14] illustrated in Fig. 5.8 for a projectile energy of 1.5 keV. (The values of the impact parameters that correspond semiclassically to the respective partial wave angular momenta are also shown).

As an aid to interpretation of the quantum mechanical results, it is instructive to consider the semiclassical charge transfer deflection function, which was calculated

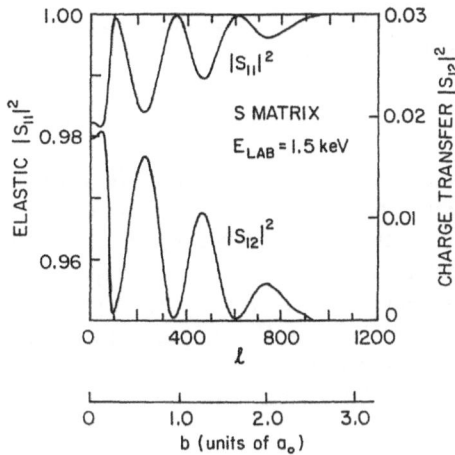

Fig. 5.8. Absolute value of the S-matrix element, derived by Kimura and Lane in *Johnson* et al. [5.3] for direct as well as charge transfer processes, as a function of partial wave l (upper abscissa) and impact parameter b (lower abscissa)

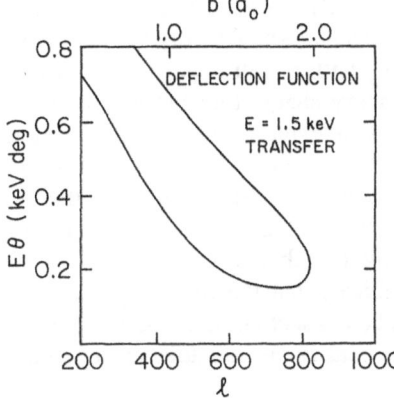

Fig. 5.9. Deflection function (in cm frame) for 1.5 keV charge transfer scattering, derived by *Kimura* and *Lane* in [5.3] as a function of partial wave l (upper abscissa) and impact parameter b (lower abscissa)

by *Kimura* and *Lane* [5.14] for the single energy 1.5 keV. The deflection function, shown in Fig. 5.9, exhibits a branch at $l_x = 820$ or $b_x = 2a_0$ (1.1 Å). Thus, for angles greater than $\theta = \theta(l_x)$, structure in the cross section arises from interference between semiclassical scattering amplitudes corresponding to two different trajectories. In Fig. 5.9 one notes a range of impact parameters, $1.5 < b < 2.0a_0$, for which the scattering takes place around $E\theta = 0.18$ keV-deg. This (lower) branch corresponds to a trajectory in which the particles follow the initial-channel potential inward to the classical turning point and then switch to the final-channel potential in the vicinity of $R = R_x$ on the outward portion of the trajectory. The upper branch of the deflection function corresponds to a trajectory in which the particles switch to the final-channel potential on the initial approach and then remain on this potential throughout the collision. The long-range repulsive wall of the final-channel potential gives rise to the larger deflection angles. Kimura and Lane indicate (*Johnson* et al. [5.3]) that since each trajectory possesses a scattering amplitude with a characteristic phase, for all scattering angles $\theta > \theta_x$ interference effects will give rise to oscillatory structure in the cross sections, which can be written in the simple form:

$$\frac{d\sigma(\theta)}{d\omega} \propto A - B\cos[\eta_1(b_1) - \eta_2(b_2)] , \qquad (5.15)$$

where A and B are weak functions of impact parameter and energy, and $\eta_1(b_1)$, $\eta_2(b_2)$ are semiclassical phase shifts at given impact parameters b_1, b_2, respectively, for trajectories denoted by 1 and 2. Variations in the difference between phase shifts $(\eta_1 - \eta_2)$ as a function of angle θ (or b) lead to the oscillatory structure apparent in the experimental data. This interference is seen in the 5.0 keV charge transfer results (Fig. 5.7) for $0.04° \leq \theta \leq 0.3°$ and in the 1.5 keV results (Fig. 5.6) at larger angles. It should be noted that, at the higher energy of 5 keV, oscillations for direct and charge transfer scattering are out of phase, as can be expected from consideration of flux loss and gain between coupled channels.

The direct scattering cross sections measured by *Johnson* et al. [5.3] correspond closely to both the theoretical molecular orbital predictions of Kimura and Lane and to single potential calculations using the ab initio Wolniewicz and Michels potentials. The ground state potential obtained from the molecular orbital theory lies very close to the results of the large ab initio calculations, with a well depth value between those of Michels and Wolniewicz. The MO theory charge transfer cross sections exhibit a lesser degree of agreement to the experimental results of *Johnson* et al. [5.3]. Kimura and Lane (in [5.3]) note that since the ground state potential is well established by the direct scattering results, the deviations may be due to either the coupling or the H-He$^+$ state potential. Preliminary calculations by Kimura and Lane (private communication) using the close-coupling formalism of charge transfer cross sections for other rare gas-hydride systems indicate that the angular positions of oscillation minima in charge transfer differential cross sections are sensitive to small adjustments in the internuclear separation of the peak in the non-adiabatic coupling.

Table 5.5. Absolute integral cross sections in Å^2 from *Johnson* et al. [5.3], where the differential cross sections from experiment and *Kimura* and *Lane*'s MO results have been integrated over the experimental angular range. The letter D denotes direct scattering while CT denotes charge-transfer. Comparisons to other published data have been made where available

Process	θ-range	Experiment	Theory int.	tot.	Literature
H$^+$(0.5 keV)-He D	0.08°–1.2°	7.2	7.1	10.7	
H$^+$(1.5 keV)-He D	0.040°–0.88°	4.0	3.4	4.7	
H$^+$(5.0 keV)-He D	0.02°–1.0°	2.3	1.6	2.2	
H$^+$(0.5 keV)-He CT	0°–1.0°		9.9×10^{-5}	3.4×10^{-4}	
H$^+$(1.5 keV)-He CT	0°–1.0°		0.014	0.022	0.02[a]
H$^+$(5.0 keV)-He CT	0°–1.2°	0.29	0.30	0.30	0.30[a], 0.35[b], 0.37[c]

[a] [5.27]
[b] [5.29]
[c] [5.20]

The measured differential cross sections of *Johnson* et al. [5.3] and *Gao* et al. [5.1] have been integrated over the observed angular range and compared in Table 5.5 to total charge transfer cross sections measured by *Stedeford* and *Hasted* [5.27], *Stier* and *Barnett* [5.28], and *Becker* and *Scharmann* [5.29]. Such comparisons are meaningful because the differential cross sections are so strongly peaked in the forward direction that the bulk of the total cross section results from scattering inside the range of the measurements. The integrated experimental cross sections so obtained are in reasonable agreement with Kimura and Lane's MO calculated results.

5.4 Summary

The experimental results reviewed here underscore the utility of high-resolution absolute angular differential scattering measurements. The level of agreement between experiment and theory indicates that the two-state close-coupling method is a practical and accurate technique for the collision system and conditions discussed, and it is possible that this method will prove useful to describe other collision systems.

Acknowledgements. The authors wish to acknowledge useful discussion with N. F. Lane and M. Kimura concerning the details of the theoretical methods. The work described was carried out with the support of the National Science Foundation, the National Aeronautics and Space Administration, and the Robert A. Welch Foundation.

References

5.1 R. S. Gao, L. K. Johnson, K. A. Smith, R. F. Stebbings: Phys. Rev. A **40**, 4914 (1989)
5.2 J. H. Newman, Y. S. Chen, K. A. Smith, R. F. Stebbings: J. Geophys. Res. **91**, 8947 (1986)
5.3 L. K. Johnson, R. S. Gao, R. G. Dixson, K. A. Smith, R. F. Stebbings, N. F. Lane, M. Kimura: Phys. Rev. A **40**, 3626 (1989)
5.4 D. E. Nitz, R. S. Gao, L. K. Johnson, K. A. Smith, R. F. Stebbings: Phys. Rev. A **35**, 4541 (1987)
5.5 R. S. Gao, L. K. Johnson, K. A. Smith, R. F. Stebbings: Phys. Rev. A **36**, 3077 (1987)
5.6 L. D. Landau: Z. Phys. Sov. Union **2**, 46 (1932); C. Zener: Proc. Roy. Soc. (London) A**137**, 696 (1932); E. C. G. Stuckelberg: Helv. Phys. Acta **5**, 370 (1932); Y. N. Demkov: Sov. Phys. JETP **18**, 138 (1964)
5.7 H. F. Helbig, D. B. Millis, L. W. Todd: Phys. Rev. A **2**, 771 (1970)
5.8 R. S. Gao, P. S. Gibner, J. H. Newman, K. A. Smith, R. F. Stebbings: Rev. Sci. Instrum. **55**, 1756 (1984)
5.9 R. S. Gao, L. K. Johnson, J. H. Newman, K. A. Smith, R. F. Stebbings: Phys. Rev A **38**, 2789 (1988)
5.10 S. W. Nagy, S. M. Fernandez, E. Pollack: Phys. Rev. A **3**, 280 (1971)
5.11 L. K. Johnson, R. S. Gao, C. L. Hakes, K. A. Smith, R. F. Stebbings: Phys. Rev. A **40**, 4920 (1989)
5.12 J. H. Newman, K. A. Smith, R. F. Stebbings, Y. S. Chen: J. Geophys. Res. **90**, 11045 (1985)

5.13 C. L. Hakes, J. H. Newman, D. A. Schafer: private communication (1989)
5.14 M. Kimura: Phys. Rev. A **31**, 2158 (1985); see also the AO-MO matching results in M. Kimura, C. D. Lin: Phys. Rev. A **34**, 176 (1986)
5.15 F. T. Smith: Phys. Rev. **179**, 111 (1969)
5.16 B. R. Johnson: J. Comput. Phys. **13**, 445 (1973)
5.17 L. S. Rodberg, R. M. Thaler: *Introduction to the Quantum Theory of Scattering* (Academic, New York 1967) Chap. 3
5.18 H. Faxén, J. Holtzmark: Z. Physik **45**, 307 (1927)
5.19 C. J. Joachain: *Quantum Collision Theory* (North-Holland, Amsterdam 1975), Chap. 9
5.20 H. S. W. Massey, C. B. O. Mohr: Proc. Roy. Soc (London) **A144**, 188 (1934)
5.21 H. H. Michels: J. Chem. Phys. **44**, 3834 (1966)
5.22 L. Wolniewicz: J. Chem. Phys. **43**, 1087 (1965)
5.23 Yu. N. Belyaev, V. B. Leonas: Sov. Phys. Dokl. **12**, 233 (1967); D. J. O'Connor, J. P. Biersack: Nuc. Instrum. Meth. B **15**, 14 (1986)
5.24 G. Theodorakopoulos, S. C. Farantos, R. J. Buenker, S. D. Peyerimhoff: J. Phys. B **17**, 1453 (1984)
5.25 H. J. Beier: J. Phys. B **6**, 683 (1973)
5.26 R. L. Fitzwilson, E. W. Thomas: Phys. Rev. A **6**, 1054 (1972)
5.27 J. B. H. Stedeford, J. B. Hasted: Proc. Roy. Soc. (London) **A227**, 466 (1955)
5.28 P. M. Stier, C. F. Barnett: Phys. Rev. **103**, 896 (1956)
5.29 M. Becker, A. Scharmann: Z. Naturforsch. **24**, 854 (1969)
5.30 B. Van Zyl, M. W. Gealy: Phys. Rev. A **35**, 3741 (1987)
5.31 J. H. Birely, R. J. McNeal: Phys. Rev. A **5**, 257 (1972)
5.32 B. Van Zyl, M. W. Gealy, H. Neumann: Phys. Rev. A **28**, 176 (1983)
5.33 B. Van Zyl, T. Q. Le, R. C. Amme: J. Chem. Phys. **74**, 314 (1981)
5.34 I. Amdur, E. A. Mason: J. Chem. Phys. **25**, 630 (1956)

6. High-Resolution Translational Energy Spectrometry of Molecular Ions

M. Hamdan[1] and A. G. Brenton[2]

With 23 Figures

Over the last twenty years, the technique of translational energy spectrometry (TES) has yielded considerable information on electronic transitions, excited electronic states and minimum lifetimes of a variety of singly and multiply charged atomic and molecular ions [6.1–7, 16–20]. Despite its inherent lack of energy resolution, compared with optical spectroscopic techniques, TES possesses a number of advantages. Firstly, the spectra of multiply charged ions may be acquired with almost the same sensitivity and resolution as that of singly charged ions. Secondly, due to the nature of the excitation process, the change in the orbital angular momentum accompanying the observed transitions is not necessarily restricated to 0, ± 1; therefore, intercombinations such as $\Delta \leftrightarrow \Sigma$ may be observed. Thirdly, this technique is inherently free from the problems of low detection efficiency and poor spectral transmission.

Briefly, the TES technique utilises a monoenergetic beam of mass-selected ions which are focused into a collision cell containing a suitable gas maintained at a pressure low enough to ensure single collision conditions. Collision induced transitions between the various energy levels of the incident ion and/or the neutral target manifest themselves as discrete changes in the translational energy of the incident ion. At keV energies, and small scattering angles, these measured energy changes can be related to excitation and de-excitation amongst the low-lying electronic states of the collision partners.

Consider, for example, a projectile ion of mass m_1 and velocity v_1^0 approaching a stationary target of mass m_2, as shown in Fig. 6.1. Let v_1 and v_2 be the velocities of m_1 and m_2 after the collision at angles Θ_1 and Θ_2 to the original direction of motion of m_1. Let E_1^0 be the translational energy of m_1 before collision, E_1 and E_2 the translational energies of m_1 and m_2 after collision, and let ΔE be the fraction of the translational energy of the projectile which is deposited as internal excitation energy in the collision partners. Five relationships defining E_1, E_2, Θ_1, Θ_2, and ΔE are theoretically possible from the laws of conservation of energy and momentum and the equations of motion. However, the absence of a specified impact parameter (the perpendicular distance between the asymptotic line of approach and a parallel line

Springer Series in Chemical Physics, Vol. 54

Deepak Mathur (ed.): Physics of Ion Impact Phenomena

© Springer-Verlag Berlin Heidelberg 1991

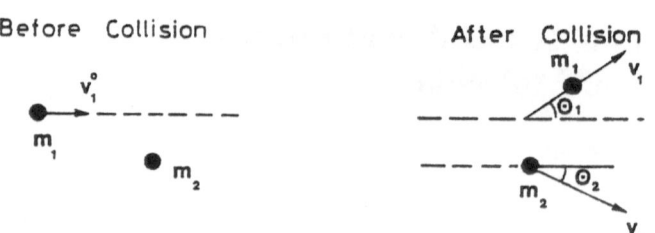

Fig. 6.1. Schematic representation of collision between a keV ion and a neutral target molecule m_2

drawn through the centre of the target atom) and unknown potential energy surface describing the interaction (except for simple systems) leaves us with three conservation equations to deduce as much as possible about the post-collision situation. These relationships can be expressed as

$$E_1^0 = E_1 + E_2 + \Delta E \,, \tag{6.1}$$

$$(m_1 E_1^0)^{1/2} = (m_1 E_1)^{1/2} \cos \Theta_1 + (m_2 E_2)^{1/2} \cos \Theta_2 \,, \tag{6.2}$$

$$0 = (m_1 E_1)^{1/2} \sin \Theta_1 - (E_2 m_2)^{1/2} \sin \Theta_2 \,. \tag{6.3}$$

By algebric elimination of the two normally inaccessible parameters, (E_2, Θ_2), and making approximations appropriate for small angle of scattering ($\Theta_1 \approx 0$) and small translational energy loss [$(E_1^0 - E_1) \ll E_1^0$], it can be shown that

$$\Delta E = (E_1^0 - E_1) - \frac{m_1}{m_2} E_1^0 \Theta_1^2 \tag{6.4}$$

where Θ_1 is in radians. The second term on the right represents the recoil energy of the target; under experimental conditions for which $\frac{m_2}{m_1} > 20$ or so, the correction to ΔE for the translational energy imparted to the target becomes negligible. Therefore, the first term on the right, which is experimentally measured, can be directly related to collisional excitation transitions between the low-lying electronic states and/or vibrational levels of the collision partners.

The applications of translational energy spectrometry (TES) discussed in this chapter will concentrate on the following collision processes

$$AB^+ + G \rightarrow AB^{+*} + G - \Delta E \,, \tag{6.5}$$

$$AB^{+*} + G \rightarrow AB^+ + G + \Delta E \,, \tag{6.6}$$

$$AB^+(AB^{+*}) + G \rightarrow AB^+(AB^{+*}) + G, \quad \Delta E = 0 \,, \tag{6.7}$$

where AB^+ and AB^{+*} are the respective ground and excited states of the projectile ion, G represents the target gas and ΔE is the overall energy change of the collision process. Equation (6.5) refers to inelastic scattering processes where part of the translational energy is converted into internal excitation energy, while (6.6) describes superelastic scattering in which internal excitation energy of the projectile ion is converted into translational motion of the fast projectile; (6.7), on the other hand, refers to an elastic scattering in which no interchange between translational and internal excitation energy occurs.

This chapter will be mainly concerned with experimental data on non-dissociative electronic states of doubly charged diatomic as well as some atomic ions which have been investigated by means of translational energy spectrometry. Of course, a full description and interpretation of data necessitates reference to theoretical calculations and other techniques which have been used.

6.1 Some Typical Experimental Arrangements

Figure 6.2 illustrates a high resolution ion impact spectrometer used by *Moore* and *Doering* [6.19] in the early 1970s to investigate electronic and vibrational transitions in collision of H_2^+ and H^+, of translational energies 150–500 eV, with a variety of neutral molecular targets. The hydrogen ions were formed in a duoplasmatron ion source, then mass analyzed by a magnetic sector and energy selected by an electrostatic deflection type of analyzer before impinging on a static target gas contained in a collision chamber. Ions scattered through a given angle were allowed to enter a second energy analyzer where their post-collision energy was determined. Thus, TES spectra were recorded as a function of scattering angle.

To increase the low intensity ion signals associated with this type of experimental arrangement, the ions were extracted at energies of $\sim$ 10 keV. Upon reaching the first energy analyzer, the ion beam was de-accelerated to less than 100 eV and allowed to pass through an aperture that defined the object for the subsequent lens system. This aperture was chosen to limit the angular divergence of the incident ion beam to less than 1.5° FWHM.

The excitation cross section, Q, for a particular transition was determined using (6.8) and (6.9)

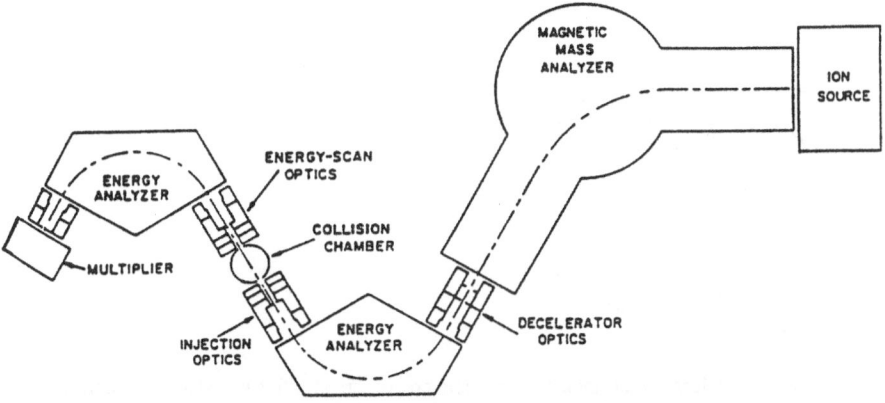

Fig. 6.2. High resolution ion impact spectrometer of *Moore* and *Doering* [6.19]. Reproduced by permission of the American Chemical Society from the Journal of Chemical Physics, Vol. 52 (1970)

$$QNl = \frac{I^*(\Theta = 0)}{I_0(\Theta = 0)} \frac{\int \frac{I^*(\Theta=0)}{I_0(\Theta=0)} \Theta \, d\Theta}{\int \frac{I_0(\Theta)}{I_0(\Theta=0)} \Theta \, d\Theta} \qquad (6.8)$$

with

$$I = I_0 \exp(-Q_0 N l), \qquad (6.9)$$

where $I^*(\Theta)$ is the intensity integrated over the transition at a scattering angle Θ, $I_0(\Theta)$ is the primary beam intensity profile with gas in the collision cell, Q_0 is the beam attenuation cross section, N the target gas density and l the path length within the collision cell.

The most notable characteristic of the scattering processes investigated by *Moore* and *Doering* [6.19] was the specific nature of the observed excitation transitions. Proton impact on molecular targets was found to promote only singlet-singlet transitions (due to the absence of electrons in the proton, spin exchange is, of course, forbidden). Furthermore, within the isoelectronic series N_2, CO and C_2H_2 the same class of transitions, $^1\Pi \leftarrow X\,^1\Sigma$, was observed even though the energy of this transition varies from 4.5 to 7.5 eV for these three molecules. In the case of H_2^+, both singlet-singlet transitions were observed.

Another experimental arrangement used to investigate ion-neutral collisions at energies extending up to 200 keV is shown in Fig. 6.3 [6.20]. The main innovative feature of this ion energy-loss spectrometer is the stability of the accelerating voltage power supply, which allowed high energy experiments without compromising energy resolution. To achieve this end, the ion source was connected to the positive terminal of an accurate power supply which provided a positive offset voltage V_1.

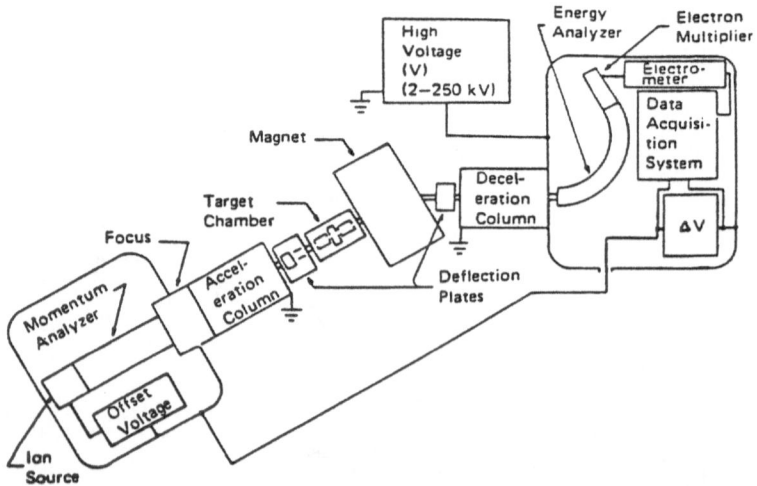

Fig. 6.3. Ion energy loss spectrometer of *Park* and coworkers [6.20]. Reproduced by permission of the American Physical Society from the Review of Scientific Instruments, Vol. 40 (1969)

The negative terminal of this supply was connected to variable power supply providing a smaller voltage ΔV. The high voltage power supply (V_0) was connected to the common of the ΔV power supply and to the de-accelerating column. The ion beam issuing from the source was momentum analyzed and passed through an einzel lens into the main accelerating column where it was accelerated by a total voltage of ($V_1 + V_0 + \Delta V$). Following collisions with the target gas, the emerging ion beam was magnetically analyzed, de-accelerated through a voltage V_1 and focused into a post-collision energy analyzer. In this experimental set-up, an ion carrying a charge q, and having an average kinetic energy E_a (within the ion source), accelerated through the voltage ($V_1 + V_0 + \Delta V$) and de-accelerated through a voltage V_1, can only be detected if its post-collision energy satisfies the relationship given in (6.12) below. This experimental condition can be understood by considering the following expressions

$$E_1 = q(V_1 + V_0 + \Delta V) + E_2 , \tag{6.10}$$

$$E_p = q(V_1 + V_0 + \Delta V) + E_a - qV_1 - \Delta E , \tag{6.11}$$

where E_1 and E_p are the respective energies of the incident ion prior to and after de-acceleration and ΔE is the ion energy loss due to collision with the target. If this energy loss happens to match the product $q\Delta V$, then (6.11) would yield

$$E_p = qV_0 + E_a . \tag{6.12}$$

In experimental apparatus used by *Park* and coworkers [6.20], the energy loss spectrum of the incident ion was acquired by recording the ion current as a function of the offset voltage, ΔV. This method had the advantage of eliminating difficulties associated with varying de-accelerating conditions and ensured that all detected ions passed through the magnet, the de-accelerating column and the electrostatic analyzer with the same energy regardless of their energy losses within the stopping cell.

6.1.1 Double-Focusing Arrangements

High resolution mass spectrometry, for the precise measurement of mass and the determination of molecular empirical formulae, utilised the combination of magnetic and electrostatic analyzers to obtain simultaneous velocity and direction focusing. This combination has been the foundation of high resolution mass spectrometry for over fifty years, following the classic studies of *Mattauch* and *Herzog* [6.9, 10] in 1934. The sensitivity afforded by the precise control and reduction of image aberrations in such devices has enabled resolving powers in excess of 10^5 to be readily achieved. *Illies* and *Bowers* [6.4] demonstrated how a reversed geometry mass spectrometer, normally used for organic chemical analysis, might be used for high resolution TES measurements. Prior to their experiments in 1982 on the energy loss of Kr^+, the energy resolving capabilities of conventional mass spectrometers had not been fully appreciated. This is probably due to the fact that double-focusing mass spectrometers are specifically designed to have zero net energy dispersion between the source slit (object) and collector slit (image).

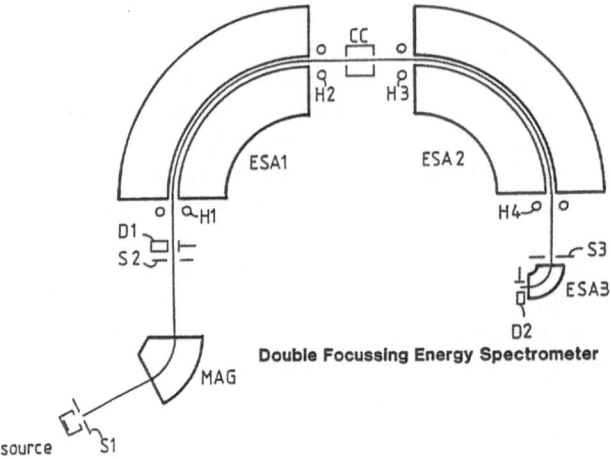

Double Focussing Energy Spectrometer

Fig. 6.4. Schematic diagram of the high resolution energy loss spectrometer in Swansea. S_1 – source slit; S_2 and S_3 – object and image resolving slits; D_1 and D_2 – electron multiplier detectors; FFR – field free regions; H_{1-4} – hexapoles; YZ – Y and Z lens assemblies; CC – collision cell

The principle of TES applied to such equipment lies in a collision cell being positioned midway along the spectrometer, between the two analyzing sectors. In this location, if an ions' translational energy is changed (positively or negatively), then that ion will be refocused at a slightly different position on the image focal plane, thus producing a net energy dispersion, which is the TES spectrum. However, there are certain deficiencies with this design of instrument, some of which can be serious, limiting ultimate performance and sensitivity. These were realised and led to a new design of spectrometer [6.8], the design and advantages of which are described below.

Figure 6.4 shows a schematic diagram of a double-focusing energy loss spectrometer designed and constructed at Swansea. The main feature of the instrument is a pair of identical cylindrical, electrostatic analyzers (ESA) of 15 inch radius and 90° deflection angle. The combination of two identical sectors has important advantages in design. Similar arguments to those presented here would also apply to a combination of two magnetic sectors but this combination has the difficulty that magnetic fields are inherently more difficult to control and measure than electric fields, and small variations in the properties of the magnetic material can also cause slight differences in performance from one sector to another.

The method of combining the two sectors is illustrated in Fig. 6.5 which shows a beam of perfectly monoenergetic ions with a divergence angle $\pm \alpha$ entering the first sector. A first-order focus is achieved between the two sectors at A. The focus of the two regions of the beam of greatest divergence can be seen to be displaced from the position at which the centre of the beam reaches A. This second-order distortion of the image is the well-known $\alpha^2 R$ term. Were it possible to reverse the direction of motion of all the ions in the beam at A, the ions would all follow exactly the same paths in the reverse direction and come to a perfect focus at the object slit. Passing

170

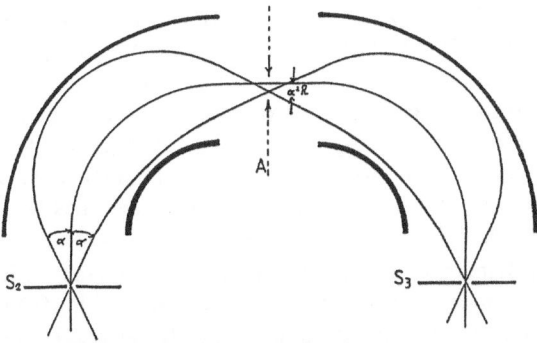

Fig. 6.5. Schematic representation of the ion trajectories of a monoenergetic beam, of angular divergence ± α, through a pair of cylindrical electrostatic analysers arranged in double focus

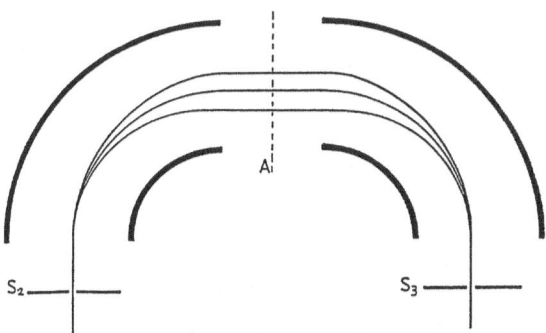

Fig. 6.6. Schematic representation of the ion trajectories of a beam containing ions of translational energy spread ± δE through a pair of cylindrical analysers arranged in double focus

the beam into an identical electric sector, as shown in Fig. 6.5, closely approximates such beam reversal and so a high order angular focus is achieved at the image slit.

Considering the energy focus, Fig. 6.6 shows a perfectly collimated beam, but one which also contains a range of translational energies ($E \pm \Delta E$) entering the sector. An energy dispersion is produced at A. Again, reversing the direction of motion of the ions in the beam and returning them through the sector with the same individual energies with which they arrived at A would produce a perfect focus.

A detailed study of the imaging properties of double-focusing designs for energy loss studies, consisting of symmetric arrangements of electrostatic analyzers, has been made [6.4, 8] using the third order ion optics (TRIO) computer programme developed at Osaka [6.10–13]. By this method, image aberrations can be calculated to third order and a variety of electrostatic field lenses (for example, quadrupole, hexapole, ...) can be incorporated and assessed. A number of suitable arrangements were found but for pragmatic reasons as to the accessibility of standard equipment, the design shown in Fig. 6.4 was chosen. Various auxilliary lenses were incorporated for specific functions. For the control of first order angular focus, a pair of crossed Y and Z deflector plates were used at positions close to H_1 and H_4. Such a lens system is simple to construct, acts like a quadrupole providing first order focus, allows for beam deflection in two orthogonal directions and occupies less than an inch in space. Again, to maintain 'symmetry', two lenses were used; one at the entry and another at the exit of the double-electric sector system. For control of second order aberrations, in particular image rotation, for slight misalignment of the collector slit (S_3), hexapole lenses H_1 and H_4 were incorporated. Additionally, a pair of hexapoles (H_2 and H_3) were positioned on either side of the collision cell (CC) where it has been calculated [6.14] that such devices could help to refocus the scattered ion beam.

Since this arrangement is non-dispersing in mass, it is preceeded by a magnetic sector (5 inch radius, 60° deflection angle) for mass analysis of the ions issuing from the source. This has another advantage in that the ion source is well separated from the high resolution object slit (S_2) from which it is essential to minimise sample

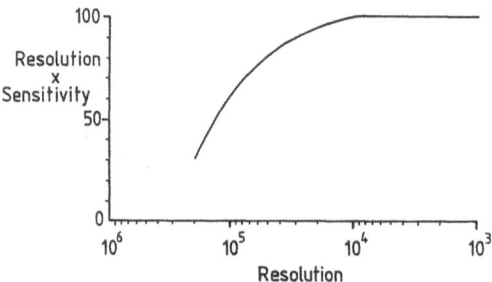

Fig. 6.7. Resolution (R) × sensitivity (S) of the Swansea high resolution energy loss spectrometer

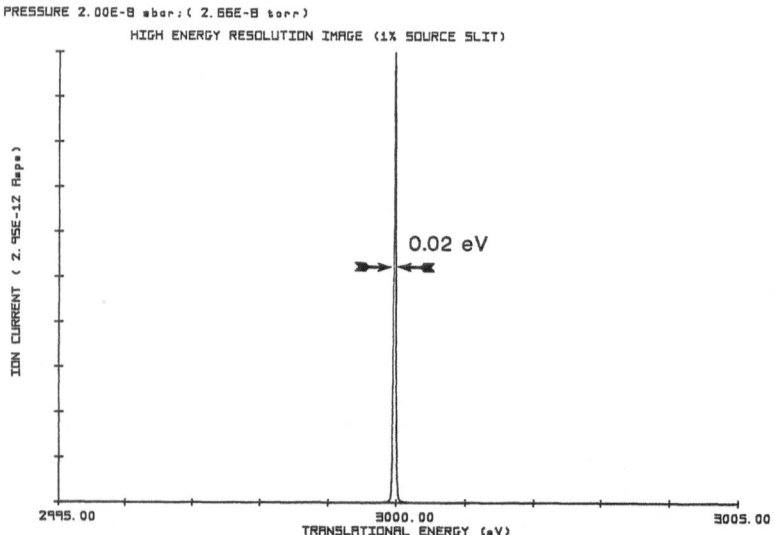

Fig. 6.8. Illustration of a high resolution ion beam obtained on the Swansea high resolution energy loss spectrometer

contamination and ion beam damage of the slits which have to close to a width of ca. 5 μm over a length of 8 mm for ultimate resolving power to be attained.

Another feature of this novel design is the inclusion of a small radius electrostatic analzser (ESA3) positioned after the collector slit (S_3). This performs the function of preventing artefact peaks arising in TES spectra which may be caused by the scattering of neutrals and ions off the walls of ESA2 and from the edges of the slit S_3, as was noted in earlier investigations [6.15].

The performance of this instrument and its utility for translational energy spectrometry is amply demonstrated by examples cited later in this chapter. However, a generalization of its performance can be appreciated from Fig. 6.7 which shows the product of resolution (R) and sensitivity (S) plotted against resolution. This curve is virtually flat up to resolutions of 25000 and shows that ultimate resolutions of 300000 can be achieved; this is shown by the parent ion beam profile in Fig. 6.8 which has a width of 0.02 eV (FWHM) for a translational energy of 3000 eV. To illustrate that this resolving power is achievable (in favourable circumstances) Fig. 6.9 shows a high resolution energy loss spectra of 3 keV Ar$^+$ on helium for inelastic $^2P^0_{1/2} \leftarrow {}^2P^0_{3/2}$ and superelastic $^2P^0_{1/2} \rightarrow {}^2P^0_{3/2}$ transitions exhibiting spin-orbit splitting of the ground state of Ar$^+$.

6.2 Results and Discussion

6.2.1 TES of keV Atomic Ions

These collision processes are fully discussed in Chap. 4 of this volume; therefore, this section will only deal with a limited number of ions which have been investigated using the high energy resolution spectrometer described in the previous section.

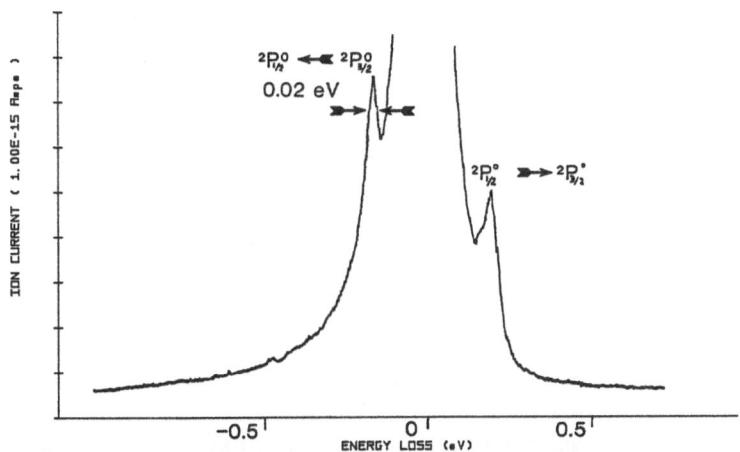

Fig. 6.9. Energy loss spectrum of Ar$^+$ (3 keV) colliding with He

It is well known that singly and multiply charged atomic ions formed in conventional electron impact ion sources are produced both in their ground and low-lying excited electronic states. Interest in the population of states in ion beams stems primarily from well documented effects that metastable excited states have on charge transfer cross sections. A number of publications [6.21–25] clearly demonstrated that this role is completely disproportionate to the fractional population of these species.

Hamdan et al. [6.16], using high resolution translational energy spectrometry, were able to identify a number of long-lived electronic excited states as well as their fractional populations in multiply charged Xe ion beams formed by electron impact. Typical translational energy spectra of $Xe^{2+,3+,4+}$ scattered off He at collision energies qV ($V = 3\,\text{kV}$ and q is the number of charges) are shown in Fig. 6.10. The observed excitation and de-excitation transitions and energy losses are compared with spectroscopically measured energies [6.26] which are summarised in Table 6.1.

From the TES spectra recorded at different ionizing electron energy and using a simple procedure first proposed by *Moore* [6.27, 28] and *Kobayashi* et al. [6.29], *Hamdan* et al. [6.16] determined the fractional populations of various electronic states within the incident Xe^{3+} ion beams. It was proposed that the observed intensities of the excitation and de-excitation transitions between an upper state, i, and a lower state, k, could be expressed as

$$\frac{I(i \rightarrow k)}{I(i \leftarrow k)} = \frac{\sigma(i \rightarrow k)}{\sigma(i \leftarrow k)} \frac{F(i)}{F(k)} , \tag{6.13}$$

where $F(i)$ and $F(k)$ are the fractional populations of both states, σ is the cross section for the indicated process and $I(i \rightarrow k)$, $I(i \leftarrow k)$ are the intensities measured

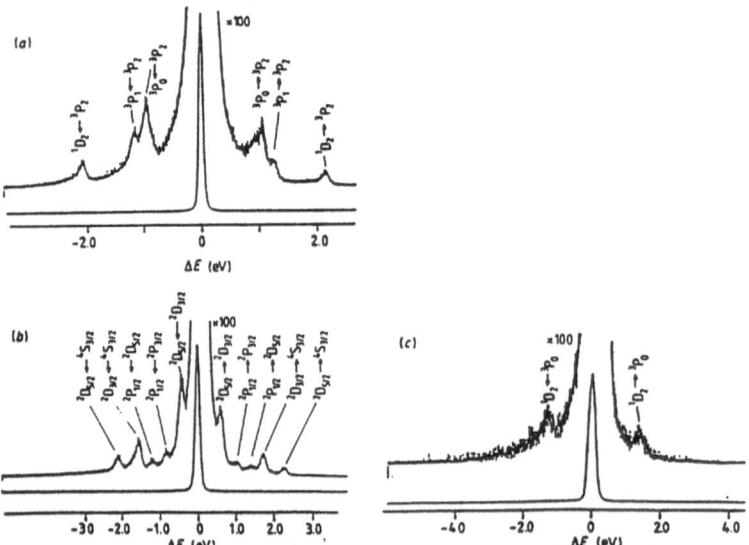

Fig. 6.10 a–c. Translational energy spectra of Xe ions scattered off He: (a) 6 keV Xe^{2+}, $E_e = 100\,\text{eV}$; (b) 9 keV Xe^{3+}, $E_e = 100\,\text{eV}$; (c) 12 keV Xe^{4+}, $E_e = 150\,\text{eV}$

Table 6.1. Experimentally observed transitions and corresponding energies of $3q$ keV Xe^{2+}, Xe^{3+} and Xe^{4+} scattered off He [6.16]. Spectroscopic energies are taken from [6.26]

Ion	Transition	Energy [eV]	
		measured	spectroscopic
Xe^{2+}	$(5p^4)^3P_1 \leftrightarrow {}^3P_2$	1.2 ± 0.05	1.21
	$(5p^4)^3P_0 \leftrightarrow {}^3P_2$	1.0 ± 0.05	1.0
	$(5p^4)^1D_2 \leftrightarrow {}^3P_2$	2.1 ± 0.05	2.12
Xe^{3+}	$(5p^3)^2D_{5/2} \leftrightarrow {}^2D_{3/2}$	0.5 ± 0.1	0.46
	$(5p^3)^2P_{3/2} \leftrightarrow {}^2P_{1/2}$	1.0 ± 0.1	1.71
	$(5p^3)^2P_{1/2} \leftrightarrow {}^2D_{5/2}$	1.4 ± 0.1	1.37
	$(5p^3)^2D_{3/2} \leftrightarrow {}^4S_{3/2}$	1.7 ± 0.1	1.64
	$(5p^3)^2D_{5/2} \leftrightarrow {}^4S_{3/2}$	2.2 ± 0.1	2.1
Xe^{4+}	$(5p^2)^1D_2 \leftrightarrow {}^3P_j$	1.3 ± 0.2	...

for de-excitation and excitation between the two states. On the basis of the principle of 'detailed balance' [6.30] the ratio of the cross sections can be written in terms of the statistical weights of the final states

$$\frac{\sigma(i \rightarrow k)}{\sigma(i \leftarrow k)} = \frac{(2J_k + 1)}{(2J_i + 1)} , \qquad (6.14)$$

where J is the total angular momentum; substituting (6.14) in (6.13) and rearranging gives

$$\frac{F(i)}{F(k)} = \frac{I(i \rightarrow k)}{I(i \leftarrow k)} \frac{(2J_i + 1)}{(2J_k + 1)} . \qquad (6.15)$$

Using this relationship and the measured intensities of excitation and de-excitation transitions recorded at different ionizing electron energy values, the fractional populations of $Xe^{2+}({}^3P_j, {}^1D_2)$ and $Xe^{3+}({}^4S, {}^2D, {}^2P)$ have been determined (see Fig. 6.11).

6.2.2 TES and the Spin-Conservation Rule

Conservation rules are important concepts and guides for the analysis of data, and for TES the *Wigner* [6.31] spin-conservation rule is important. It requires that the total electron spin angular momentum of a pair of atoms or molecules does not change in the course of the collision event.

Moore [6.28], in a series of TES measurements, demonstrated that spin-changing transitions are nearly three orders of magnitude less probable than their spin-conserving counterparts. However, the same author pointed out that the rule did not hold as well for the collision system, $N^+ + O_2$. This deviation was attributed to the fact that both N^+ and O_2 possess two unpaired orbital electrons in their respective ground

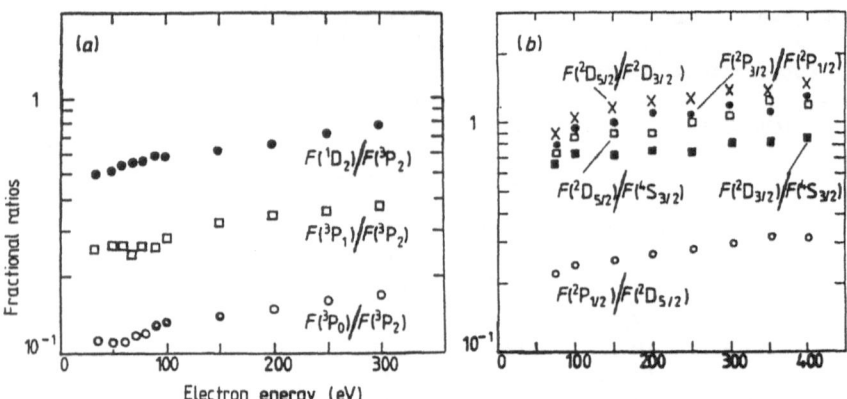

Fig. 6.11 a,b. Measured fractional ratios of Xe^{2+} and Xe^{3+} low-lying excited states. **(a)** Xe^{2+}: $\bigcirc - F(^3P_0)/F(^3P_2)$; $\blacksquare - F(^3P_1)/F(^3P_2)$; $\bullet - F(^1D_2)/F(^3P_2)$. The statistical weight ratios are 0.19, 0.58 and 1.0, respectively. **(b)** Xe^{3+}: $\bigcirc - F(^2P_{1/2})/F(^2D_{5/2})$; $\bullet - F(^2P_{3/2})/F(^2P_{1/2})$; $\blacksquare - F(^2D_{3/2})/F(^4S_{3/2})$; $\square - F(^2D_{5/2})/F(^4S_{3/2})$; $\times - F(^2D_{5/2})/F(^2D_{3/2})$. The ratios expected from the statistical weight model are 0.33, 2.0, 1.0, 1.5 and 1.5, respectively

states as well as several low-lying electronic states which may correlate to produce a large number of low-lying states in the N^+-O_2 collision complex. The crossings and pseudo-crossings of the potential energy curves associated with these states could provide for the spin-orbit coupling which may lead to spin non-conserving transitions. A recent study by *Lee* et al. [6.32] on a series of atomic ions colliding with CO confirmed that in every instance the selection rule is rigorously obeyed.

The promotion of spin-nonconserving electronic transitions by using molecular, rather than atomic targets, has also been demonstrated by *Hamdan* and *Brenton* [6.33]. Their results for 3 keV O^+ scattered off He and O_2 are shown in Fig. 6.12. The two peaks at ± 1.65 eV observed in the TES of O^+-He were assigned to excitation and de-excitation transitions between the two electronic states, $O^+(^2P)$ and $O^+(^2D)$. On the other hand, the TES of the collision system, O^+-O_2 displayed a number of additional transitions. The most prominent and interesting was that assigned to the excitation transition, $O^+(^4S \leftarrow {}^2D)$. In collisions with He this transition is spin-nonconserving; however, the use of an O_2 target, whose ground electronic state is $^3\Sigma_g^-$, allows the overall transition to become spin-conservative, as is expressed by the two collision processes possible in their experiment

$$O^+(^2D) + O_2(X^3\Sigma_g^-) \rightarrow O^+(^4S) + O_2(a^1\Delta_g) + 2.34\text{eV} , \qquad (6.16)$$

$$O^+(^2P) + O_2(X^3\Sigma_g^-) \rightarrow O^+(^4S) + O_2(a^1\Delta_g) + 4.03\text{eV} . \qquad (6.17)$$

Despite the limited amount of experimental data so far presented, there are a number of unambiguous deductions which can be safely made. Firstly, the excellent agreement between spectroscopically measured energies and those obtained from translational energy spectrometry demonstrates the important role of the latter technique, not only in confirming existing energies of atomic ions but also in the determi-

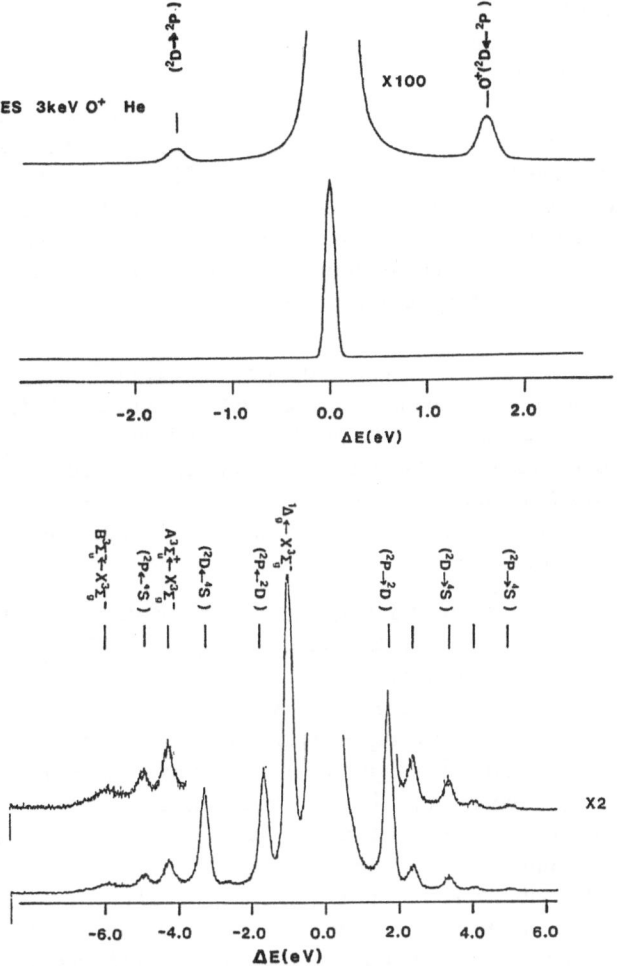

Fig. 6.12. Translational energy spectra of 3 keV O^+ scattered off He and O_2

nation of hitherto unknown energy values. Secondly, spin-nonconserving transitions, which are difficult to observe in photoelectron and/or in conventional optical spectroscopy, might be investigated by TES in systems which exhibit a non-conserving tendency.

6.2.3 Doubly Charged Diatomic Molecules

The existence of doubly charged diatomic ions which are stable (or metastable) on the microsecond time scale has been established for over fifty years [6.34]. Despite this, spectroscopic information on this class of ions remains very sparse. This is partially due to the widely held belief that doubly charged molecular ions tend to behave like two singly charged atoms interacting along a Coulombic repulsive potential energy curve resulting in spontaneous dissociation.

The first quantitative insight into this problem was provided by *Pauling* [6.35] for the simple two electron case of He_2^{2+}. It was proposed that resonance between the valance bond configurations He^+-He^+ and He^{2+}+He would lead to a ground state which is quasibound. Recent advances in quantumchemical calculations involving large basis set configuration interaction procedure [6.46–49, 65, 66, 69] have provided substantial evidence that dissociation processes encounter activation barriers deep enough to permit the existence of one or more bound electronic states of doubly-charged molecular ions.

This section will be mainly concerned with non-dissociative electronic states of doubly-charged diatomic ions investigated by means of high resolution TES. Relevant data regarding these observations, which have been obtained by means of other experimental techniques and/or by theoretical calculations, will also be discussed where necessary. More detailed descriptions of experimental techniques and contemporary quantumchemical methods employed in studies of multiply charged molecular ions can be found in Chap. 8.

a) NO²⁺. The NO^{2+} ion is isoelectronic with N_2 and the CN radical, a fact that was used in a semi-empirical method by *Hurley* [6.36, 37] to construct the first set of potential energy curves for this doubly charged ion. These calculations showed a $^3\Sigma^+$ ground state having an appearance energy of 38:1 eV and two bound electronic excited states, $A^2\Pi$ and $B^2\Sigma^+$, at energies of 0.6 eV and 4.8 eV above the ground state. Calculations using a multiconfigurational method by *Thulstrup* et al. [6.38] gave the same ordering of bound states but predicted a dissociative $A^2\Pi$ excited state. Discrepancy between these two sets of calculations and the absence of any reliable data motivated a number of experimental investigations which included Auger spectrometry [6.39], double charge transfer [6.41, 42], translational energy spectrometry [6.5, 7, 44] and photoion-photon of fluorescence concidence [6.45]. Table 6.2 gives a summary of the calculated and experimentally measured energies of the three lowest electronic states of NO^{2+} which have been reported over the last 25 years.

None of the listed energies in Table 6.2 fully account for the value of 3.9 eV which *Besnard* et al. [6.45] have attributed to the optical emission, $NO^{2+}(X^2\Sigma^+ \leftarrow B^2\Sigma^+)$. The calculated as well as the measured energy values for this transition, which are listed in Table 6.2, range from 4.3 to 4.96 eV. These discrepancies regarding the only known emission of NO^{2+} may be partially explained by considering the latest MC/SCF calculations of *Cooper* [6.46] and the experimental results of two recent high resolution TES investigations carried out by *Jonathan* et al. [6.44] and *Hamdan* and *Brenton* [6.7].

The potential energy curves for the low-lying electronic states of NO^{2+}, which are depicted in Fig. 6.13, show the vertical energy difference between the minima of the $X^2\Sigma^+$ and $B^2\Sigma^+$ states at an internuclear distance of $r = 1.16$ Å to be 4.96 eV.

Table 6.2. Calculated and experimentally measured appearance energies [eV] of the three lowest states of NO^{2+}

$X^2\Sigma^+$	$A^2\Pi$	$B^2\Sigma^+$	Method	Authors and Reference
–	(0.7)	(3.6)		Thulstrup et al. [6.38]
38.1	38.7	42.9		Hurley [6.36]
38.6 ± 0.1	40.0	42.5 ± 0.5	PIFCO	Besnard et al. [6.45]
38.3 ± 0.5	–	–		Newton and Sciamanna [6.77]
39.3 ± 0.3	–	–		Dorman and Morrison [6.41]
37.1	40.1	–	Auger Spectroscopy	
36.8	40.7	–	Auger Spectroscopy	
39.3 ± 0.5	–	42.4 ± 1.0	DEC	Appell et al. [6.41]
39.3	41.2	–		Kim et al. [6.43]
0	$(1.75)^a$	(4.85)	TES	O'Keefe et al. [6.5]
0	(1.7 ± 0.1)	(4.8 ± 0.1)	TES	Jonathan et al. [6.44]
39.4 ± 0.12	–	–		Samson et al. [6.43]
	$(1.30)^b$	(4.96)	MC/SCF	Cooper [6.46]
	(1.5)	(4.9)	TES	Hamdan and Brenton [6.7]

[a] Values in parantheses refer to energies measured with respect to the indicated ground state
[b] The energies are for $v = 0$ at internuclear separation of 1.15 Å

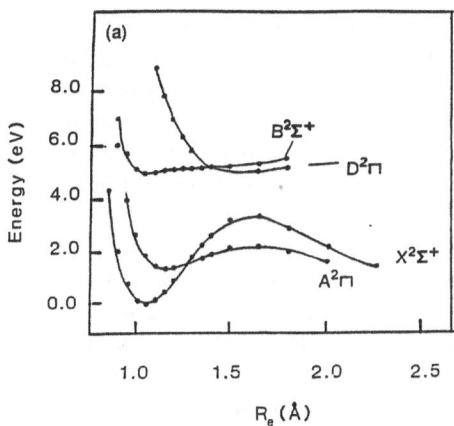

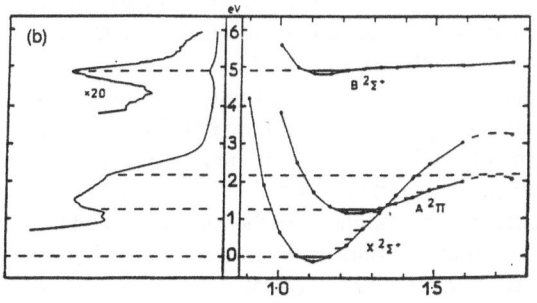

Fig. 6.13 a,b. (a) Potential energy curves for low-lying $^2\Sigma^+$ and $^2\Pi$ states of NO^{2+}. (b) Correlation of experimental data for $\Delta E < 0$ and theoretical data for the $X^2\Sigma^+$, $A^2\Pi$ and $B^2\Sigma^+$ states of NO^{2+}

179

The same curves also show that the three lowest electronic states are deep enough to support a number of vibrational levels. *Cooper* [6.46] has calculated the average vibrational spacing for the first six vibrational levels of $X^2 \Sigma^+$ to be 0.24 eV, while the average vibrational spacing for the first five vibrational levels of $B^2 \Sigma^+$ was found to be 0.16 eV. These high-level calculations were followed by two high resolution TES investigations [6.7, 44]. In both experiments, the following scattering processes were investigated

$$NO^{2+} + He \rightarrow NO^{2+*} + He - \Delta E \,, \tag{6.18}$$

$$NO^{2+*} + He \rightarrow NO^{2+} + He + \Delta E \,, \tag{6.19}$$

$$NO^{2+}(NO^{2+*}) + He \rightarrow NO^{2+}(NO^{2+*}) + He, \quad \Delta E = 0 \,. \tag{6.20}$$

Here NO^{2+} and NO^{2+*} are the respective ground and excited states of the incident ion and ΔE is the net energy gain/loss of the indicated process. *Jonathan* et al. [6.44] have interpreted their TES data by constructing a correlation diagram (Fig. 6.13b) between *Cooper's* [6.46] potential energy curves and their experimental data. The correlation diagram shows that the elastically scattered NO^{2+} projectile ion beam coincides with the zeroth vibrational level of the calculated $NO^{2+}(X^2 \Sigma^+)$ ground state. The onset of the first broad peak is in good agreement with the calculated energy for a Franck-Condon transition from $X^2 \Sigma^+$, $\nu = 0$ to $A^2 \Pi$, $\nu = 0$, whereas the sharp cut-off point of the same peak corresponds to a vertical transition of the dissociative threshold of the $A^2 \Pi$ upper electronic state. The energy of the excitation transition, which is centered at 4.9 eV, and its unusual sharpness, infer that the excitation is between $X^2 \Sigma^+$, $\nu = 0$ and $B^2 \Sigma^+$, $\nu = 0$. This deduction is in accord with Cooper's calculations which predict a shallow potential energy well for the $B^2 \Sigma^+$ state which is not expected to support more than one bound vibrational level.

The main deductions made on the basis of the experimental results of *Jonathan* et al. [6.44] and the calculations performed by Cooper regarding the three lowest electronic states of NO^{2+} have been supported by a more recent TES experiment by *Hamdan* and *Brenton* [6.7]. Their spectra for 6 keV, NO^{2+} scattered off He, acquired at two different values of energy resolution, are shown in Fig. 6.14a and b. Three excitation transitions which are evident in the spectrum of Fig. 6.14a were assigned to excitation transitions between the $X^2 \Sigma^+$ ground state and the two lowest electronic excited states, $A^2 \Pi$ and $B^2 \Sigma^+$. The TES of the same collision system, acquired at higher energy resolution and a narrower energy range, shows a number of inelastic and superelastic features which the authors attributed to excitation and de-excitation between the vibrational levels of the two electronic states, $X^2 \Sigma^+$ and $A^2 \Pi$. The measured energies and tentative assignments of these transitions are summarised in Table 6.3. The observation of the five superelastic submaxima within the energy range $\Delta E = 1.6$ eV was interpreted as a clear indication that a number of vibrational levels of the $X^2 \Sigma^+$ ground state are populated in the ion source and these may have a minimum lifetime of few microseconds (the ion transit time between the source and the collision cell). Although the measured vibrational spacing is in

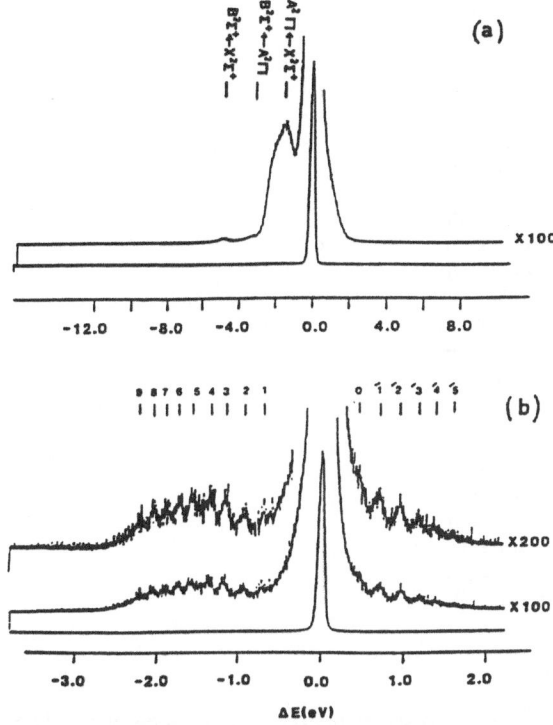

Fig. 6.14a,b. Translational energy spectra of 6 keV NO^{2+} scattered off He, taken at energy resolutions (FWHM) of (a) 0.35 eV and (b) 0.08 eV

reasonable agreement with the calculations of *Cooper* [6.46], the possibility of level mixing and internal conversion between the levels of the two states cannot be ruled out. This assumption is based on the potential energy curves (see Fig. 6.13a) which show part of the attractive region of the $A^2 \Pi$ curve to be within the potential energy well of the $X^2 \Sigma^+$ ground state.

As pointed out earlier, the energy separation between $X^2 \Sigma^+$, $\nu = 0$ and $B^2 \Sigma^+$, $\nu = 0$ was measured by *Besnard* et al. [6.45] to be 3.9 ± 0.5 eV, whereas the technique of double electron capture [6.41] yielded the value of 3.1 ± 0.5 eV. Both values are lower than the value of 4.8 to 4.96 eV reported in three different TES measurements [6.5, 7, 44] as well as the value obtained in Cooper's MC/SCF calculations. The low value obtained by means of double electron capture may be simply due to poor energy resolution which did not allow the separation of $X^2 \Sigma^+$ and $A^2 \Pi$ electronic states. It is worthwhile to point out that the energy differences between these two states is of a similar magnitude to the discrepancy between the energy measured by means of TES and that measured in a double electron capture experiment [6.41].

The discrepancy of ~ 1 eV between the TES and the photoion-photon of fluorescence concidence (PIFCO) technique regarding the energy of the transition, $NO^{2+}(B^2 \Sigma^+ \rightarrow X^2 \Sigma^+)$ can be tentatively attributed to either of the following pos-

Table 6.3. Observed vibrational transitions and corresponding energies involving excited NO^{2+} $(X^2\Sigma^+, A^2\Pi)$

Peak	ΔE [eV]	Transition
0[a]	–	$X^2\Sigma^+(\nu=2 \leftarrow \nu=0)$
1	–0.69	$X^2\Sigma^+(\nu=3 \leftarrow \nu=0)$
2	–0.95	$X^2\Sigma^+(\nu=4 \leftarrow \nu=0)$
3	–1.15	$X^2\Sigma^+(\nu=5 \leftarrow \nu=0)$
4	–1.36	$X^2\Sigma^+(\nu=6 \leftarrow \nu=0)$
5	–1.58	$X^2\Sigma^+(\nu=7 \leftarrow \nu=0)$
6	–1.75	$A^2\Pi^+(\nu=3 \leftarrow \nu=0)$
7	–1.92	$A^2\Pi^+(\nu=4 \leftarrow \nu=0)$
8	–2.07	$A^2\Pi^+(\nu=5 \leftarrow \nu=0)$
9	–2.23	$A^2\Pi^+(\nu=6 \leftarrow \nu=0)$
1'	0.69	$X^2\Sigma^+(\nu=3 \rightarrow \nu=0)$
2'	0.95	$X^2\Sigma^+(\nu=4 \rightarrow \nu=0)$
3'	1.15	$X^2\Sigma^+(\nu=5 \rightarrow \nu=0)$
4'	1.36	$X^2\Sigma^+(\nu=6 \rightarrow \nu=0)$
5'	1.60	$X^2\Sigma^+(\nu=7 \rightarrow \nu=0)$

[a] The transition is not sufficiently resolved to allow accurate measurement of ΔE

sibilities. First, in PIFCO, the excitation process is nearly vertical, whereas in TES, the excitation can occur at an NO^{2+} geometry which will probably differ from that of the neutral NO molecule. However, this hypothesis does not account for the difference between the energy measured by PIFCO and that calculated by Cooper, despite the fact that both values refer to vertical excitation. Another possible, and more plausible, explanation for the low value measured in the PIFCO experiment is that the observed emission is between $B^2\Sigma^+$, $\nu=0$ and $X^2\Sigma^+$, $\nu \geq 0$. The population of a number of vibrational levels of this state has been clearly demonstrated in the high resolution TES measurements carried out by *Hamdan* and *Brenton* [6.7]. Although the ionization technique is not the same in both experimental arrangements, there is no reason to suggest that the same levels cannot be populated when photons rather than electrons are used to form NO^{2+}.

b) CO^{2+}. Despite a relatively large volume of theoretical and experimental data concerning the doubly charged molecular ion of carbon monoxide, there persists a considerable ambiguity regarding the symmetries and energies of its low-lying electronic states. The question whether the ground state of this ion has $^3\Pi$ or $^1\Sigma^+$ character is still not fully settled.

The first set of potential energy curves of CO^{2+} was constructed by *Hurley* [6.36, 37] who used a semiempirical procedure relating the potential energy curves of this ion to the isoelectronic neutral, BN. Hurley's calculations yielded an appearance energy of 41.17 eV and $^3\Pi$ was assigned as the lowest electronic state of CO^{2+}. *Wetmore* et al. [6.47], using a restricted (3s,2p,1d) contracted Guassian basis set and

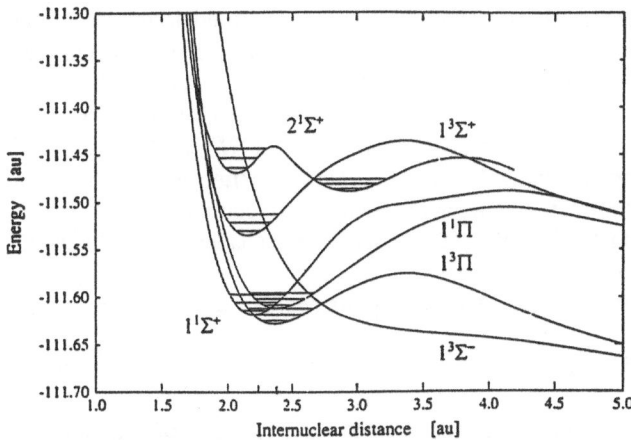

Fig. 6.15. Potential energy curves for the lowest singlet and triplet states of CO^{2+} [6.49]

a multi-reference configuration interaction treatment with self consistent field (SCF) molecular orbitals, calculated a large set of potential energy curves for CO^{2+}. They calculated a double ionization energy of 41.1 eV and also ascribed $^3\Pi$ character to the lowest electronic state. *Ågren* et al. [6.48] employed a multireference contracted configuration interaction (CCI) treatment with molecular orbitals obtained by complete active space SCF(CASSCF) calculations and basis sets larger than those used by *Wetmore* et al. [6.47]; it was concluded that the lowest electronic state of CO^{2+} is $^1\Sigma^+$ and not $^3\Pi$ as was suggested by *Hurley* [6.36] and *Wetmore* et al. The most recent high quality potential energy curves for a number of electronic states of CO^{2+} have been calculated with the complete active space SCF(CASSCF) and multireference contracted CI(MRCCI) method with a 1-particle basis set of a medium size (8s,6p,2d) [6.49]. These curves, which are depicted in Fig. 6.15, are used here to interpret experimental data acquired by various experimental techniques, including TES and single electron capture spectra.

The most detailed experimental results regarding the low-lying electronic states of CO^{2+} result from Auger spectroscopy [6.39, 40, 50–52]. A remarkably sharp feature was observed in the Auger spectrum at 45.4 eV above the CO ($^1\Sigma^+$, $\nu = 0$) ground electronic state and identified as the $2^1\Sigma^+$ state of the doubly charged molecule. The logical explanation for this sharp feature has been discussed by *Correi* et al. [6.50]; the $2^1\Sigma^+$ state has equilibrium internuclear distance, r_e, of 1.095 Å, very similar to that of the carbon core hole state of CO^+ ($r_e = 1.073$ Å). From analysis of band shapes, the energies of the $^1\Sigma^+$ and $^1\Pi$ states were determined to be at 41.5 and 41.9 eV above the CO $^1\Sigma^+$ ground state. *Marathe* and *Mathur* [6.53] have recently reported the results of ab initio all-electron molecular orbital calculations on the low-lying electronic states of CO^{2+} and concluded that the lowest electronic state is $^1\Sigma^+$. A fuller discussion of recent high-level theoretical efforts at quantally identifying the low lying states of this doubly charged ion is presented in Chap. 8.

Mass spectrometry and photoionization experiments have also yielded considerable information on the energies and lifetimes of a number of dissociative and non-dissociative electronic states of this dication. In a recent combined double photoionization and kinetic energy spectrometric study, *Dujardin* et al. [6.55] determined the following CO^{2+} state energies: 41.25 ± 0.05, 41.3 ± 0.1, 41.8 ± 0.1 and 43.8 ± 0.1 eV. On the basis of the ab initio calculations of *Larsson* et al. [6.49], these were assigned to the four low-lying electronic states, $(^3\Pi, \, ^1\Sigma^+)$, $(^3\Pi)$, $(^1\Pi)$ and $(^3\Sigma^+)$. The $^3\Pi$ and $^1\Sigma^+$ are grouped together because the energy resolution was insufficient to separate them. Furthermore, the same authors have deduced that the two lowest electronic states were metastable on the μs time scale, whereas the two higher states were rapidly dissociating states.

Utilizing the technique of double electron capture represented by

$$H^+ + CO \rightarrow H^- + CO^{2+} \, , \tag{6.21}$$

Lablanquie et al. [6.56] observed two reaction channels at energies of 41.3 and 45.4 eV above the CO $^1\Sigma^+$ ground state. These energies are in good agreement with the results of Auger spectroscopy [6.50] and were assigned to CO^{2+} $1^1\Sigma^+$ and $2^1\Sigma^+$ states. *Lablanquie* et al. [6.56] also reported energies of 40.75 ± 0.5 and 38.5 ± 0.5 eV for the production of C^+ and O^+ by means of single photon dissociative ionization of CO $^1\Sigma^+$. Assuming the former value to be the appearance energy for CO^{2+} $^1\Sigma^+$ or $^3\Pi$, which is in reasonable agreement with Auger results, the lower data value of 38.5 eV cannot be assigned to any of the known electronic states of CO^{2+}. It has been suggested [6.56] that this low-energy value may result from a highly-excited electronic state of CO^+, lying at $\sim 38\,eV$ above the CO $^1\Sigma^+$, which autoionises to CO^{2+} $^3\Sigma^-$ and an electron. Two double electron capture investigations carried out by *Mazumdar* et al. [6.57, 58] have yielded appearance energies of 39.45 ± 0.2 and $39.6\,eV$ which they assigned to $^1\Sigma^+$.

Translational energy spectra of 6 keV CO^{2+} scattered off He reported by *Hamdan* and *Brenton* [6.59] offered no evidence of long-lived, excited electronic states. However, because of resolution limitations, the above statement does not include electronic states which may have energies within 0.3 eV above the ground state of CO^{2+}.

The potential energy curves of CO^{2+} are predicted to have several triplet and singlet electronic states which can be described as quasibound. In order to explain the absence of excitation/de-excitation transitions in the TES spectrum shown in Fig. 6.16, the latest, and in our opinion one of the most accurate, calculations of potential energy curves [6.49] of CO^{2+} are considered. These are depicted in Fig. 6.15 and are summarised in Table 6.4.

Electron impact ionization used in the TES measurements may lead to the population of the states listed in Table 6.4. Earlier TES measurements [6.4–6, 19] demonstrated that significant excitation and/or de-excitation transitions can only occur within the same spin-manifold (that is singlet $\leftrightarrow$ singlet or triplet $\leftrightarrow$ triplet). Therefore, in the experiment of *Hamdan* and *Brenton* [6.59] the population of the three lowest electronic states, $^3\Pi$, $^1\Sigma^+$ and $^1\Pi$, in the ion source will not guarantee the observation of energy gain/loss transitions among the listed states. This is because the transition $^1\Pi \leftrightarrow {}^3\Pi$ is spin forbidden, whilst $^1\Pi \leftrightarrow {}^1\Sigma^+$ requires an energy

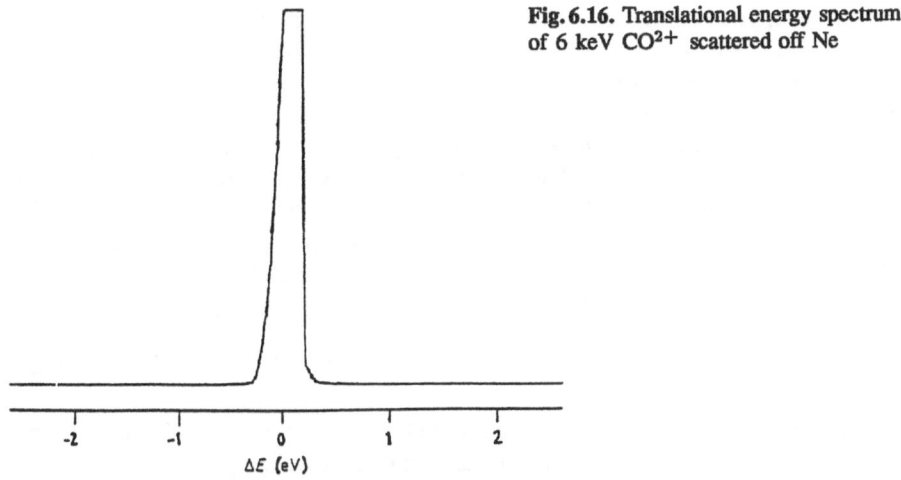

Fig. 6.16. Translational energy spectrum of 6 keV CO^{2+} scattered off Ne

ΔE (eV)

resolution higher than that used in their measurements. This leaves three possible transitions, $^3\Sigma^+ \leftrightarrow {}^3\Pi$, $2^1\Sigma^+ \leftrightarrow 1^1\Sigma^+$ and $2^1\Sigma^+ \leftrightarrow {}^1\Pi$ at energy losses of 2.52, 4.07 and 3.88 eV, respectively. The absence of peaks in TES spectra at these energies (see Fig. 6.16) can be partially rationlized by considering the lifetime and stability towords dissociation of the two upper electronic states, that is, $2^1\Sigma^+$ and $1^3\Sigma^+$. The potential energy curve of the first state has a double minima [6.49], the inner one only supporting three vibrational levels of which $\nu = 1$ and 2 have tunnelling lifetimes of 10^{-11} s and 10^{-13} s, respectively. This is a very short time to allow their detection in a mass spectrometer. The $\nu = 0$ level, to which 75 % of the population is expected to go, has a tunnelling lifetime of 10 ns [6.60] and transitions to the $1^1\Sigma^+$ and $^1\Pi$ states cannot be ruled out. The $1^3\Sigma^+$ state is expected to be populated entirely in the $\nu = 0$ level. A possible decay route for this level is a transition to the $^3\Pi$ state via fast radiative decay. Short lifetimes associated with the upper electronic states CO^{2+} $2^1\Sigma^+$ and $1^3\Sigma^+$ have also been reported in a PIFCO experiment [6.55] in which attempts to observe emission from these states yielded negative results.

The argument of a short lifetime may explain the absence of de-excitation peaks (which would be observed at positive values of ΔE). However, the same argument

Table 6.4. Minima (r_e) and energies (T_e) of some low-lying states of CO^{2+}

State	r_e [Å]	T_e [eV]
$^3\Pi$	1.253	0.0
$1^1\Sigma^+$	1.167	0.25
$^1\Pi$	1.260	0.44
$^3\Sigma^+$	1.140	2.52
$2^1\Sigma^+$	1.095	4.32

cannot be applied to explain the absence of excitation transitions (which would be observed at negative ΔE). Observation of TES spectra is dependent on the product ion not dissociating and is not necessarily dependent on the lifetime of an upper state, if dissociation does not result. Therefore, our tentative explanation for non-observation of excitation in the spectrum of Fig. 6.16 is a fast dissociative excitation via the repulsive $1^3\Sigma^-$ state. This is a drawback of the TES methodology; it does not currently yield high resolution data on dissociative states. Further evidence that electronically excited states of CO^{2+}, originating from the ion source do not survive long enough to be detected has come from a mass spectrometric investigation by *Hamdan* and *Brenton* [6.59] of single electron capture of the type

$$CO^{2+} + Ne \rightarrow CO^+ + Ne^+ \pm \Delta E , \tag{6.22}$$

which yielded the energy-change spectrum shown in Fig. 6.17; utilising data from this spectrum, and spectroscopically measured ionization energies of Ne (21.56 eV) and CO (14.01 eV), the appearance energy for CO^{2+} was calculated to be 40.6 eV. This is in reasonable agreement with the appearance energy of CO^{2+} measured by Auger spectroscopy but not to the value of 41.25 ± 0.05 eV measured by PIFCO and assigned to CO^{2+} $^3\Pi, {}^1\Sigma^+$.

An important use of the technique of single electron capture is that the appearance energy can often be estimated directly from a particular electron capture channel. However, a disadvantage, as in other TES measurements, is that the state composition of the projectile ion beam is not known and has to be calculated or deduced from TES measurements.

The sharpness of the peak observed at ~ 2.6 eV in Fig. 6.17a follows from the two Π states having almost identical equilibrium internuclear distances ($r_e = 1.243$ Å) [6.61] for the $A^2\Pi$ state of CO^+ and $r_e = 1.253$ Å for the $^3\Pi$ state of CO^{2+}. Electron capture by CO^{2+} ($^3\Pi, \nu = 0$) yielding CO^+ ($X^2\Sigma^+, r_e = 1.115$ Å) [6.61] is expected to populate more than one vibrational level. This is clearly evident in the spectrum (see Fig. 6.17a). It is also likely that the peak observed at $\Delta E \sim 5.3$ eV is enhanced by contributions from the electronic excited state, CO^{2+} ($^1\Sigma^+, \nu = 0$), predicted by the MRCCI [6.60] calculations to be at 0.25 eV above the CO^{2+} ($^3\Pi, \nu = 0$) ground electronic state. Similar investigations of the collision system CO^{2+}-Ne by *Pedersen* and *Hvelplund* [6.62] yielded an appearance energy of 40.6 eV and was assigned to CO^{2+} ($^3\Pi, \nu = 0$).

From the above discussion it is clear that in view of the prevailing ambiguity surrounding the potential energies and the exact character of the low-lying electronic states, particularly the character of its ground state, further high resolution experimental data as well as high quality large basis quantal calculations on CO^{2+} are necessary.

c) N_2^{2+}. Doubly-charged molecular nitrogen has also been the subject of a large number of theoretical and experimental investigations. The considerable attention paid to this ionic species can be partially attributed to its potential influence on the properties of the ionosphere. Basic spectroscopic information (spectroscopic constants) on

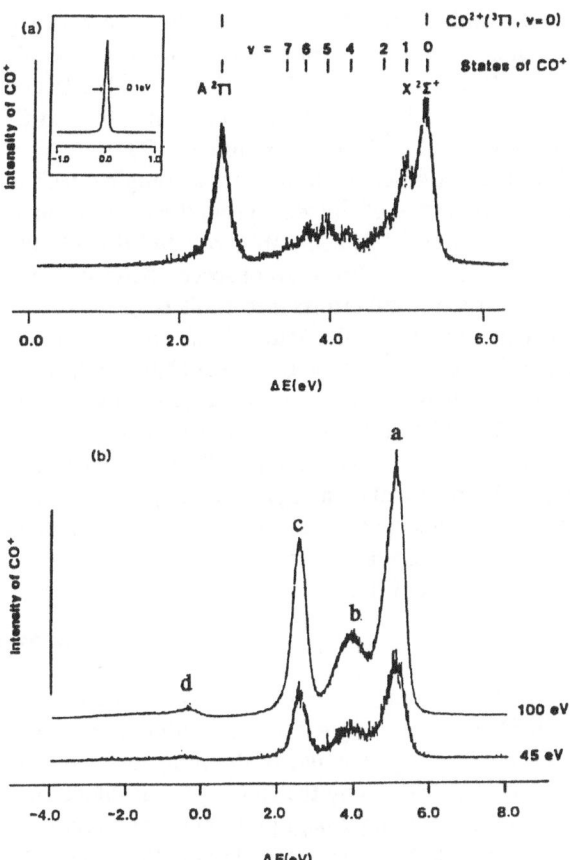

Fig. 6.17. (a) Energy change spectrum for CO^+ ions produced by 6 keV collisions of CO^{2+} $(^3\Pi$, $\nu = 0$) with Ne. The inset shows 6 keV CO^{2+} projectile ion beam having a FWHM of 0.1 eV at ionizing electron energy of 100 eV. (b) Energy change spectra of 6 keV CO^{2+}-Ne collision system taken at two ionizing electron energies of 45 eV and 100 eV and FWHM of 0.3 eV

N_2^{2+}, accurate vertical ionization energies from the N_2 ground state are all necessary for detailed modelling of its collision dynamics within the ionospheric environment.

The earliest set of potential energy curves of N_2^{2+} were constructed by *Hurley* [6.36]. These potential energy curves were used to identify the emission, N_2^{2+} $D^1\Sigma_u^+ \rightarrow X^1\Sigma_g^+$ which was first observed by *Carroll* [6.63].

To date, there exist extensive calculations extending beyond the Hartree-Fock level. *Taylor* [6.64] and *Taylor* and *Partridge* [6.65] have used the complete active space (CASSCF) approach to generate a large number of potential energy curves of moderate accuracy. *Wetmore* and *Boyd* [6.66, 69] used a variant of the SCF-CI approach, including perturbation calculations, to obtain a large number of potential energy curves for different symmetries. The SCF-CI method was used by *Cossart* et al. [6.67], with a CI expansion more limited than that used by Wetmore and Boyd. Other less accurate and/or less complete SCF calculations regarding the electronic states of N_2^{2+} have been extended by *Fraga* and *Ransil* [6.68], *Cade* et al. [6.70] and *Sahni* and *Delorenzo* [6.71]. *Thulstrup* and *Andersen* [6.72] performed limited CI calculations for several states of N_2, N_2^+ and N_2^{2+} using a minimal STO basis set.

Despite these extensive theoretical activities, there persists a complete lack of agreement regarding the character and energy of the N_2^{2+} ground electronic state, with *Taylor* and *Partridge* [6.65] predicting $^1\Sigma_g^+$ while *Wetmore* and *Boyd* [6.66] predict $^3\Pi_u$ symmetry. Differences along these lines are observed between other theoretical calculations and various experimental efforts (see Table 6.5).

Experiments in which the quasibound electronic states of N_2^{2+} have been observed include photoion-photon of fluorescence concidence (PIFCO) [6.67], laser spectroscopy [6.74], a high-frequency deflection technique [6.75], Auger spectroscopy [6.39, 40], electron impact [6.76, 77], collision spectrometry [6.78], double-charge transfer [6.41, 42] and photoion-photoion coincidence (PIPICO) [6.73]. In view of the diverse experimental and theoretical data regarding the potential energies of the low-lying electronic states of this species, a brief summary of the main conclusions of the diffrent experimental investigations is in order.

In a double charge transfer measurement of the type

$$H^+ + N_2 \rightarrow H^- + N_2^{2+} - \Delta E , \qquad (6.23)$$

where ΔE is the energy defect of the reaction, two energies, 43.1 ± 0.5 eV and 45.2 ± 0.5 eV, were measured [6.41, 42] and assigned to the vertical appearance energies of N_2^{2+} a$^1\Sigma_g^+$ and b$^1\Pi_u$, respectively. The use of the term 'vertical' and the ascribed assignments can be partially justified by the following considerations. For first-row elements in which spin-orbit coupling is negligible, spin conservation is expected to hold, as was demonstrated by earlier experimental work. Since both H^+ and H^- are singlet electronic states, the states of the N_2^{2+} product ion are expected to

Table 6.5. Measured appearance energies of N_2^{2+} ground electronic state

Technique	Energy [eV]	Assignment	Authors and Reference
Electron impact	42.7	$^3\Pi_u$	Dorman and Morrison [6.77]
Electron impact	42.9	$^3\Pi_u$	Mark [6.76]
Auger spectroscopy	43.2 ± 0.5	$^1\Sigma_g^+$	Siegbahn et al. [6.51]
Auger spectroscopy	43.3	$^1\Sigma_g^+$	Stolheim et al. [6.39]
Auger spectroscopy	43.4 ± 0.4	–	Moddeman et al. [6.40]
Double charge transfer	43.1	$^1\Sigma_g^+$	Appell et al. [6.41]
Double charge transfer	43.1 ± 0.2	$^1\Sigma_g^+$	Fournier and Aouchiche [6.42]
Laser spectroscopy	43.6 ± 0.2	$^1\Sigma_g^+ (\nu = 0)$	Cosby et al. [6.74]
Photoion-Photoion coincidence	43.1 ± 0.1	$^1\Sigma_g^+$	Besnard et al. [6.73]
Single-electron capture	42.6 ± 0.2	$^3\Pi_u$	Hamdan and Brenton [6.78]

have the same multiplicity as that of the N_2 precursor, that is, singlets. Furthermore, the interaction time between a 4 keV H^+ projectile and the N_2 target is sufficiently short ($\sim 10^{-16}$ s) that transitions between N_2 and N_2^{2+} may be considered vertical.

An experimental investigation of N_2^{2+} by photofragment spectroscopy [6.74] has yielded valuable information regarding the two N_2^{2+} electronic states $X^1 \Sigma_g^+$ and $1^1 \Pi_u$. In these experiments, N_2^{2+} ions were produced by electron impact, accelerated to 5000 eV, mass selected and collimated. The ion beam was subsequently coaxially merged, using an electrostatic quadrupole, with a laser beam over a length of 60 cm. The fragment ions produced in the interaction region were separated from the primary ion beam by a second quadrupole and then either detected or passed through a hemispherical electrostatic energy analyzer and detected by an electron multiplier. In this experiment, the photodissociation of N_2^{2+} yielded a series of structural bands in the photon energy dependence for the production of N^+ photofragments. Rotational analysis of five observed bands and comparison with earlier emission spectra indicated that the absorption was occurring from $X^1 \Sigma^+$ ($\nu = 0, 1, 2$) into three predissociated levels of the $1^1 \Sigma_g^-$ state. Furthermore, the two highest levels of this state could predissociate by tunnelling, while the lowest level could predissociate via the $1^1 \Sigma_g^-$ state. Comparison with existing experimental data also inferred that the lowest energy state of this dication may not be the $^1 \Sigma_g^+$ state.

Accurate fluorescence lifetime data for the zeroth vibrational level of N_2^{2+} $D^1 \Sigma_u^+$ has been obtained by *Olsson* et al. [6.75] using what is known as the high frequency deflection technique (HFD). In this, a 20 keV electron beam was focused and deflected over the entrance slit of a collision chamber containing nitrogen gas. The fluorescence following the pulsed excitation was dispersed with a monochromator followed by a photomultiplier. The monochromator was evacuated in order to avoid absorption by the O_2 Schumann-Runge bands. Further experimental details of this interesting technique have been provided by *Erman* [6.79]. *Olsson* et al. [6.75] reported a value of 6 ± 0.5 ns for the lifetime of the zeroth vibrational level of N_2^{2+} $D^1 \Sigma_u^+$. This conclusion infers that fast tunnelling through the potential energy barrier of this state can be ruled out. However, there are a number of electronic states which cross the Franck-Condon zone of this state; thus predissociation cannot be dismissed as an alternative route of decay.

A number of low-lying electronic states of N_2^{2+} have also been investigated by the photoion-photoion coincidence (PIPICO) technique [6.73]. Briefly, dispersed synchrotron radiation acts as a photon source of variable energy in the range 40–58 eV. The doubly-charged N_2^{2+} cations were formed by double photoionization of neutral N_2 in a gas jet at a pressure of $\sim 10^{-4}$ Torr. The two resulting N^+ photofragment ions were accelerated and detected in coincidence by a single detector arrangement. The coincidence count rate was measured as a function of the difference of time-of-flight of the N^+ fragment pair. Further details of the experimental arrangement and interpretation of the PIPICO curves have been given by *Dujardin* et al. [6.80] and *Winkoun* and *Dujardin* [6.81]. Apart from the measurement of a number of excitation energies and their tentative assignments to a number of electronic states of N_2^{2+} the PIPICO experiment has highlighted two other important aspects. Firstly, the $^1\Sigma_g^+$ state is extremely stable against predissociation and may have a minimum lifetime of 15 μs. Secondly, the $^1\Sigma_u^+$ state radiatively decays to the $^1\Sigma_u^+$ state.

In a recent mass spectrometric study, the non-dissociative electronic states of $^{29}N_2^{2+}$ have been investigated by *Hamdan* and *Brenton* [6.78]. Two collision processes were studied

$$^{29}N_2^{2+} + He \rightarrow {}^{29}N_2^{2+*} + He - \Delta E \,, \tag{6.24}$$

$$^{29}N_2^{2+} + He(Ne) \rightarrow {}^{29}N_2^{+} + He^{+}(Ne^{+}) - \Delta E \,. \tag{6.25}$$

The translational energy spectrum of 6 keV $^{29}N_2^{2+}$ scattered off He (process (6.24)) is shown in Fig. 6.18. The fact that no de-excitation peaks are observed implies that the projectile ion beam is predominantly in the ground electronic state. The wide and partially stuctured inelastic peak centered at $-\Delta E \sim 1.8$ eV can be assigned to collisional excitation between the low-lying electronic states of the N_2^{2+} projectile ion. Tentative assignments of the states involved were arrived at by considering the following points. First, the absence of any de-excitation transitions implies that the observed excitation transitions originate from either $^3\Pi_u$ and/or $a^1\Sigma_g^+$. Specific assignment into one state or another is not possible since the energy separation of only 0.1 eV is beyond the energy resolution employed in this investigation. Second, the energy range over which the observed transition extends, 1.1–2.0 eV, is expected, according to various theoretical calculations, to contain a number of quasibound electronic states of N_2^{2+} which may include the $B^3\Sigma_u^-$, $b^1\Pi_u$, $C^1\Delta_g$, $C^3\Sigma_u^+$ and $d^1\Sigma_g^+$ states. Taking these two points into account, together with spin and symmetry conservation, the observed excitation transitions are tentatively assigned

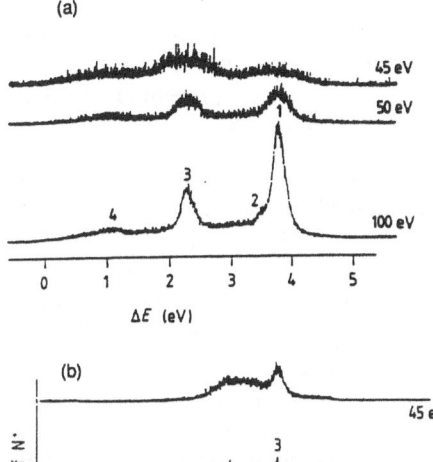

Fig. 6.19. Energy change spectra of 6 keV N_2^+ product ion following electron capture by N_2^{2+} from (a) He and (b) Ne

to the following excitations: $(B^3 \Sigma_u^- \leftarrow X^3 \Pi_u)$, $(c^1 \Delta_g, b^1 \Pi_u \leftarrow a^1 \Sigma_g^+, X^3 \Pi_u)$ and $(C^3 \Sigma_u^+ \leftarrow X^3 \Pi_u)$.

The energy change spectra resulting from single-electron capture by N_2^{2+} from He and Ne are shown in Fig. 6.19a and b. Figure 6.19a exhibits a well-resolved spectrum comprising four electron capture channels with exoergicities ranging from 1.3–4.2 eV. To establish the reaction channel associated with the ground state of N_2^{2+}, resulting in ground state N_2^+, the energy-change spectra were recorded at different electron ionizing energies. Considering the dependence of the intensity of the observed electron capture channels as a function of ionizing electron voltage, together with the measured exoergicities, it was deduced that peak 3 in both spectra (Fig. 6.19a, b) was due to capture by ground state N_2^{2+} forming N_2^+ ground state. Thus, using an exoergicity of 2.4 eV, measured from the energy-change spectrum, together with the ionization energies of He (24.58 eV) and N_2^+ (15.58 eV), the double-ionization energy of N_2 was found to be 42.6 ± 0.2 eV. The two reaction channels observed at 1.8 eV and 1.4 eV, above the ground state reaction channel, were tentatively assigned to electron capture by the dication excited states $C^3 \Sigma_g^+$ and $c^1 \Sigma_g,^1 \Pi$, respectively. Observation of these states in single electron capture collisions but not in the TES (refer to Fig. 6.18) could arise for the following reasons.

Firstly, due to interference by the electronic states of N^+ which has the same mass-to-charge ratio as N_2^{2+} ($m/z = 14$), the TES of N_2^{2+} was acquired using the natural abundance of $^{15}N^{14}N^{2+}$ (0.365 %); this resulted in a very weak projectile ion beam with consequential loss of sensitivity in the product TES spectrum. Single electron capture is free from this type of interference and therefore $^{28}N_2^{2+}$ was used and a stronger product ion spectrum obtained. Secondly, the collision de-excitation cross section may have a considerably lower magnitude than the cross section for electron capture; this may be another plausible reason for not observing long-lived electronic excited states in the TES spectrum (refer to Fig. 6.19).

d) Singly and doubly-charged ions of diatomic hydrides. TES has been used by *Hamdan* et al. [6.6] to investigate non-dissociative electronic states of the singly and doubly charged ions of the diatomic hydrides CH, NH, SH and OH. For the latter molecule, only its singly charged ion has been investigated. All measurements were performed on the high resolution double focusing translational energy mass spectrometer described above. Because of the absence of other experimental data relevant to the energies and spectroscopic properties of the electronic states of these doubly charged ions, interpretation of the present TES spectra is based on comparison with sparse existing theoretical calculations [6.6, 82–88]. In the case of the singly charged ions, there exists a large volume of data, both theoretical and experimental, on the energies, lifetimes and spectroscopic constants of low-lying electronic states. Agreement between the TES data and such existing data would, to a certain extent, lend an element of confidence to the measured excitation energies and assignments regarding the electronic states of their doubly charged counterparts.

The TES data of 6 keV CH^+, NH^+, SH^+, and OH^+ scattered off He, reported by *Hamdan* et al. [6.6], are given in Fig. 6.20. Two prominent peaks are observed in

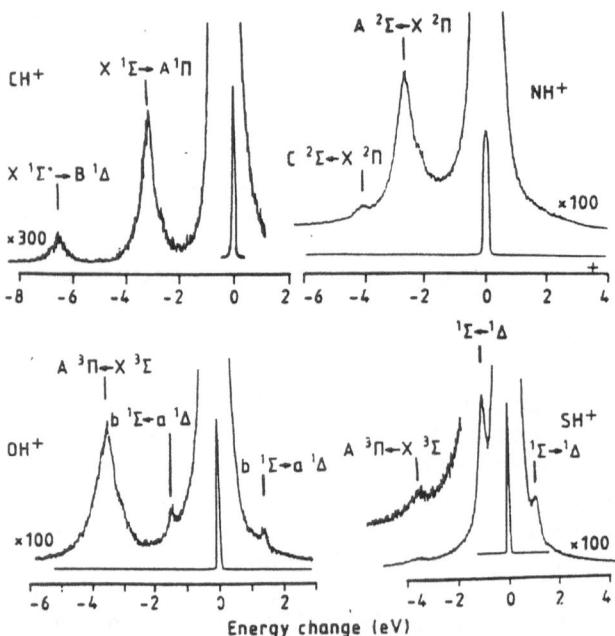

Fig. 6.20. Translational energy spectra of singly-charged hydride molecules CH^+, NH^+, OH^+ and SH^+

the TES of CH^+ at energy losses of 3.08 and 6.58 eV. These peaks were attributed to collisional excitation transitions from the $X^1\Sigma^+$ ground electronic state to the $A^1\Sigma$ and $B^1\Delta$ excited states; the corresponding spectroscopically determined energies for the two transitions are known to be 2.99 and 6.51 eV, respectively [6.89]. Absence of superelastic peaks infer that the incident CH^+ ion beam, which arrives at the collision cell a few microseconds after its formation in the ion source, is predominantly in the ground electronic state.

The first excited state of OH^+ that is accessible to optically allowed transitions from the $X^3\Sigma$ ground state is $^3\Pi$, whose excitation energy is 3.5 eV [6.89]. This transition is clearly evident in the TES of OH^+, at an energy loss of 3.5 eV. The absence of an equivalent superelastic peak at an energy gain of 3.5 eV was again interpreted as an indication that the OH^+ $^3\Pi$ state was not populated by electron impact. This suggestion is in accord with the results of an earlier crossed-beam experiment by *Kimura* and *Nishitani* [6.95] which showed that the transition OH^+ $A^3\Pi \rightarrow X^3\Sigma$ was observed when a 3 keV beam of He^+ ions collided with an H_2O beam, but the same transition was absent when the H_2O beam was crossed by a 50 eV electron beam. Both the results of TES and the crossed-beam experiment infer that the $(A^3\Pi)$ state of OH^+ can be populated in heavy particle collisions but not by electron impact.

The TES spectrum of OH^+ also shows a partially resolved structure at an energy loss of ~ 0.4 eV, measured from the centre of the 3.5 eV energy loss peak, which compares favourably with the vibrational spacing of 0.38 eV of the $X^3\Sigma^-$ state [6.89]; hence excitation from $\nu > 0$ levels of the OH^+ $^3\Sigma^-$ ground state cannot be ruled out. Two more transitions, centered at energy changes of ± 1.4 eV are assigned to $^1\Delta \leftrightarrow b^1\Sigma^+$. This type of transition, which violates the conservation of angular momentum, is difficult to observe in photoelectron and/or optical spectroscopy.

The translational energy spectrum of 3 keV NH^+ scattered off He shows two inelastic peaks centered at energy losses of 2.7 and 4.1 eV which are assigned to excitation transitions between the NH^+ $X^2\Pi$ ground state and the two electronic excited states, $A^2\Sigma^-$ and $C^2\Sigma^+$, respectively. The measured energies are in good agreement with spectroscopically measured data for these two transitions. Furthermore, the peak at 4.1 eV energy loss shows some indication of a partially resolved structure at ~ 0.35 eV from its centre. It has been suggested that the agreement between this energy and the vibrational spacing (0.36 eV) [6.74] of the $X^2\Pi$ state indicated that transitions from the $\nu > 0$ levels of the NH^+ ground state cannot be ruled out.

The TES spectrum pertaining to 3 keV SF^+ scattered off He shows a weak inelastic peak centered at an energy loss of 3.6 eV which is in reasonable agreement with the energy of 3.7 eV assigned to the $X^3\Sigma \rightarrow A^3\Pi$ transition in emission spectroscopy [6.61]. The same spectrum shows two narrow peaks, with energy changes of ± 1.1 eV, which have been assigned to dipole-forbidden transitions between the $^1\Sigma^+$ and $^1\Delta$ states. The measured energies of both transitions were found to be in accord with the energy separation between the zeroth vibrational levels of the two states, as determined from the photoelectron spectrum of *Dunlavey* et al. [6.90]. Furthermore, if it is asssumed that the cross section for excitation is the same as that for

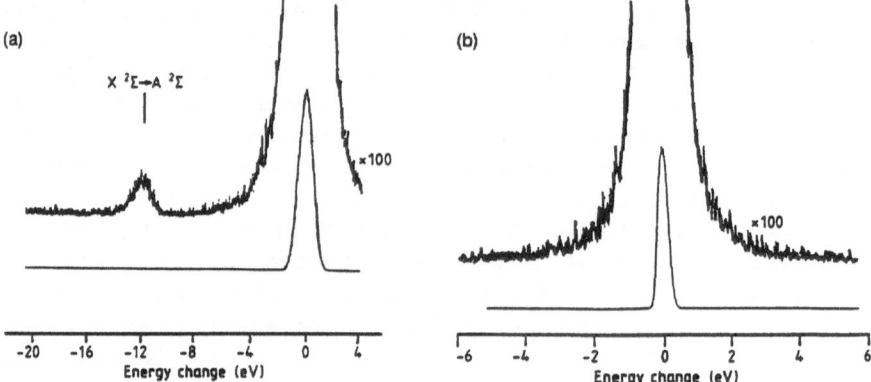

Fig. 6.21 a,b. Translational energy spectra of (a) CH^{2+} and (b) NH^{2+}

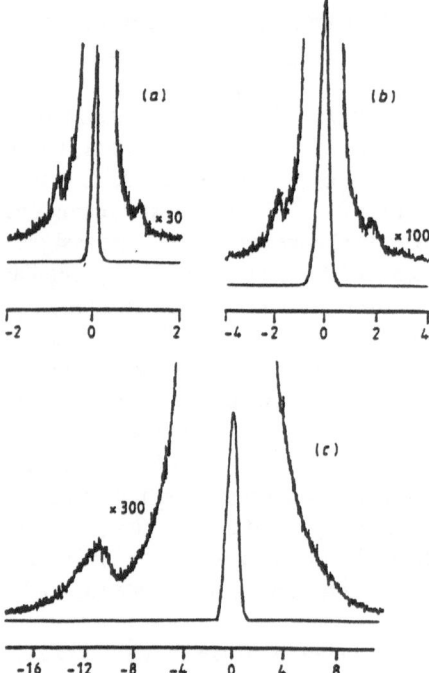

Fig. 6.22 a–c. Translational energy spectra of SH^{2+} measured at (a) high energy resolution and increased sensitivities (b) and (c)

de-excitation, then the relative heights of the two observed peaks imply that the $^1\Delta$ and $^1\Sigma^+$ states are populated in the ratio 3:1.

TES spectra of 6 keV CH^{2+}, NH^{2+} and SH^{2+} scattered off He, measured by *Hamdan* et al. [6.6], are shown in Figs. 6.21 and 6.22. The authors have interpreted these spectra with the aid of ab initio and multireference CI calculations. These the-

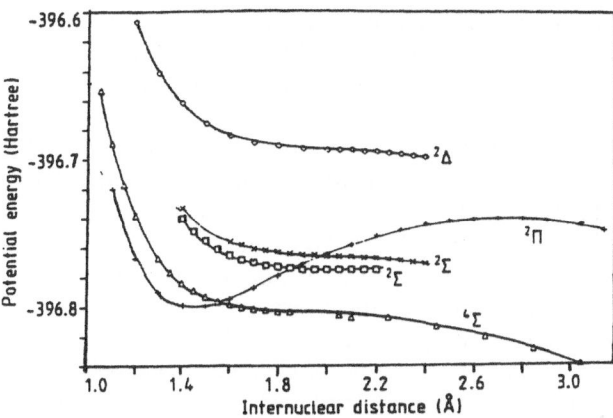

Fig. 6.23. Computed potential energy curves of the ground and low-lying states of SH^{2+}

oretical results [6.6, 87] have demonstrated that the ground state of SH^{2+} has $^3\Pi$ character and its potential energy curve is deep enough to prevent spontaneous dissociation. The metastability of this state and two low-lying excited electronic states is clearly evident in the TES of SH^{2+} scattered off He. The elastic peak centered at $\Delta E = 0$ is attributed to the ground state of the incident SH^{2+} ion, whereas the pairs of peaks observed at ± 0.9 eV and ± 2.0 eV are tentatively assigned to transitions between the ground state and the mixed $^2\Sigma$, $^2\Pi$ state and the upper excited state $^2\Delta$. In view of the mixed character of the ground state (see Fig. 6.23), the above assignments have to be tentative. More definite conclusions require calculations which incorporate the effect of spin-orbit coupling. Despite the preliminary nature of the given assignments, the TES spectra clearly demonstrate that SH^{2+} ions formed by electron impact can populate the ground and two metastable electronic excited states which survive the few microseconds necessary to reach the collision region.

For CH^{2+} scattered off He, one inelastic peak at an energy loss of 12.3 ± 0.5 eV is attributed [6.6] to a transition of CH^{2+} to an excited state, possibly of $^2\Sigma^+$ symmetry. The existence of this transition is still disputed on experimental and theoretical grounds [6.85]. The earliest ab initio calculations on CH^{2+}, carried out by *Pople* et al. [6.85] in 1982, indicated that the Hartree-Fock potential energy function for the ground state was purely repulsive; later, more refined, configuration interaction calculations, performed by *Heil* et al. [6.91] and *Wetmore* et al. [6.84], have indicated the existence of a potential minimum. *Friedman* et al. [6.82] have carried out semi-empirical adjustments to the earlier potential energy curve of *Heil* et al. in order to deduce a value of the potential well depth which would account for some of the experimental observations of CH^{2+}.

Disagreement regarding the stability of this doubly-charged ion is not only amongst theoretical calculations, but also between experimental measurements. Three experimental investigations [6.6, 92, 93] have reported the observation of a metastable ion, whereas a recent mass spectrometric investigation by *Koch* et al. [6.94] disputed its stability and suggested that earlier data pertains to $^{13}C^{2+}$ rather than CH^{2+}. Although this conjecture has been disputed by *Mathur* [6.96], prevailing uncertainty regarding the stability and energies of CH^{2+} electronic states require further high-resolution experiments and more accurate high level quantumchemical calculations.

The stablity of NH^{2+} has been theoretically investigated by *Pople* et al. [6.85] and by *Koch* and *Schwarz* [6.87]. Both sets of calculations indicated that NH^{2+} ions could not exist in a metastable state. However, a recently calculated potential energy curve for ground state NH^{2+} appears to contain a shallow minimum at an equilibrium N–H distance of 1.28 Å. A TES study [6.6] of NH^{2+} confirms the existence of a metastable ground state, but no electronically excited states.

As has been pointed out earlier, the sparse experimental data on the stability and energies of the electronic states of these ions hinders rational assessment of existing theoretical calculations; further experimental investigations of this class of ions would definitely resolve a number of fundamental questions.

Acknowledgements. We wish to thank the Royal Society (Paul Instrument Fund), University College Swansea and the Science and Engineering Research Council for financial support. We are also indebted to our colleagues and collaborators, particularly Dr. D. Mathur, Dr. A. R. Lee and Dr. D. P. Almedia, for their contributions to the developments of TES over the last few years and for their many helpful discussions.

References

6.1   J. H. Moore Jr., J. P. Doering: Phys. Rev. **177**, 218 (1969)
6.2   J. H. Moore Jr.: J. Chem. Phys. **55**, 2760 (1971)
6.3   U. Thielmann, J. Krutein, M. Barat: J. Phys. B. **13**, 421 (1980)
6.4   A. J. Illies, M. T. Bowers: Chem. Phys. **65**, 281 (1982)
6.5   A. O'Keefe, A. J. Illies, J. R. Gilbert, M. T. Bowers: Chem. Phys. **82**, 471 (1988)
6.6   M. Hamdan, S. Mazumdar, V. R. Marathe, C. Badrinathan, A. G. Brenton, D. Mathur: J. Phys. B **21**, 2571 (1988)
6.7   H. Hamdan, A. G. Brenton: Chem. Phys. Lett. **155**, 321 (1989)
6.8   A. G. Brenton, M. Hamdan, J. H. Beynon, P. Jonathan: Int. J. Mass Spectrom. Ion Proc. (1990) submitted.
6.9   R. F. K. Herzog: Z.Physik **89**, 447 (1934)
6.10 J. Mattauch, R. F. K. Herzog: Z. Physik **89**, 786 (1934)
6.11 T. Matsuo, H. Matsuda, Y. Fujita, H. Wollnik: Mass Spectrosc.(Japan)**24**, 19 (1975)
6.12 H. Matsuda: Mass Spectrom. Reviews **2**, 299 (1983)
6.13 Int. J. Mass Spectrom. Ion Proc., (Special Issue) 'Aspects of Ion Optics and High Mass Studies: A series of papers from the Japanese School Founded by Professor H. Matsuda', **91**, 1 (1989)
6.14 P. Jonathan, A. G. Brenton: Int. J. Mass Spectrom. Ion Proc. (1990) submitted.
6.15 M. G. Guilhaus, R. K. Boyd, A. G. Brenton, J. H. Beynon: Int. J. Mass Spectrom. Ion Proc. **68**, 91 (1986)
6.16 M. Hamdan, A. R. Lee, A. G. Brenton: J. Phys. B **21**, L561 (1988)
6.17 T. Nakamura, N. Kobayashi Y. Kaneko: J. Phys. Soc. Japan **54**, 2774 (1985)
6.18 N. Kobayashi, T. Nakamura, Y. Kaneko: J. Phys. Soc. Japan **52**, 1581 (1983)
6.19 J. H. Moore, Jr., J. P. Doering: J. Chem. Phys. **52**, 1693 (1970)
6.20 J. T. Park, F. D. Schowengerdt: Rev. Sci. Instrum. **40**, 753 (1969)

6.21 K. Tanaka, J. Durup, T. Kato, T. Koyano: J. Chem. Phys. **74**, 5561 (1981)
6.22 T. T. Jones, K. Birkinshaw, J. D. C. Jones, N. D. Twiddy: J. Phys. B **15**, 2439 (1982)
6.23 M. Hamdan, K. Birkinshaw, N. D. Twiddy: Int. J. Mass Spectrom. Ion Proc. **57**, 225 (1984)
6.24 N. Matic, V. Sidis, M. Viljovic, B. Cobic: J.Phys. B **13**, 3665 (1980)
6.25 F. T. Moran, J. B. Wilcox: J. Chem. Phys. **68**, 2855 (1978)
6.26 C. E. Moore: *Atomic Energy Levels, NBS circular number 467*, vol. 3 (U.S. Govt. Printing Office, Washington,D.C.,1957)
6.27 J. H. Moore: Phys. Rev. A **10**, 724 (1971)
6.28 J. H. Moore: Phys. Rev. A **8**, 2359 (1973)
6.29 N. Kobayashi, T. Nakamura, Y. Kaneko: J. Phys. Soc. Japan. **50**, 3541 (1983)
6.30 J. B. Hasted: *Physics of Atomic Collisions* (London, Butterworth, 1964)
6.31 E. Wigner: Nach. Akad. Wiss. Göttingen, Math. Physik. Kl. **11A**, 375 (1927)
6.32 A. R. Lee, C. S. Enos, A. G. Brenton: Rapid Commun. Mass Spectrom. **4**, 256 (1990)
6.33 M. Hamdan, A. G. Brenton: J. Phys. B **22**, 2289 (1989)
6.34 E. Friedlander, H. Kallman, W. Lasereff, B. Rosen: Z. Phys. **76**, 60 (1932)
6.35 L. Pauling: J. Chem. Phys. **1**, 56 (1933)
6.36 A. C. Hurley: J. Mol. Spectroscopy **9**, 18 (1962)
6.37 A. C. Hurley: J. Chem. Phys. **54**, 3656 (1971)
6.38 P. W. Thulstrup, E. W. Thulstrup, A. Andersen, Y. Ohrn: J. Chem. Phys. **60**, 3975 (1974)
6.39 D. Stolherm, B. Cleff, H. Hillig, W. Melhorn: Z. Naturforsch. **24**, 1728 (1969)
6.40 W. E. Moddeman, T. A. Carlson, M. A. Krause, B. P. Pullen, W. E. Bull, G. K. Schweitzer: J. Chem. Phys. **55**, 2317 (1971)
6.41 J. Appell, J. Durup, F. C. Fehsenfeld, P. G. Fournier: J. Phys. B **6**, 197 (1973)
6.42 P. G. Fournier, H. Aouchiche: CNRS Report (1987).
6.43 J. A. R. Samson, T. Masuoka, P. N. Pareek: J. Chem. Phys. **83**, 5531 (1985)
6.44 P. Jonathan, Z. Herman, M. Hamdan, A. G. Brenton: Chem. Phys. Lett. **141**, 511 (1987)
6.45 M. J. Besnard, L. Hellner, Y. Malinovich, G. Dujardin: J. Chem. Phys. **85**, 1316 (1986)
6.46 D. L. Cooper: Chem. Phys. Lett. **132**, 377 (1986)
6.47 R. W. Wetmore, R. J. Le Roy, R. K. Boyd: J. Phys. Chem. **88**, 6318 (1984)
6.48 H. Ågren: J. Chem. Phys. **75**, 1267 (1981)
6.49 M. Larsson, B. J. Olsson, P. Sigray: Chem. Phys. **139**, 457 (1989)
6.50 N. Correia, A. Flores-Riveros, H. Ågren, K. Helenelund, L. Asplund, U. Gelius: J. Chem. Phys. **89**, 2035 (1985)
6.51 K. Siegbahn, C. Nording, G. Johansson, J. Hedman, P. F. Heden, K. Harmin, U. Gelius, T. Bergmark, L. O. Werme, R. Manne, Y. Baer: *ESCA Applied to Free Molecules* (North-Holland, Amsterdam 1969)
6.52 J. A. Kelher, D. R. Jennison, R. R. Rye: Phys. Rev. A **75**, 652 (1981)
6.53 V. R. Marathe, D. Mathur: Chem. Phys. Lett. **163**, 189 (1989)
6.54 A. S. Newton, A. F. Sciamanna: J. Phys. Chem. **50**, 4868 (1969)
6.55 G. S. Dujardin, L. Hellner, M. Hamdan, A. G. Brenton, B. J. Olsson, M. J. Besnard-Ramage: J. Phys. B **23**, 1165 (1990)
6.56 P. Lablanquie, J. Delwiche, M.-J. Hubin-Franskin, I. Nenner, P. Morin, K. Ito, J. H. D. Eland, J.-M. Robbe, G. Gandara, J. Fournier, P. G. Fournier: Phys. Rev. A **40**, 5673 (1989)
6.57 S. Mazumdar, F. A. Rajgara, V. R. Marathe, C. Badrinathan, D. Mathur: J. Phys. B. **21**, 2815 (1988)
6.58 D. Mathur, V. R. Marathe, S. Mazumdar, J. Phys. B. **22**, L385 (1989)
6.59 M. Hamdan, A.G. Brenton: J. Phys. B **22**, L45 (1989)
6.60 M. Larsson, B. J. Olsson, P. Sigray: to be published
6.61 K. P. Huber, G. Herzberg: *Molecular Spectra and Molecular Structure*, vol. 4, (Van Nostrand Reinhold, New York, 1979)
6.62 J. O. K. Pedersen, P. Hvelplund: J. Phys. B **20**, L117 (1987)
6.63 P. K. Carroll: Can.J. Phys. **36**, 1585 (1958)
6.64 P. R. Taylor: Mol. Phys. **49**, 1297 (1983)
6.65 P. R. Taylor, H. Partridge: J. Phys. Chem. **91**, 6148 (1987)

6.66 R. W. Wetmore, R. K. Boyd: J. Phys. Chem. **90**, 5540 (1986)
6.67 D.Cossart, F. Launay, J. M. Robbe, J. Gandara: J. Mol. Spectrosc. **113**, 142 (1985)
6.68 S. Fraga, B. J. Ransil: J. Chem. Phys. **35**, 669 (1961)
6.69 R. W. Wetmore, R. K. Boyd: J. Phys. Chem. **90**, 6091 (1986)
6.70 P. E. Cade, K. D. Sales, A. C. Wahl: J. Chem. Phys. **44**, 1973 (1966)
6.71 R. C. Sahni, A. J. Delorenzo: J. Chem. Phys. **42**, 3612 (1965)
6.72 E. W. Thulstrup, A. Andersen: J. Phys. B **8**, 968 (1975)
6.73 M. J. Besnard, L. Hellner, G. Dujardin, D. Winkoun: J. Chem. Phys. **88**, 1732 (1988)
6.74 P. C. Cosby, R. Moller, H. Helm: Phys. Rev. A **28**, 766 (1983)
6.75 B. J. Olsson, C. Kindvall, M. Larsson: J. Chem. Phys. **88**, 7501 (1988)
6.76 T. D. Mark: J. Chem. Phys. **63**, 3731 (1975)
6.77 F. H. Dorman, J. D. Morrison: J. Chem. Phys **39**, 1906 (1963)
6.78 M. Hamdan, A. G. Brenton: J. Phys. B **22**, L9 (1989)
6.79 P. Erman: *Molecular Spectroscopy* (The Chemical Society, London, 1979)
6.80 G. Dujardin, S. Leach, O. Dutuit, P. M. Guyon, M. Richard-Viard: Chem. Phys. **88**, 339 (1984)
6.81 D. Winkoun, G. Dujardin: Z. Phys. D **4**, 57 (1986)
6.82 R. Friedman, S. Preston, A. Dalgarno: Chem. Phys. Lett. **4**,469 (1987)
6.83 S. A. Pope, I. Hillier, M. F. Guest, J. Kendric: Chem. Phys. Lett. **95**, 247 (1983)
6.84 R. W. Wetmore, R. K. Boyd, R. J. LeRoy: Chem. Phys. **89**, 329 (1984)
6.85 J. H. Pople, B. Tidor, P. v. R. Schleyer: Chem. Phys. Lett. **88**, 533 (1982)
6.86 S. A. Pope, I. H. Hillier, M. F. Guest: Faraday Symp. Chem. Soc. **19**, 109 (1984)
6.87 W. Koch, H. Schwarz: Int. J. Mass Spectrom. Ion Proc. **68**, 49 (1986)
6.88 W. Koch, N. Heinrich, H. Schwarz, F. Maquin, D. Stahl: Int. J. Mass Spectrom. Ion Proc. **67**, 305 (1985)
6.89 K. Huber, G. Herzberg: *Constants of Diatomic Molecules* (van Nostrand, New York, 1979) p. 516
6.90 S. J. Dunlavey, J. M. Dyke, N. K. Fayed, N. Jonathan, A. Morris: Mol. Phys. **44**, 265 (1981)
6.91 T. G. Heil, S. E. Butler, A. Dalgarno: Phys. Rev. A **27**, 2365 (1983)
6.92 T. Ast, C. J. Porter, C. J. Proctor, J. H. Beynon: Chem. Phys. Lett. **78**, 439 (1981)
6.93 D. Mathur, C. Badrinathan: J. Phys. B **20**, 1517 (1987)
6.94 W. Koch, B. Liu, T. Weiske, C. B. Lebrilla, T. Drewello, H. Schwarz: Chem. Phys. Lett. **142**, 147 (1987)
6.95 M. Kimura, T. Nishitani: J. Phys. Soc. Japan **39**, 551 (1975)
6.96 D. Mathur: Chem. Phys. Lett. **150**, 547 (1988)

7. Molecular Ionization Energies by Double Charge Transfer Spectrometry

F.M. Harris

With 5 Figures

Theoretical and experimental studies of the electronic states of doubly charged molecular ions have increased considerably in recent years. Experimental research is often difficult because doubly charged ions are, in general, unstable to dissociation. Two experimental techniques have, however, been widely used because the energy information concerning formation and excitation of the dications is carried away by electrons (Auger electron spectroscopy [7.1]) and by negative ions (double charge transfer spectrometry [7.2]). The more recent photoion-photoion coincidence (PIPICO) technique [7.3] makes use of the dissociation process by recording the time difference in the arrival at a detector of pairs of singly charged fragment ions. In this chapter, double charge transfer (DCT) spectrometry will be described, and an overview presented of its use to study a wide variety of ions.

In DCT spectrometry, fast-moving singly charged positive ions (usually having several keV of translational energy) undergo double electron capture (DEC) reactions in collisions with molecules. As will be seen below, these reactions are endoergic, and the energy to drive them comes from the translational energy of the projectile ion. Thus, the negative ions formed have a lower translational energy than the incident positive ions, and this energy reduction is measured in the experiment. If the projectile ion is denoted by $\underset{\rightarrow}{A}^+$, the arrow indicating that it has a high velocity, and the target molecule by M, the $\overrightarrow{\text{DEC}}$ reaction can be represented by

$$\underset{\rightarrow}{A}^+ + M \rightarrow \underset{\rightarrow}{A}^- + M^{2+} . \tag{7.1}$$

If the reduction in the translational energy of A^+ when it converts to A^- is denoted by ΔE_1, then the energy balance equation for reaction (7.1) is

$$\Delta E_1 = IE_2(M) - E(A^+ \rightarrow A^-) , \tag{7.2}$$

where $IE_2(M)$ is the double ionization energy of M (assuming M to be initially in its ground state), and $E(A^+ \rightarrow A^-)$ is the energy released when A^+ converts to A^-. The latter quantity can in many cases be equated to $IE_1(A) + E_a(A)$, that is, the sum of the single ionization energy and electron affinity of A. Another term which could be included in (7.2) is the recoil energy of M^{2+}, but under appropriate experimental

Springer Series in Chemical Physics, Vol. 54
Deepak Mathur (ed.): Physics of Ion Impact Phenomena
© Springer-Verlag Berlin Heidelberg 1991

conditions (high translational energy for A^+, and a geometry such that the resulting A^- ions are transmitted only at small scattering angles) it can be shown [7.4] to be negligibly small except when A^+ is significantly heavier than M. For most of the experiments described in this chapter, these conditions apply, and the recoil energy of M is not considered.

$IE_2(M)$ in (7.2) represents the double ionization energy required to populate the ground state of M^{2+} in reaction (7.1). The energies to populate higher-lying states of M^{2+}, that is, $IE_2'(M)$, $IE_2''(M)$, etc., will necessitate larger translational energy losses, $\Delta E'$, $\Delta E''$, etc., and these will give extra peaks in the measured energy loss spectrum. Thus, it can be seen that by determining the energy losses associated with the peaks in DCT spectra, the double ionization energies to ground and excited states of M^{2+} can be evaluated.

When DCT spectra are recorded, peaks often appear in them which are not due to A^- ions generated by reaction (7.1). If the target-gas pressures are high enough, two sequential single electron capture reactions can take place which also generate A^- ions. These reactions can be represented by

$$\underset{\rightarrow}{A^+} + M \rightarrow \underset{\rightarrow}{A} + M^+ \tag{7.3}$$

and

$$\underset{\rightarrow}{A} + M \rightarrow \underset{\rightarrow}{A^-} + M^+ . \tag{7.4}$$

In this case also, A^- has a lower translational energy than A^+, the difference being given by

$$\Delta E_2 = IE_1'(M) + IE_1''(M) - E(A^+ \rightarrow A^-) . \tag{7.5}$$

$IE_1'(M)$ and $IE_1''(M)$ represent the single ionization energies to different electronic states of M, since it is not necessary that the same state be populated in reactions (7.3) and (7.4).

The peaks in the DCT spectra due to reaction (7.1) and those due to the sequential reactions (7.3) and (7.4), are readily distinguished by varying the target-gas pressure. Since only one target molecule is involved in reaction (7.1), the resulting negative ion currents are linearly dependent on the pressure; the formation of A^- by reactions (7.3) and (7.4), however, requires two target molecules, so the resulting negative ion currents are quadratically dependent on pressure.

When information is sought about the properties of M^{2+}, the peaks of interest are those that result from reaction (7.1). The A^- ion currents are, in general, much smaller than those of A^+ but it is difficult to be categoric about the difference because the cross section for reaction (7.1) depends on several factors. The endoergicity of the reaction is one, and the way in which it affects the cross section will be discussed in Sect. 7.4. It is possible, however, to get some idea of the fate of an H^+ projectile ion on the target molecule H_2 from previously published cross section data [7.5, 6]. Assuming the proton has a translational energy of 6 keV, the relevant cross sections, read off the published graphs, are for $H^+ \rightarrow H$, 8.2×10^{-16} cm^2, for $H \rightarrow H^-$, 1.5×10^{-17} cm^2, and for $H^+ \rightarrow H^-$, 1×10^{-18} cm^2/gas molecule. Thus, only a small fraction of the incident H^+ ions which undergo reactions with H_2 give H^- by double electron capture, the reaction of interest in DCT spectrometry.

7.1 Apparatus and Experimental Techniques

7.1.1 Double Charge Transfer Spectrometry Prior to 1977

The first report in the literature of a DCT spectrometry experiment was an abstract of a paper presented at the Twenty-third Annual Gaseous Electronics Conference by *Witteborn* and *Ali* [7.7]. They measured the energies of formation of H^- by H^+ impact at 4 keV on various gases. No details of the apparatus appear in the abstract and the authors appear not to have continued with this type of research. The group working at Paris published at about the same time a paper [7.8] in which the translational energy spectra of H^- ions arising from single and double charge transfer of 4 keV protons on H_2 were presented. The apparatus used was essentially a mass spectrometer equipped with a collision gas cell within which the protons collided with the target molecules. The resulting H^- ions were energy analysed by the magnetic analyser. A full description of this apparatus is contained in a detailed review article by *Appell* [7.4] who also describes most of the work carried out with it up to 1977. In the present chapter, mainly work since 1977 will be described since the work prior to that is adequately covered by Appell.

7.1.2 Double Charge Transfer Spectrometer Used at Paris

Following the earlier work in Paris, *Fournier* and his co-workers developed and used an apparatus having a higher energy resolution. Information concerning the apparatus appears in several publications but most detail is to be found in one published in 1985 [7.9]. The projectile ions are formed in a modified plasma ion source. Most of the Paris work has been with H^+ projectile ions formed in the source by dissociative ionization of H_2. The ions are accelerated to several keV translational energy and mass-analysed by a Wien filter with a resolution of 300. The beam is angularly defined by two 30 μm diameter holes, 30 cm apart, before entering an electrically floating collision gas chamber of 10 cm length. The negative ion beam exiting the cell is also angularly selected using a hole of 30 μm diameter situated 20 cm from the cell. Energy dispersion of the negative ions is achieved with a 127° electrostatic analyser whose resolving power is 3000. The DCT spectra are accumulated in a multichannel analyser.

7.1.3 Double Charge Transfer Spectrometry at Swansea

Most of the DCT spectrometry experiments at Swansea have been carried out using a commercial, double focusing, high resolution mass spectrometer, the ZAB-2F [7.10]. It was manufactured in 1974 and has been considerably modified over the

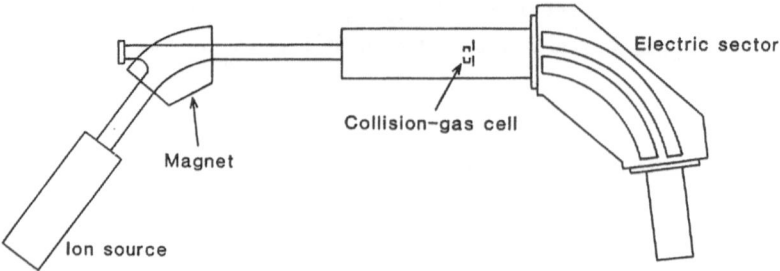

Fig. 7.1. Schematic representation of the ZAB-2F spectrometer used for DCT spectrometry at Swansea

years for specific research experiments. For example, the flight tube through the magnet has been modified [7.11] to allow the introduction of laser radiation for studies of photodissociation of ions. For DCT spectrometry, however, the various modifications are of no importance and any commercial ZAB-type spectrometer could be used provided it is fitted with a collision gas cell in the second field-free region of the spectrometer. The ZAB-2F spectrometer is shown schematically in Fig. 7.1. The projectile ions are generated in the source and accelerated to a translational energy of 6 keV before being mass-selected by the magnet and transmitted into the collision gas cell in the second field-free region which contains molecules of the sample under investigation. The negative ions generated by reactions (7.1), (7.2) and (7.3) are transmitted by scanning the voltages applied to the electric sector. These voltages are of opposite polarity to those normally applied when transmitting positive ions. The ion currents are measured using an electron multiplier/amplifier system and the spectra recorded using a computer-based data acquisition system.

The electric sector voltages are scanned using a digital scanner which relates the transmission energy of the ions to the source voltage. Voltage offsets can occur, however, so the translational energy loss scale has to be calibrated. This is done by carrying out DCT spectrometry with a calibrant gas in the cell, that is, one for which the relevant ionization energies are known accurately. If for a DEC reaction with a calibrant molecule C the value of translational energy loss is ΔE_c and the corresponding double ionization energy of the molecule is $IE_2(C)$, then

$$\Delta E_c = IE_2(C) - E(A^+ \rightarrow A^-) . \tag{7.6}$$

Subtracting (7.6) from (7.2) gives

$$\Delta E_1 - \Delta E_c = IE_2(M) - IE_2(C) , \tag{7.7}$$

from which $IE_2(M)$ is readily evaluated since $\Delta E_1 - \Delta E_c$ is determined experimentally, and $IE_2(C)$ is known. This demonstrates that the energy $E(A^+ \rightarrow A^-)$ need not be known but it does affect the endoergicity of the reaction and as will be seen in Sect. 7.4, the endoergicity must fall within a 'reaction window' for the cross section for the reaction to be non-zero.

Another spectrometer used at Swansea for DCT spectrometry is a ZAB-E, a much more modern mass spectrometer. This has an extended geometry which allows the analysis of masses up to 12,000 u. Other than that, the principle of operation is the same as that described above for the ZAB-2F.

The gas lines to the collision gas cells in both the ZAB-2F and ZAB-E spectrometers are unheated since they were originally installed by the manufacturer to allow the introduction of gas for collision induced dissociation of mass-selected projectile ions. Thus, the DCT spectrometry experiments carried out to date with these spectrometers have been with gaseous atoms and molecules, and with molecules of liquids having relatively high vapour pressures. To work with liquids having low vapour pressures requires heated lines to the cells; to provide those would have involved considerable changes to the ZAB-2F or ZAB-E spectrometer. To extend the range of molecules which can be studied, it was decided to modify [7.12] an MS9 spectrometer at Swansea for DCT spectrometry. This spectrometer is well known to the mass spectrometry community and is double focusing with a Nier-Johnson geometry in which the electric sector precedes the magnet (see, also, Chap. 8). Molecules under investigation are introduced into a collision gas cell located close to the intermediate focal point between the electric sector and magnet. It is connected via a heated line to a reservoir which can be heated to 150° and into which the sample under study is introduced. A separate line to the cell allows the introduction of a gas such as xenon or argon for calibration purposes. The scanning procedure with this spectrometer is different from that with the ZAB spectrometers. The magnetic field is held constant to transmit A^- ions of one momentum (that is, one energy, say E^-). The energy of the A^+ ions is scanned so that a peak in the A^- ion spectrum appears when the energy of A^+ reaches a critical value, say E_M^+ with molecules M in the cell, when the energy balance equation

$$E_M^+ - E^- = IE_2(M) - E(A^+ \rightarrow A^-) \tag{7.8}$$

is satisfied. When the molecules are replaced by, say, Ar atoms, the energy of A^+ to ensure transmission of A^- is E_{Ar}^+ given by

$$E_{Ar}^+ - E^- = IE_2(Ar) - E(A^+ \rightarrow A^-) . \tag{7.9}$$

It can be seen that $IE_2(M)$ is readily evaluated from (7.8) and (7.9) since E^- and $E(A^+ \rightarrow A^-)$ are common to both, E_M^+ and E_{Ar}^+ are measured in the experiment, and $IE_2(Ar)$ is known.

7.1.4 Double Charge Transfer Spectrometer Used at Bombay

Several DCT spectrometry experiments have been carried out by Mathur's group in Bombay. A laboratory-built translational energy spectrometer [7.13] has been used which has a low-voltage arc source from which ions are extracted, accelerated to 3 keV and focused on the entrance plane of a Wien filter. The mass-selected ions are allowed to pass through a collision gas cell containing the molecules under investigatons, and the resulting negative ions analysed using a high resolution, electrostatic energy analyser. The ion source is generally operated under relatively high pressure conditions (10^{-1}–1 Torr) which ensure that collisional deactivation of excited electronic states proceeds efficiently and, consequently, the projectile ion beam comprises essentially ground state ions.

7.2 Studies of Electronic States of Doubly Charged Ions

7.2.1 Spin Conservation

The Wigner spin-conservation rule [7.14] requires that the total electron spin angular momentum of a pair of atoms or molecules does not change in the course of a collision. In the early development of DCT spectrometry, Fournier and his co-workers examined several (H^+, noble gas) spectra to search for evidence of spin conservation. In the (H^+, Ar) spectrum [7.4], the singlet states $^1D_2$ and $^1S_0$ were populated. Since H^+ and H^- ions both have an electronic spin equal to zero, the doubly charged ion will have the same spin state as the neutral molecule if the spin-conservation rule applies. For a heavy target atom such as Xe, however, the (H^+, Xe) spectrum [7.15] had within it a peak which was evidence of population of the $^3P_2$ state. This effect of a collision with a change in the total spin of the colliding system was also observed weakly in the (H^+, Kr) spectrum [7.4]. The validity of spin conservation for DEC reactions involving a moleculer target was found in (H^+, O_2) spectra [7.2].

Most of the DCT spectrometry research carried out at Swansea has been with OH^+ as the projectile ion. It has a triplet ground state and the resulting OH^- ion is in a singlet state. It follows that collisions with molecules M, which are usually in singlet states, will populate triplet states of M^{2+}. On the other hand, the use of H^+ as a projectile ion will result in the population of singlet states of M^{2+}. These predictions are borne out in studies of the molecules listed below.

a) CO_2, OCS and CS_2. In a detailed paper, *Millie* et al. [7.16] have described previous studies of CO_2^{2+}, OCS^{2+} and CS_2^{2+}, presented new experimental data, and have also presented the results of ab initio SCF-CI calculations of electronic states up to an excitation energy of 10–12 eV above each ground state. A very large basis set was used in the computations, and they estimated the uncertainity of the calculated relative energies of the lower states to be less than 0.2 eV while for the highest states it was about 0.5 eV. They found, however, that the absolute values for vertical double ionization energies to the ground states were lower by about 1 eV than those measured experimentally. In order to define an absolute scale for the calculated data, the authors matched the energies of certain ground and excited states with those measured experimentally.

Consider first the situation for CO_2^{2+}. The appearance energy for the formation of metastable CO_2^{2+} was determined as 37.7 eV in a photoionization experiment [7.16]. This was in agreement with an earlier similar measurement [7.17] which yielded a value of 37.9 eV. *Millie* et al. [7.16] matched 37.7 eV with the energy of the lowest calculated CO_2^{2+} state which is $^3\Sigma_g^-$. The first state observed in a DCT spectrometry experiment reported in the same paper was found at 38.9 eV. Since H^+ was the projectile ion used in that experiment, it follows from the above discussion of the spin-conservation rule that a singlet state of CO_2^{2+} was being populated. The first singlet state is $^1\Delta_g$ and the calculated energy difference between it and the $^3\Sigma_g^-$ state is 1.2 eV [7.16], which is in exact agreement with the energy gap between the two experimental values. *Millie* et al. [7.16] therefore took these two energies to define an absolute scale for their calculated data. DCT spectrometry experiments (using H^+ projectile ions) and Auger electron spectroscopy experiments provided information on the energies of singlet states only and results from both [7.16, 18] were matched with eight calculated energies below 48 eV with reasonable certainty. Both singlet and triplet states are likely to be populated in double photoionization but the experimental results [7.16, 17, 19] could not be matched satisfactorily to the calculated values. Experiments carried out at Swansea [7.20], in which single electron capture by doubly charged ions was studied, provided information on the energies of the first three singlet states of CO_2^{2+} and these too agree well with the calculated values [7.6]. It can be seen from the above description that little, if any, information about triplet states of CO_2^{2+} was available up to 1986 when *Millie* et al. published their paper [7.16].

The same was true for OCS^{2+} and CS_2^{2+}. In 1989, however, the results of a DCT spectrometry experiment were reported [7.21] in which OH^+ was the projectile ion and, because of spin conservation, triplet states were populated. CO_2 and OCS were studied, and the position of the peaks in the spectra corresponded well with calculated double ionization energies to certain triplet states [7.16]. More recently, in the light of information on a 'reaction window' (see Sect. 7.4), F^+ projectile ions have also been used [7.22] which effectively populate triplet states of higher energies than those populated with the OH^+ projectile ion. The results from this investigation also correlate well with calculated triplet state energies.

b) CCl_4. Singly and doubly charged ions of CCl_4 are not readily observed in mass spectrometric studies. Evidence for the existence of the monocation has been obtained in recent years but it is still uncertain whether the dication exists for times long enough to be observed in a mass spectrometer. The single ionization energy and information about excited states of CCl_4 have been obtained in photoelectron spectrometry experiments [7.23]. However, the double ionization energy had not been measured until a DCT spectrometric study was carried out at Swansea [7.24]. It was stressed by *Appell* [7.4] that DCT spectrometry yields double ionization energies of molecules regardless of the stability of the doubly ionized molecule with respect to dissociation. In effect, the energy information concerning the formation of the doubly charged ion is contained in the reduction of translational energy of A^+ in acquiring two electrons from the molecule.

The DCT spectrum obtained [7.24] with OH^+ ions incident on CCl_4 molecules is shown in Fig. 7.2. Peaks A and B were identified as due to double collision processes [reactions (7.3) and (7.4)] and peaks C and D to the single-collision reaction (7.1) in which the doubly charged ions is formed. Information from the photoelectron spectrometry experiment [7.23] readily allows the positions of A and B to be

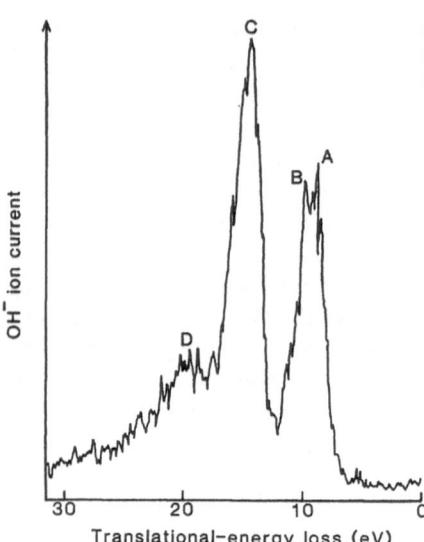

Fig. 7.2. Translational energy loss spectrum of OH$^-$ ions formed when 6 keV OH$^+$ ions collide with CCl$_4$ molecules

interpreted. The position of peak C on the energy loss scale indicates that the double ionization energy of CCl$_4$ is 29.3 ± 0.5 eV, which is the first recorded measurement of this quantity. A broad peak D indicates that an excited state of CCl$_4^{2+}$ exists at about 34 eV. Although the main reason for this investigation [7.24] was to determine the double ionization energy to the ground state of CCl$_4^{2+}$, it has been pointed out in a subsequent study [7.25] that spin conservation may well apply to DEC reactions of OH$^+$ with CCl$_4$, and that the states at 29.3 and 34 eV may be triplets. In that study, CCl$_4$ was throughly investigated by DCT spectrometry using the H$^+$ projectile ion, and by Auger electron spectroscopy. The energies of numerous states were measured and singlet/triplet assignments made. The data from both investigations [7.24, 25] are shown in Table 7.1.

c) **NH$_3$.** In a recent collaborative project, a DCT spectrometric study of NH$_3$ was carried out [7.26], using H$^+$ projectile ions in the apparatus at Paris, and OH$^+$ and F$^+$ projectile ions in the ZAB-2F spectrometer at Swansea. If spin conservation holds for DEC reactions of these ions with NH$_3$, singlet and triplet states of NH$_3^{2+}$ would be populated depending on whether the projectile ion was in a singlet state or a triplet state. The results obtained are shown in Table 7.2 together with the assignments of the states likely to have been populated. These results illustrate quite clearly that, for certain molecules, energies to singlet and triplet states of the corresponding dications are readily measured by DCT spectrometry using appropriate projectile ions.

d) **CO.** The doubly charged molecular ion of carbon monoxide has been the subject of numerous recent experimental and theoretical studies. All-electron ab initio molecular orbital calculations of the potential energy curves of several low-lying singlet, triplet and quintet state of CO^{2+} using unrestricted Hartree-Fock techniques showed [7.27] that the sequence of energy levels for the ion was $^3\Pi < {}^1\Sigma^+ <$

Table 7.1. Energies and assignments of electronic states of CCl_4^{2+}

Energies of states [eV]			Assignments [7.25]
DCT spectrometry		Auger-electron spectroscopy [7.25]	
OH^+ projectile ion [7.24]	H^+ projectile ion [7.25]		
29.3	–	29.8	Triplet
–	30.4	30.5	Singlet
–	–	31.4	Triplet
–	33.0	32.9	Singlet
34.0	–	33.7	Triplet
–	–	34.5	Triplet
–	35.8	36.1	Singlet
–	–	36.4	Singlet
–	–	39.1	
–	–	40.7	
–	–	42.3	

Table 7.2. Energies of NH_3^{2+} states measured in a DCT spectrometric investigation [7.26] using H^+, OH^+ and F^+ projectile ions

Measured energies [eV] and reacting projectile ion	H^+	35.6	–	39.5	–	44.7
	OH^+	–	37.7	–	–	–
	F^+	–	38.0	–	43.7	–
State assignment		$\tilde{X}^1A_1$	$\tilde{A}^3E$	$\tilde{A}^1E$	$\tilde{B}^3A_2$	$\tilde{B}^1E$

$^3\Sigma_-$. In a DCT spectrometry experiment carried out in the same investigation, 39.45 ± 0.2 eV was measured for the double ionization energy which was assigned to the $^3\Pi$ state. However, *Fournier* [7.28] pointed out that, since the DCT spectrometry experiment [7.27] was carried out with the H^+ projectile ion, population of the $^3\Pi$ state appears to violate the Wigner spin-conservation rule. He also raised doubts about the method used to calibrate the energy loss scale. In a subsequent experiment [7.29] in which the energy loss scale was determined making use of a (H^+, Ar) spectrum, the double ionization energy of CO was determined to be 39.6 eV, in agreement with the earlier results of 39.45 ± 0.2 eV. However, more rigorous configuration interaction all-electron ab initio calculations using a VAX/VMS version of the GAUSSIAN 86 series of programs indicated [7.29, 30] that the sequence of the first two energy levels is reversed to $^1\Sigma^+ < {}^3\Pi < {}^3\Sigma^-$, with an energy gap between $^1\Sigma^+$ and $^3\Pi$ of 1.4 eV. Thus, the lowest energy electronic state of CO^{2+} was then designated as $^1\Sigma^+$ which is accessible in (H^+,CO) DCT spectrometry experiments without violation of the Wigner spin-conservation rule. Subsequent studies on this dication (see Chap. 8) appear to indicate that the lowest $^1\Sigma^+$ and $^3\Pi$ states lie very close to each other in the Franck-Condon region accessible from neutral CO.

7.2.2 Studies of Small Molecules

Fournier and co-workers have used the high resolution DCT spectrometer at Paris to study doubly ionized states of CO_2, OCS and CS_2 [7.16], H_2O [7.31], Cl_2 [7.32], H_2 [7.33], HCl [7.34], N_2O [7.35] and NO_2 [7.36]. These were particularly detailed studies, often involving the use of other experimental techniques, such as Auger electron spectroscopy and/or double photoionization, in conjunction with DCT spectrometry. In addition, calculation of the energies of the states of the doubly charged ion were carried out which were correlated, where possible, with the energies measured. In this way, the investigators were able to confirm that for collisions of H^+ projectile ions with these molecules, the Wigner spin-conservation rule applied and singlet states of the doubly charged ions were populated. The results for CO_2, OCS and CS_2 have already been described and an outline will now be given of the results for some other molecules.

a) H_2. The DCT spectrometric study of H_2 [7.33] was carried out with protons having translational energies of 3, 5, 7 and 9 keV. This is essentially a repeat of an earlier study by *Fournier* et al. [7.8] but with better energy resolution and improved signal-to-noise ratio, using both H_2 and D_2 targets. The authors attempted to interpret the measured H^- energy distributions in terms of Franck-Condon vertical electronic transitions from H_2 to H_2^{2+}. Theoretical distributions were calculated using the well-known formula

$$F(E) = \rho(E)|\langle X_e(R)X_v(R)\rangle|^2 , \qquad (7.10)$$

where $\rho(E) = (2\mu h)^{-1}(2/W)^{1/2}$ is the state density in the dissociation continuum of H_2^{2+} and μ is the H_2 or D_2 reduced mass. W is the centre-of-mass kinetic energy of $H^+ + H^+$ fragments for a given electronic transition. The vibrational wave function $X_v(R)$ for H_2 (v = 0, J = 1) is space normalized and the amplitude of the continuum wave function $X_e(R)$ is taken equal to 1 for large values of the internuclear distance. It was found [7.33] that the experimental distributions for H_2 and D_2, including instrumental effects, were remarkably close in shape to the theoretical distribution. However, a systematic energy shift of about 0.7 eV was present. This small discrepancy corresponds to a shortening of the H_2 internuclear distance by about 0.05 Å.

b) H_2O. Experimentally, both DCT spectrometry and the PIPICO technique were used to obtain information about the energies of H_2O^{2+} states [7.31]; these were compared with values calculated by several procedures. By adding measured kinetic energy releases determined from the PIPICO spectra to the thermodynamic limits for each set of ion-pair products, an energy of 36.5 eV was found for the reaction giving $H^+ + OH^+$ products and one of 41 eV for that giving the more abundant $H^+ + O^+ +$ H products. This latter energy is close to 41.8 eV from an earlier Auger electron spectroscopy experiment [7.37] and to 41.4 eV found in the DCT spectrometry part of the investigation. The lower energy of 36.5 eV corresponds to a state of H_2O^{2+} not seen in the DCT or Auger electron spectroscopy experiments. However, the results of a charge stripping experiment [7.38], in which the energy to ionize H_2O^+ was

Table 7.3. Calculated and measured double ionization energies of various states of Cl_2^{2+} [7.32]

$IE_2(Cl_2)$ [eV]		Attributed state
Experimental	Calculated	
	31.24 ± 0.3	$1^1\Delta_g$
32.0 ± 0.2	31.58 ± 0.3	$1^1\Sigma_g^+$
33.1 ± 0.3	32.50 ± 0.3	$1^1\Sigma_u^-$
33.5 ± 0.4	34.79 ± 0.3	$1^1\Pi_u$
33.6 ± 0.2	36.07 ± 0.3	$1^1\Delta_u$
37.8 ± 0.8	37.63 ± 0.3	$1^1\Sigma_u^+$
39.2 ± 0.3	39.15 ± 0.3	$2^1\Sigma_u^-$
	39.39 ± 0.3	$2^1\Pi_u$

measured, suggest a state exists at 36.1 eV. The authors interpret the low energy state near 36.5 eV as the ground state, $^3B_1$ of H_2O^{2+}, while the states seen between 41 and 42 eV are unresolved singlet states $^1A_1$ and $^1B_1$. The experimental results are consistent with their SCF-CI calculations.

c) Cl_2. Energies of singlet states of Cl_2^{2+} ions were determined [7.32] using the proton as projectile ion, and were compared with the results of multireference single and double excitation configuration interaction calculations. Six peaks were identified in the DCT spectrum from which the double ionization energies shown in Table 7.3 were determined. The calculated data and state attributions are also shown in the table.

d) N_2O. The doubly charged cation of nitrous oxide has been known for many years as a short-lived species. Because of this, the states of N_2O^{2+} can best be studied by examining the energetics of its formation or decay. Gas-phase Auger electron spectra have been measured and analysed [7.39] and subsequently reanalysed theoretically [7.40]. A DCT spectrum was first measured [7.41] in 1974, and a PIPICO spectrum was one of the first to be obtained [7.42] by double photoionisation. *Price* et al. [7.35] report new results obtained with more recently developed PIPICO equipment. Also, new calculations on the electronic structure of N_2O^{2+} are presented in this study. The results obtained are summarised in Table 7.4.

7.3 Studies of Larger Molecules

7.3.1 CH₃OH

Methanol is one of a group of molecules represented by the general formula CH_3X (X = F, Cl, Br, I, OH, SH, NH_2, PH_2) which are known to have ylide isomers CH_2XH. In general, the ylide isomers are unstable, having small or no barriers to re-

Table 7.4. Electronic states of N_2O^{2+} [7.35]

State	Energy [eV]			
	Calculated		Measured	
	Triplets	Singlets	DCT Spectrometry	PIPICO
$^1\Sigma^-$	36.0[a]			$36.0^{+0.1}_{-0.5}$
$^1\Delta$		37.21	37.0 ± 0.2	
$^1\Sigma^+$		37.82		
$^3\Pi$	38.47			
$^1\Pi$		39.64	39.6 ± 0.3	
$^1\Sigma^-$		40.54		
$^3\Delta$	41.2			
$^3\Sigma^+$	41.32			
$^3\Sigma^-$	41.49			
$^3\Pi$	43.16			
$^1\Delta$		43.19	42.8 ± 0.4	
$^3\Pi$	43.32			
$^5\Pi$	43.45			
$^1\Sigma^-$		43.49		
$^3\Delta$	43.55			
$^3\Sigma^+$	43.64			
$^1\Sigma^+$		44.32	44.1 ± 0.4	
$^1\Pi$		44.44		44.3 ± 0.1
$^1\Sigma^+$		46.1	45.5 ± 0.4	
$^1\Pi$		49.5	47.7 ± 0.7	

[a] The calculated energy difference were referenced to 36.0 eV as zero

arrangement or dissociation. In contrast, ylidions (CH_2XH^+) have been calculated to lie in moderately deep wells in potential energy surfaces and are of comparable stability to their conventional isomers, CH_3X^+. Recently, interest has centred on the doubly charged cations CH_3X^{2+} ions and CH_2XH^{2+}. Theoretical results predict that CH_2XH^{2+} exist in relatively deep potential wells whilst, in the case of CH_3X^{2+}, little or no barriers exist to dissociation and/or rearrangement. Experimental evidence for this was obtained in charge stripping energy loss experiments in which the energy Q_{min} required to ionize the singly charged ions was measured. For CH_2XH^+ ions, the values of Q_{min} [7.43] agree with those calculated [7.43, 44]. For CH_3X^+ ions, however, the values of Q_{min} were found to be approximately the same as those for the corresponding ylidions, and are markedly different from those calculated [7.43, 44]. This led to the suggestion [7.43] that charge stripping reactions involving CH_3X^+ ions result in the formation of CH_2XH^{2+} ions and not CH_3X^{2+} ions.

The use of DCT spectrometry to study CH_3X^{2+} ions is particularly appropriate since the technique does not require that the doubly charged ions formed are stable. Thus, using the ZAB-2F spectrometer at Swansea, the double ionization energy to the ground state of CH_3OH^{2+} was determined to be 32.4 ± 0.5 eV [7.45]. Q_{min} can be determined from this by subtracting from it the single ionization energy of CH_3OH (10.96 eV). The value obtained is 21.4 ± 0.5 eV, which is substantially higher than that measured in the charge stripping experiment [7.43], 16.34 ± 0.05 eV, but close to each calculated value (22.05 eV [7.43] and 21.6 eV [7.44]). The DCT spectrometry investigation thus confirmed that charge stripping energy loss experiments do not give the correct ionization energy of the CH_3OH^+ ion. Similar DCT spectrometric studies of CH_3NH_2 and CH_3SH were subsequently carried out [7.46].

7.3.2 SF$_6$

Sulphur hexafluoride is widely used in the electrical transmission industry as a highly efficient insulator of high voltage equipment. It was established over forty years ago in a mass spectrometric study that electron-impact ionization of SF_6 does not result in the formation of stable SF_6^+ ions. The double ionization energy of SF_6 had not been measured prior to a DCT spectrometric investigation at Swansea [7.47]; the technique of charge stripping energy loss spectrometry could not be applied because of the instability of SF_6^+. By examining the spectrum obtained [7.47] as the result of DEC reactions of OH^+ with SF_6, it was established that the double ionization energy of SF_6 is 38.9 ± 0.5 eV. There was also evidence of one, or possibly two, electronic states of SF_6^{2+} between 38.9 and 44 eV.

7.3.3 CH$_4$

Considerable experimental and theoretical effort has been expended in investigating the properties of the methane dication [7.48]. The experimental techniques employed have been Auger electron spectroscopy [7.49, 50], electron energy loss spectroscopy [7.51, 52], DCT spectrometry [7.53, 9, 54] charge stripping energy loss spectrometry [7.55, 56, 57], PIPICO [7.58] and PEPIPICO [7.59], the last technique being a modification of PIPICO in that an electron provides the start pulse for the timing sequence. The double ionization energies determined in the Auger electron spectroscopy investigations are about 40 eV, but in a charge-stripping experiment [7.55, 56] a value of 30.6 eV was determined indirectly by adding the single ionization energy of CH_4 to the measured Q_{min} value. An explanation for this low value was put forword by *Hanner* and *Moran* [7.60]. Although CH_4^{2+} is not observed in the electron impact mass spectrum of CH_4 the charge stripping experiments apparently produced methane dications which were stable for times of at least a few µs. Calculations showed [7.60] that vertical transitions from CH_4^+ produced CH_4^{2+} in the bound portion of the ground state potential energy surface. A direct transition from CH_4, however, was predicted to give CH_4^{2+} ions lying above the ion-pair dissociation asymptote. The energy calculated [7.60] for the transition $CH_4^+ \rightarrow CH_4^{2+}$ is the same value as that measured [7.55, 56]. The vertical double ionization energy calculated for the transition $CH_4 \rightarrow CH_4^{2+}$ is 35.1 eV, markedly lower than the results from Auger electron spectroscopy [7.49, 50].

The DCT spectrometry measurements on methane yield values of 38.9 ± 0.7 eV [7.53], 38.6 ± 0.3 eV [7.9] and 35.6 ± 0.25 eV [7.54] for the double ionization energy. The first two values were obtained using the proton projectile ion, and probably represent ionization to a singlet states of CH_4^{2+} [7.9]. The third value was obtained using projectile ions which should result in the population of a triplet state, if spin conservation applies. From the double photoionization experiments the lowest double ionization energies measured are 35.0 [7.58], 33.9 ± 0.4 [7.9] and 36.5 ± 0.5 [7.59].

The lack of a totally unambiguous picture of the double ionization process in methane is further discussed in Chap. 8.

7.3.4 Fluoromethanes, Chloromethanes and Bromomethanes

The halogenated methane molecules have been investigated at Swansea [7.12, 61–63] by DCT spectrometry using the OH^+ projectile ion in the ZAB-2F spectrometer. For many of these molecules, the double ionization energies were measured for the first time in these investigations. The results obtained are shown in Table 7.5 which also has incorporated into it the single ionization energies of the molecules determined in previous photoelectron spectroscopy experiments. The uncertainties in the double ionization energies are not greater than ± 0.5 eV. In general, they are considerably less in the single ionization energies from photoelectron spectra.

Looking down each column of Table 7.5, it is seen that the single and double ionization energies decrease, reflecting the decrease in the electronegativity of the halogen in the molecules. The trends in the single and double ionization energies are similar since all the ratios $IE_2(M)/IE_1(M)$ are within 6% of the mean value of 2.64. *Tsai* and *Eland* [7.64], when they examined the single and double ionization

Table 7.5. Single and double ionization energies [eV] of halogenated methane molecules [7.12, 61, 62, 63]

M	CF_4	CF_3H	CF_2H_2	CFH_3
$IE_1(M)$	16.3	15.1	13.6	13.4
$IE_2(M)$	41.5	39.0	37.0	36.0
$IE_2(M)/IE_1(M)$	2.5	2.6	2.7	2.7

M	CCl_4	CCl_3H	CCl_2H_2	$CClH_3$
$IE_1(M)$	11.69	11.48	11.40	11.31
$IE_2(M)$	29.3	29.4	29.8	31.5
$IE_2(M)/IE_1(M)$	2.5	2.6	2.6	2.8

M	CBr_4	CBr_3H	CBr_2H_2	$CBrH_3$
$IE_1(M)$	10.39	10.47	10.61	10.53
$IE_2(M)$	27.9	28.9	27.0	28.8
$IE_2(M)/IE_1(M)$	2.7	2.8	2.5	2.7

energies of 34 different aromatic molecules, found the average values of such ratios to be 2.69; the average value for the halogenated methane molecules is in good accord with this earlier finding. This provides a useful 'rule-of-thumb' if an estimate of a double ionization energy of a molecule is needed from its single ionization energy.

7.3.5 Perhalomethanes

Members of the perhalomethane family of compounds are chemically inert and non-toxic, and some are widely used as aersol propellants, in fire extinguishers and refrigeration system, and in the manufacture of rigid foams. The two most extensively used are CF_2Cl_2 and $CFCl_3$; their release into the atmosphere and their possible involvment in the reduction of the stratospheric ozone layer have been the subjects of widespread recent discussion. These and other perhalomethane molecules have been studied by DCT spectrometry at Swansea [7.65–68] and the results obtained are summarized in Tables 7.6 and 7.7. Included in these tables are the ionization energies to the ground and first excited states of the singly charged ion obtained from photoelectron spectroscopy. From Table 7.6, it can be seen that, in general, as an atom in the molecule is replaced with one of larger electronegativity, the single and double ionization energies increase. In fact, it can be seen from the ratios $IE_2(M)/IE_1(M)$ and $IE_2'(M)/IE_1'(M)$ that the trends are markedly similar for double and single ionization energies. A similar trend is seen in the data of Table 7.7.

In filling in many of the gaps which previously existed in a compilation similar to those of Tables 7.6 and 7.7, DCT spectrometry experiments have provided valuable new information for perhalomethane molecules.

7.3.6 Fluoroethanes

Six fluoroethane molecules were studied [7.69] by DCT spectrometry using the OH^+ projectile ion in the ZAB-2F spectrometer at Swansea. For some of these molecules,

Table 7.6. Ionization energies to ground states, $IE_1(M)$, and first exited states, $IE_1'(M)$, of singly charged ions, and ground states, $IE_2(M)$, and first exited states, $IE_2'(M)$, of doubly charged ions for various perhalomethane molecules

M	CF_3Cl	CF_2Cl_2	CF_2ClBr	CF_2Br_2	$CFCl_3$	CBr_2FCl	$CFBr_3$
$IE_1(M)$ [eV]	13.0	12.6	11.83	11.18	11.9	11.8	10.67
$IE_1'(M)$ [eV]	15.0	13.5	12.9	12.02	13.0	–	11.24
$IE_2(M)$ [eV]	35.1[a]	31.3[a]	28.9[b]	28.4[b]	30.0[a]	29.0[c]	27.8[b]
$IE_2'(M)$ [eV]	40.3[a]	33.6[a]	30.0[b]	30.4[b]	32.3[a]	30.8[c]	30.0[b]
$IE_2(M)/IE_1(M)$	2.7	2.5	2.4	2.5	2.5	2.5	2.6
$IE_2'(M)/IE_1'(M)$	2.7	2.5	2.4	2.5	2.5	–	2.7

[a] From [7.65]
[b] From [7.66]
[c] From [7.67]

Table 7.7. Ionization energies to ground states, $IE_1(M)$, and first exited states, $IE_1'(M)$, of singly charged ions, and ground states, $IE_2(M)$, and first exited states, $IE_2'(M)$, of doubly charged ions for various hydrofluorocarbon and perhalomethane molecules

M	CF_3H	CF_2H_2	CF_2HCl	CFH_3	$CFHCl_2$	CBr_2FH	CH_2Br_2
$IE_1(M)$ [eV]	14.8	13.29	12.6	13.05	12.0	10.8	10.61
$IE_1'(M)$ [eV]	15.5	15.5	14.0	17.0	12.2	12.7	11.28
$IE_2(M)$ [eV]	39.0[a]	37.0[a]	33.1[b]	36.0[a]	30.3[b]	26.9[c]	27.0[d]
$IE_2'(M)$ [eV]	44.0[a]	41.0[a]	38.0[b]	39.0[a]	32.4[b]	28.9[c]	28.9[d]
$IE_2(M)/IE_1(M)$	2.6	2.8	2.6	2.8	2.5	2.5	2.5
$IE_2'(M)/IE_1'(M)$	2.8	2.6	2.7	2.3	2.5	2.2	2.6

[a] From [7.61]
[b] From [7.68]
[c] From [7.67]
[d] From [7.63]

Table 7.8. Double ionization energies [eV] of six fluoroethane molecules [7.69]

M	CH_3CH_2F	CH_3CHF_2	CH_3CF_3	CH_2FCHF_2	CHF_2CF_3	CF_3CH_3
$IE_2(M)$	32.5	35.0	36.4	33.3	35.8	38.7

information was obtained from the spectra of the ionization energies to excited states of the corresponding doubly charged ions. For all, however, the double ionization energy to the ground states was measured and these are tabulated in Table 7.8. The uncertainities in the values are not greater than ± 0.5 eV.

For the first three, it can be seen that as the fluorine content of the molecule increases so does $IE_2(M)$. However, replacing two further H atoms with F atoms to give the molecule CHF_2CF_3 results in a double ionization energy lower than that for CH_3CF_3. Clearly, it is not only the number of fluorine atoms in the molecule which govern the double ionization energy but also their positions relative to other atoms. It was shown [7.69] that the single ionization energies of the molecules in Table 7.8 follow the same patterns as the double ionization energies with values of $IE_2(M)/IE_1(M)$ being either 2.6 or 2.7.

7.4 Reaction Window for Double Electron Capture

7.4.1 Endoergicity of DEC Reactions

Reaction (7.1), which is central to DCT spectrometry, is endoergic since, in general, $IE_2(M) > E(A^+ \to A^-)$. For many DEC reactions, the endoergicity can be quite large. Consider the DCT spectrometry experiment [7.21] in which OH^+ was converted to OH^- in collisions with CO_2 molecules. The double ionization energy to the ground triplet state of CO_2^{2+} was measured to be 37.7 ± 0.5 eV. The single ionization energy and electron affinity of OH are 13.01 eV and 1.83 eV, respectively, so that the endoergicity of the DEC reaction of OH^+ in collisions with CO_2 is about 23 eV. Thus, DEC reactions are markedly different from single electron capture by single, doubly and triply charged positive ions which are known to be nearly resonant (in the case of singly charged projectile ions) or slightly exoergic for ions in charge state greater than or equal to 2+.

7.4.2 Theoretical Predictions of a Reaction Window

Mathur [7.70] was the first to employ a Landau-Zener type treatment to study double electron capture. He postulated that the DEC process can be considered in terms of the long-range potential energy function of two quasimolecules, A^+M and A^-M^{2+}, representing, respectively, the reactant and product channels in reaction (7.1). In the case of the A^+M quasimolecule, the potential energy varies as a function of the internuclear separation, r, of A^+ and M according to

$$V_{A^+M}(r) = -\frac{Z^2 e^2 \alpha(M)}{2r^4} , \tag{7.11}$$

in which $\alpha(M)$ represents the dipole polarizability of the target molecule. In the case of the A^-M^{2+} quasimolecule, the potential energy function is dominated by the extremely strong Coulombic attraction between the A^- and M^{2+} products but also has a weaker attractive component due to the polarizabilities of A^- and M^{2+}. The potential energy function takes the form

$$V_{A^-M^{2+}}(r) = -\frac{2Ze^2}{r} - \frac{Z^2 e^2}{2r^4}[\alpha(A^-) + \alpha(M^{2+})] . \tag{7.12}$$

The two potential energy functions are shown schematically in Fig. 7.3 and illustrate that although the DEC reaction is strongly endoergic, avoided crossing of the potential energy functions can occur which can result in fairly large cross sections for the reactions [7.70]. A Landau-Zener approach to a reaction occurring at a crossing point was applied by *Mathur* [7.70] and, later, by *Almedia* and *Langford* [7.71]. In both studies, it was shown that the cross section for DEC reactions is dependent on a quantity β which is a complex function, strongly dependent on the endoergicity of the reaction. Of particular importance in those studies was the prediction that the cross section passed through a maximum at a particular value of β, that is, that a 'reaction window' exists in which β, and hence the endoergicity, must fall in order that a DEC reaction takes place efficiently. To date, two experimental investigations have taken place in which evidence for the existence of a reaction window was sought. These are described below.

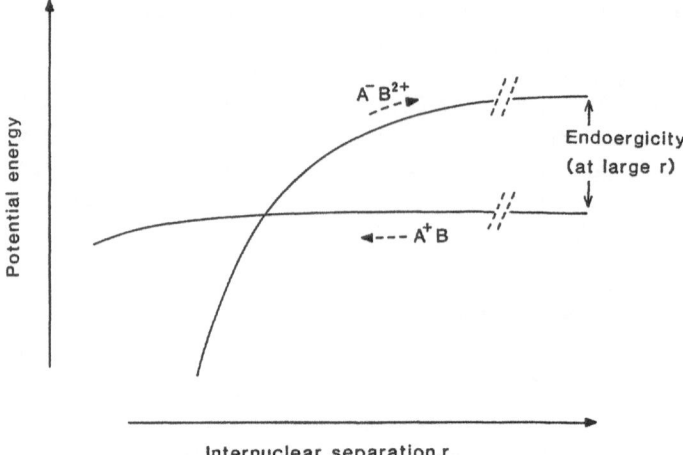

Fig. 7.3. Schematic representation of the potential energy curves of (A^+B) and (A^-B^{2+}) quasi-molecules

7.4.3 Relative Cross Sections for DEC Reactions Measured as a Function of Endoergicity

DEC reactions were studied [7.72] using 6 keV OH^+ projectile ions incident on numerous atomic and molecular target gases. Areas under the OH^- product ion peak as well as the elastically scattered OH^+ peak were measured as a function of pressure in the collision gas cell, and relative cross section values were deduced by means of the initial growth curve technique [7.13] widely used in single electron capture cross section determinations. The atoms and molecules used were Ne, Ar, Kr, N_2, CH_4, H_2S, CO_2, N_2O, NH_3 and C_2H_6. The endoergicities $IE_2(M) - E(OH^+ \rightarrow OH^-)$ ranged from 48 eV to 18.4 eV. The relative cross sections determined in this way are plotted against endoergicity in Fig 7.4 and provide clear evidence that a reaction window for DEC reactions exists.

7.4.4 Evidence for a Reaction Window in Collisions Involving the Molecular Target CH₃Br

In this experiment [7.73], a single molecular target was used and the endoergicity of the reaction was varied by using different projectile ions for which the energy $E(A^+ \rightarrow A^-)$ is different. The projectiles used (with the relevant energies in brackets) were S^+ (12.44 eV), OH^+ (14.82 eV) and F^+ (20.82 eV). Since the endoergicity is $IE_2(M) - E(A^+ \rightarrow A^-)$, it can be seen that if it is constrained to remain within a window, and if $E(A^+ \rightarrow A^-)$ increases, then $IE_2(M)$ will increase, which physically means that more highly excited states of M^{2+} will be populated in the reaction. Thus, one can anticipate that with a projectile ion having a low value of $E(A^+ \rightarrow A^-)$ the resulting spectrum will have peaks corresponding to the population of low-lying M^{2+} states. For a projectile ion with a high $E(A^+ \rightarrow A^-)$,

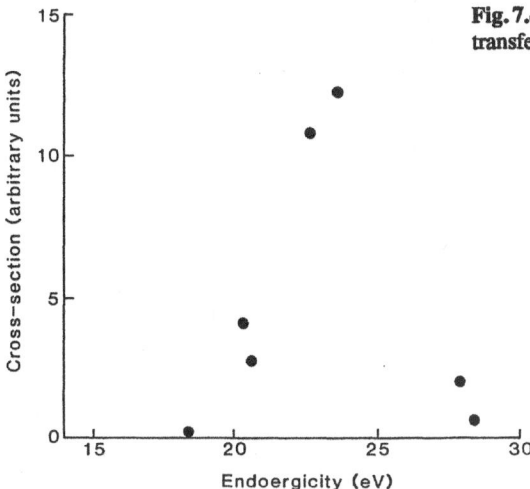

Fig. 7.4. Cross sections for double charge transfer reactions as a function of endoergicity

however, higher-lying states will be populated, that is, the spectrum profile will be changed.

The spectra obtained as a result of DEC reactions of S^+, OH^+ and F^+ with CH_3Br molecules are shown in Fig. 7.5. The measured translational energy losses have been converted into double ionization energies to give the abscissa scale. From these and similar spectra obtained at different CH_3Br pressures, it was deduced that peaks 1 and 2 were due to double electron transfer in two collisions and that peaks I–V were due to DEC reactions in single collisons. The latter peaks are of interest here. It is clear from the three spectra that the peak profile is markedly dependent on projectile ion used. With S^+, only the ground state of CH_3Br^{2+} is populated (peak I) with no peaks being apparent due to population of excited states, although a prononunced shoulder does exist at higher values of $IE_2(CH_3Br)$. With the OH^+ projectiles, the yields of ground and first excited state CH_3Br^{2+} ions are approximately equal (peaks I and II), with peak II, corresponding to the second excited state, just evident. With F^+, peak I has virtually disappeared, II is dominant and, in addition to III, peaks IV and V appear which correspond to population of higher excited states of CH_3Br^{2+}. Thus, the profiles of spectra measured agree in general with those expected on the basis of the existence of a reaction window.

A rough estimate of the width of the window was made [7.73] in the following way. The double ionization energy determined from the mean position of peak I is 29.0 eV. This is close to a previous determination [7.63] of 28.8 eV. For peak I, with the S^+ projectile ion, the endoergicity of the reaction is $29.0 - 12.44 = 16.56$ eV. The corresponding values for peak I with the OH^+ and F^+ projectile ions are 14.15 eV and 8.18 eV. Thus as the endoergicity ranges from 16.56 eV to 8.18 eV, the cross section for populating CH_3Br^{2+} in its ground state decreases from a finite value to virtually zero. This suggests that a lower limit for the reaction window is at about 8 eV. An upper limit was deduced by considering the peaks marked II in the spectra. With OH^+ and F^+, the cross sections are non-zero since the peaks are

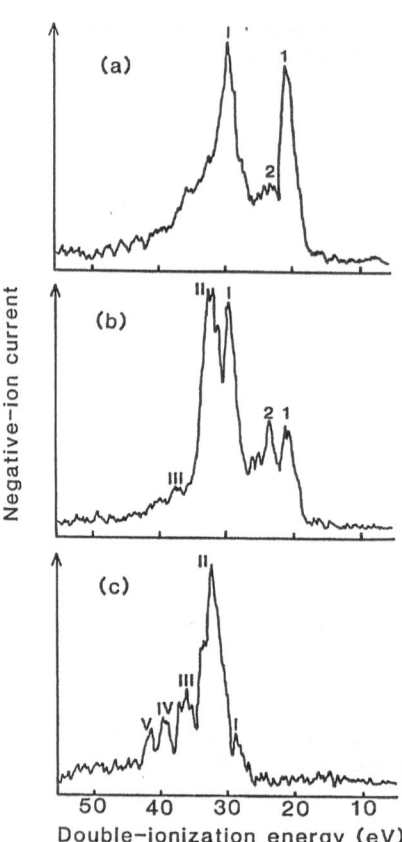

Fig. 7.5 a–c. Negative ion spectra obtained in DCT spectrometry experiments when (a) S^+, (b) OH^+ and (c) F^+ projectile ions having 5 keV translational energy reacted with CH_3Br molecules. The abscissa scale does not apply to peaks 1 and 2

large in spectra obtained with these ions. However, with S^+, peak II is not distinct but may be present in the high energy shoulder. To populate the first excited state of CH_3Br^{2+} using S^+ requires a reaction with an endoergicity of 19.56 eV which must be close to the upper limit of the reaction window, since peak II is not clearly evident. By examining other peaks in the spectra it was concluded [7.73] that the reaction window for CH_3Br exists between endoergicities of 8 eV and 22 eV.

The existence of a reaction window has important consequences for the choice of the projectile ions in DCT spectrometry. Using one with a low value of $E(A^+ \rightarrow A^-)$ causes an increase in the population of lower energy states relative to higher energy states whereas the use of a projectile ion with a high value of $E(A^+ \rightarrow A^-)$ has the opposite effect. Although in a previous DCT spectrometic study [7.63] of CH_3Br, using OH^+ as a projectile ion, the double ionization energies to the ground and first excited states were measured accurately, the third peak in the spectrum corresponding to the second excited state was small and ill-defined, making an accurate measurement of the double ionization energy to that state difficult. The use of F^+ in the present work, however, has allowed the energy of peak III to be measured accurately and also the energy to two higher states to be determined. The use of

F^+ to investigate higher lying states of CO_2, OCS, CS_2 and NH_3 has already been mentioned.

7.5 Single Ionization Energies of Radical Species

To experimentally determine the physical properties of radical species can often be difficult. Their high reactivities make it impractical to study them in solution chemistry, and generating sufficient numbers of them at a specific region of an apparatus in a gas-phase experiment can often only be done using special techniques. Some types of experiment in the gas phase can, however, be performed on ions of the radicals which, because of their charge, can be directed to an interaction region. Thus, for example, the electron affinities of many radical species have been determined in a variety of experiments with negative ions [7.74].

To measure the single ionization energies of radicals by DCT spectrometry, the positive ions of the radicals have been used as projectile ions. If the radical is denoted by R, then in a DEC reaction in a single collision the translational energy lost by R^+ in converting to R^- is given by (7.1) in which R replaces A. The method depends on equating $E(R^+ \rightarrow R^-)$ to $IE_1(R) + E_a(R)$, that is, the sum of the single ionization energy and electron affinity of R. Thus,

$$\Delta E_1 = IE_2(M) - IE_1(R) - E_a(R) . \tag{7.13}$$

For the formation of R^- in the sequential reactions (7.3) and (7.4), the translational energy loss is

$$\Delta E_2 = IE_1'(M) + IE_1''(M) - IE_1(R) - E_a(R) . \tag{7.14}$$

If the relevant ionization energies of the target molecule M are known, and ΔE_1 and/or ΔE_2 are measured, it is seen that $IE_1(R)$ can be determined from (7.13) and (7.14) if $E_a(R)$ is known. As mentioned above, the electron affinities of many radicals have been measured which opens up the possiblity of measuring the single ionization energies of those radicals by DCT spectrometry. The results of studies of several radical species by this technique are described below.

7.5.1 CH_3O and CH_3S Radicals

The target gas used in this investigation [7.75] was xenon and calibration of the energy loss scale was carried out using the Cl^+ projectile ion for which the single ionization energy and electron affinity are accurately known. The ions CH_3O^+ and Cl^+ were generated by introducing a mixture of CH_3OH and CCl_4 into the source of the ZAB-2F spectrometer. A mixture of CH_3SH and CCl_4 was used to generate CH_3S^+ and Cl^+ ions.

Only one CH_3O^- peak was observed in the DCT spectrum and, by varying the xenon gas pressure, it was established that it was due to the sequential reactions

(7.3) and (7.4). By measuring the position of the peak relative to those for Cl^- it was possible to evaluate $IE_1(CH_3O) + E_a(CH_3O)$. The average value from five of such pairs of spectra was determined to be 9.9 ± 0.3 eV. Since $E_a(CH_3O)$ is 1.570 ± 0.022 eV [7.74], the single ionization energy determined in this experiment is 8.3 ± 0.3 eV. The equivalent experiment carried out with CH_3S^+ projectile ions gave 8.1 ± 0.3 eV for the single ionization energy of CH_3S. These values are in good agreement with the results of previous investigations [7.76, 77] (8.14 ± 0.15 eV for CH_3O, 8.06 ± 0.1 eV for CH_3S) in which the radicals produced by pyrolysis of suitable derivatives were introduced into the source of a mass spectrometer, and electron impact ionization efficiency curves were measured. In a photoelectron spectroscopy study [7.78] of CH_3O, the radicals were generated by the reaction of fluorine atoms with methanol; the vertical and adiabatic ionization energies were determined to be 8.13 ± 0.2 eV and 7.37 ± 0.3 eV, respectively. The good agreement of the former energy with that measured in the present study suggests that vertical single ionization energies are determined in this type of DCT spectrometric study of radicals.

The investigation of CH_3O and CH_3S showed that DCT spectrometry gives results which agree with more direct measurements of ionization energies. Although the accuracy of the data is lower with DCT spectrometry, it may prove to be of value in investigating radicals which are produced in low abundance by pyrolysis and by chemical reactions, but for which the electron affinities are known.

7.5.2 Mercaptyl Radicals C_2H_5S and $n-C_3H_7S$

This investigation [7.79] was carried out on the ZAB-E spectrometer at Swansea using essentially the same technique as that described above. $C_2H_5S^+$ ions were generated by dissociatve ionization of molecules of either $(C_2H_5)_2S$ or C_2H_5SH. The $n-C_3H_7S^+$ ions were generated by the same process from $n-C_3H_7SH$ molecules. Xenon was used as the collision gas and experiments with OH^+ and CH_3S^+ allowed the energy loss scale to be calibrated. The single ionization energies determined are 9.6 ± 0.8 eV and 8.2 ± 0.8 eV for C_2H_5S and $n-C_3H_7S$, respectively.

7.5.3 CF_2Cl and $CFCl_2$ Radicals

Photolysis of the two most widely used chlorofluorocarbon compounds generates the free radicals CF_2Cl and $CFCl_2$ by the reactions

$$CF_2Cl_2 + h\nu \rightarrow CF_2Cl + Cl, \tag{7.15}$$

$$CFCl_3 + h\nu \rightarrow CFCl_2 + Cl. \tag{7.16}$$

These reactions are enviromentally of considerable importance since the resulting Cl atoms are thought to be involved in the catalytic destruction of ozone, thus reducing the thickness of the stratospheric ozone layer. There is also interest in the CF_2Cl and CFCl radicals formed since recent studies of the kinetics of the oxidation reactions of these radicals suggests that reaction products include Cl atoms which could also be involved in the ozone-destruction cycle. Ionization of these radicals may be an important part of ion-molecule reaction mechanisms in the upper atmoshere; so their single ionization energies were measured in a DCT spectrometry experiment [7.80]. The CF_2Cl^+ projectile ions were generated by dissociative ionization of three precursor molecules, CF_2ClBr, CF_3Cl and CF_2Cl_2. The $CFCl_2$ ions were generated by the same procedure from CF_2Cl_2 and $CF_2ClCFCl_2$. To calibrate the energy loss scale spectra were obtained with CF^+ and OH^+ as calibrant ions. The single ionization energies determined in the DCT spectrometry experiment are 9.0 ± 0.5 eV for CF_2Cl and 8.5 ± 0.5 eV for $CFCl_2$. There are no other measurements of the single ionization energies of these radicals but values have been obtained for CF_3 and CCl_3, viz. 9.25 eV [7.81] and 8.28 eV [7.82], respectively. It is seen from the four results that a decrease in the single ionization energy occurs as successive fluorine atoms are replaced by chlorine atoms. This trend is to be expected because the electronegativity of Cl is less than that of F. Thus, the results of the DCT spectrometry experiment are, in the light of the known data for CF_3 and CCl_3, of the right magnitude and follow the expected trend.

7.5.4 The SF_5 Radical

By measuring various ion-molecule reaction rates *Tichy* et al. [7.83] deduced that the single ionization energy of SF_5 is 10.0 ± 0.15 eV. This is markedly lower than the previous literature values which have an average value of 11.6 eV. Because such a large difference exists, it was decided to apply DCT spectrometry to measure the single ionization energy of SF_5 [7.84]. The SF_5^+ projectile ion was produced by dissociative ionization of SF_6 in the source of the ZAB-2F spectrometer at Swansea. The target gas used was xenon, and the calibrant ions were Cl^+, Br^+ and I^+. The opportunity was taken to study also other ions (SF_n^+, $n = 1$–4) in this experiment, and the masses of the calibrant ions were chosen to encompass the masses of all the projectile ions.

The resulting SF_5^- ion spectrum had two peaks, both of which varied in height quadratically with xenon gas pressure. They were due to the sequential single electron capture reactions (7.3) and (7.4). For the first peak, both resulting Xe^+ ions were in the $^2P_{3/2}$ state; for the second, one was in the $^2P_{3/2}$ and the other in the $^2P_{1/2}$ state. By measuring the energy losses corresponding to these peaks, it was deduced that the single ionization energy of SF_5 is 9.6 ± 0.7 eV. The relatively large uncertainty in this value is partially due to the uncertainty in the value of $E_a(SF_5)$. A figure of 3.2 ± 0.5 eV was used which is a mean of published data [7.74]. Thus, the DCT spectrometry result supports the conclusion of *Tichy* et al. [7.83] that the single ionization energy of SF_5 is 10.0 ± 0.15 eV. After the DCT spectrometric study [7.84] was completed, the results of a high-pressure mass spectrometric study [7.85] of the formation and chemistry of SF_5^+ were published which shows that $IE_1(SF_5)$ is 9.83 ± 0.18 eV, a value in agreement with those from both previous studies [7.83, 84].

References

7.1   T. A. Carlson: Photoelectron and Auger Spectroscopy (Plenum, New York, 1978)
7.2   J. Appell, J. Durup, F. C. Fehsenfeld, P. G. Fournier: J. Phys. B **6**, 197 (1973)
7.3   G. Dujardin, S. Leach, O. Dutuit, P. M. Guyon, M. Richard-Viard: Chem. Phys. **88**, 339 (1984)
7.4   J. Appell: "Double charge transfer spectroscopy", in *Collision Spectroscopy*, ed. by R. G. Cooks (Elsevier, Amsterdam, 1977) p. 227
7.5   G. W. McLure: Phys. Rev. **132**, 1636 (1963)
7.6   J. F. Williams: Phys. Rev. **153**, 116 (1967)
7.7   F. C. Witteborn, D. E. Ali: Bull. Phys. Soc. Am. **16**, 208 (1971)
7.8   P. G Fournier, J. Appell, F. C. Fehsenfeld, J. Durup: J. Phys. B **5**, L58 (1972)
7.9   P. G. Fournier, J. Fournier, F. Salama, P. J. Richardson, J. H. D. Eland: J. Chem. Phys. **83**, 241 (1985)
7.10 Manufactured by VG Analytical Limited, Manchester, UK
7.11 E. S. Mukhtar, I. W. Griffiths, F. M. Harris, J. H. Beynon: Int. J. Mass Spectrom. Ion Phys. **37**, 159 (1981)
7.12 F. M. Harris, B. C. Cooper: Int. J. Mass Spectrom. Ion Processes **97**, 165 (1990)
7.13 D. Mathur. C. Badrinathan, F. A. Rajgara, U. T. Raheja: J. Phys. B **18**, 4795 (1985)
7.14 H. S. W. Massey: Rep. Progr. Phys. **12**, 248 (1949)
7.15 P. Fournier, C. Benoit, J. Durup, R. E. March: C. R. Acad. Sci. **278**, 1039 (1974)
7.16 P. Millie, I. Nenner, P. Archirel, P. Lablanquie, P. Fournier, J. H. D. Eland: J. Chem. Phys. **84**, 1259 (1986)
7.17 G. Dujardin, D. Winkoun: J. Chem. Phys. **83**, 6222 (1985)
7.18 W. E. Moddeman, J. A. Carlson, M. O. Krause, B. P. Pullen, W. E. Bull, G. K. Schweitzer: J Chem. Phys. **55**, 2317 (1971)
7.19 D. M. Curtis, J. H. D. Eland: Int. J. Mass Spectrom. Ion Processes **63**, 241 (1985)
7.20 P. Jonathan, M. Hamdan, A. G. Brenton, G. D. Willett: Chem. Phys. **119**, 159 (1988)
7.21 W. J. Griffiths, F. M. Harris: Int. J. Mass Spectrom. Ion Processes **87**, 349 (1989)
7.22 M. L. Langford, F. M. Harris, C. J. Reid, J. A. Ballantine, D. E. Parry: Chem. Phys. **149**, 445 (1991)
7.23 A. W. Potts, H. J. Lempka, D. G. Streets, W. C. Price: Phil. Trans. Roy. Soc. Lond. **A268**, 59 (1970)
7.24 W. J. Griffiths, F. M. Harris: Rapid Commun, Mass Spectrom. **2**, 28 (1988)
7.25 P. G. Fournier, G. Comtet, J. Fournier, S. Svensson, K. Karlsson, M. P. Keane, A. Naves De Brito: Phys. Rev. A **40**, 163 (1989)
7.26 M. L. Langford, F. M. Harris, P. G. Fournier, J. Fournier: J. Phys. B – submitted (1991)
7.27 S. Mazumdar, F. A. Rajgara, V. R. Marathe, C. Badrinathan, D. Mathur: J. Phys. B **21**, 2815 (1988)
7.28 P. G. Fournier: J. Phys. B **22**, L381 (1989)
7.29 D. Mathur, V. R. Marathe, S. Mazumdar: J. Phys. B **22**, L385 (1989)
7.30 V. R. Marathe, D. Mathur: Chem. Phys. Lett. **163**, 189 (1989)
7.31 P. J. Richardson, J. H. D. Eland, P. G. Fournier, D. L. Cooper: J. Chem. Phys. **84**, 3189 (1986)
7.32 P. G. Fournier, J. Fournier, F. Salama, D. Stark, S. D. Peyerimhoff, J. H. D. Eland: Phys. Rev. A **34**, 1657 (1986)
7.33 P. G. Fournier, H. Aouchiche, V. Lorent, J. Baudon: Phys. Rev. A 34, 3743 (1986)
7.34 P. G. Fournier, M. Mousselmal, S. D. Peyerimhoff, A. Banichevich, M. Y. Adam, T. J. Morgan: Phys. Rev. A **36**, 2594 (1987)
7.35 S. D. Price, J. H. D. Eland, P. G. Fournier, J. Fournier, P. Millie: J. Chem. Phys. **88**, 1511 (1988)
7.36 P. G. Fournier, J. H. D. Eland, P. Millie, S. Svensson, S. D. Price, J. Fournier G. Comtet, B. Wannberg, L. Karlsson, P. Baltzer, A. Kaddouri, U. Gelius: J. Chem. Phys. **89**, 3553 (1988)
7.37 K. Siegbahn, L. Apslund, P. Kelfve, Chem. Phys. Lett. **35**, 330 (1975)
7.38 C. J. Proctor, C. J. Porter, T. Ast, P. D. Bolton, J. H. Beynon: Org. Mass Spectrom. **16**, 454 (1981)

7.39 J. A. Conner, I. H. Hillier, J. Kendrick, M. Barber, A. Barrie: J. Chem. Phys **64**, 3325 (1976)
7.40 F. P. Larkins, J. Chem. Phys. **86**, 3239 (1987)
7.41 J. Appell, J. Durup, F. C. Fehsenfeld, P. Fournier: J. Phys. B **7**, 406 (1974)
7.42 D. M. Curtis, J. H. D. Eland: Int. J. Mass Spectrom. Ion Processes **63**, 241 (1985)
7.43 F. Maquin, D. Stahl, A. Sawaryn, P. v. R. Schleyer, W. Koch, G. Frenking, H. Schwarz: J. Chem. Soc., Chem. Commun. 504 (1984)
7.45 W. J. Griffiths, F. M. Harris: Org. Mass Spectrom. **23**, 553 (1988)
7.46 W. J. Griffiths, F. M. Harris: Int. J. Mass Spectrom. Ion Processes **89**, 125 (1989)
7.47 W. J. Griffiths, F. M. Harris: Int. J. Mass Spectrom. Ion Processes **85**, 259 (1988)
7.48 S. Leach: J. Mol. Struct. **157**, 197 (1987)
7.49 R. Spohr, T. Bergmark, N. Magnusson, L. O. Werme, C. Nordling, K. Siegbahn: Phys. Scr. **2**, 31 (1970)
7.50 R. R. Rye, T. E. Madey, J. E. Houston, P. H. Holloway: J. Chem. Phys. **69**, 1504 (1978)
7.51 K. E. McCulloh, T. E. Sharp, H. M. Rosenstock: J. Chem. Phys. **42**, 3501 (1965)
7.52 C. Backx, M. J. Van der Wiel: J. Phys B **8**, 3020 (1975)
7.53 J. Appell, J. A. Horsley: J. Chem. Phys. **60**, 3445 (1974)
7.54 W. J. Griffiths, D. Mathur, F. M. Harris: J. Phys. B **20**, L493 (1987)
7.55 T. Ast. C. J. Porter, C. J. Proctor, J. H. Beynon: Chem. Phys. Lett. **78**, 439 (1981)
7.56 M. Rabrenovic, A. G. Brenton, J. H. Beynon: Int. J. Mass Spectrom. Ion Phys. **52**, 175 (1983)
7.57 D. Mathur, C. Badrinathan, F. A. Rajgara, U. T. Raheja: Chem. Phys. **103**, 447 (1986)
7.58 G. Dujardin, D. Winkoun, S. Leach: Phys. Rev. A **31**, 3027 (1985)
7.59 P. A. Hatherly, M. Stankiewicz, L. J. Frasinski, K. Codling, M. A. MacDonald: Chem. Phys. Lett. **159**, 355 (1989)
7.60 A. W. Hanner, T. E. Moran: Org. Mass Spectrom. **16**, 512 (1981)
7.61 W. J. Griffiths, F. M. Harris: Int. Mass Spectrom. Ion Processes. **85**, 69 (1988)
7.62 W. J. Griffiths, F. M. Harris: Rapid Commun. Mass Spectrom. **2**, 91 (1988)
7.63 W. J. Griffiths, F. M. Harris, D. E. Parry, C. J. Reid, J. A. Ballantine: Org. Mass Spectrom. **25**, 375 (1990)
7.64 B. P. Tsai, J. H. D. Eland: Int J. Mass Spectrom. Ion Phys. **36**, 143 (1980)
7.65 W. J. Griffiths, F. M. Harris: Int. J. Mass Spectrom. Ion Processes **86**, 341 (1988)
7.66 W. J. Griffiths, F. M. Harris: Org. Mass Spectrom. **24**, 323 (1989)
7.67 W. J. Griffiths, F. M. Harris: New J. Chem. **14**, 261 (1990)
7.68 W. J. Griffiths, F. M. Harris: Rapid Commun. Mass Spectrom. **3**, 130 (1989)
7.69 W. J. Griffiths, F. M. Harris: J. Chem. Soc., Faraday Trans.II **85**, 1575 (1989)
7.70 D. Mathur: Int. J. Mass Spectrom. Ion Processes **83**, 203 (1988)
7.71 D. P. Almeida, M. L. Langford: Int. J. Mass Spectrom. Ion Processes **96**, 331 (1990)
7.72 W. J. Griffiths, D. Mathur, F. M. Harris: Int J. Mass Spectrom. Ion Processes **87**, R1 (1989)
7.73 M. L. Langford, F. M. Harris: Rapid Commun. Mass Spectrom. **4**, 125 (1990)
7.74 A. A. Christodoulides, D. L. McCorkle, L. G. Christophorou: "Electron Affinities of Atoms, Molecules and Radicals" in *Electron-Molecule Interactions and their Applications*, Vol. 2, ed. by L. G. Christophorou, (Academic Press, New York 1984)
7.75 W. J. Griffiths, F. M. Harris: Chem. Phys. Lett. **142**, 7 (1987)
7.76 J. P. Fisher, E. Henderson: Trans. Faraday Soc. **63**, 1342 (1967)
7.77 T. F. Palmer, F. P. Lossing: J. Am. Chem. Soc. **84**, 4661 (1962)
7.78 J. M. Dyke: J. Chem. Soc. Faraday Trans.II **83**, 69 (1987)
7.79 W. J. Griffiths, F. M. Harris, C. J. Reid, J. A. Ballantine: Org. Mass Spectrom. **24**, 849 (1989)
7.80 W. J. Griffiths, F. M. Harris, J. D. Barton: Rapid Commun. Mass Spectrom. **3**, 283 (1989)
7.81 C. Lifshitz, W. A. Chupka: J. Chem. Phys. **47**, 3439 (1967)
7.82 J. B. Farmer, I. H. S. Hendeson, F. P. Lossing, D. G. H. Marsden: J. Chem. Phys. **24**, 348 (1956)
7.83 M. Tichy, G. Javahery, N. D. Twiddy: Int. J. Mass Spectrom. Ion Processes **79**, 231 (1987)
7.84 M. L. Langford, D. P. Almeida, F. M. Harris: Int. J. Mass Spectrom. Ion Processes **98**, 147 (1990)
7.85 J. A. Stone, W. J. Wytenberg: Int. J. Mass Spectrom. Ion Processes **94**, 269 (1989)

8. Studies of Multiply Charged Molecules by Ion Collision Techniques and Ab Initio Theoretical Methods

V.R. Marathe and D. Mathur

With 10 Figures

Experimental developments over the last two to three decades have made feasible detailed studies on the structure and spectroscopy of singly charged molecular ions, and a substantial body of information now exists on positively charged diatomic and polyatomic molecules covering the spectral region from the microwave to the infrared (see [8.1], and references therein). However, despite the fact that multiply charged molecular ions were first observed in a mass spectrometer as long ago as 1932 [8.2], there continues to be an acute paucity of experimental and theoretical information on such species. Even in the case of relatively simple species, such as doubly charged diatomic ions (like CO^{2+}) and highly symmetrical polyatomic ions (like CH_4^{2+}), the information that does exist can, in the main, be considered to be somewhat ambiguous.

Studies of multiply charged molecules continue to pose a serious theoretical and experimental challenge. The high density of electronic states in such species makes accurate calculations difficult and demand the use of ab initio quantum-chemical methods well beyond the simple Hartree-Fock level. The relatively high energies required for their formation and the intrinsic instability of the ions complicate experiments. Nevertheless, there are several compelling fundamental reasons for carrying out studies on such molecular species.

Determination of the structures and bonding arrangements of metastable multiply charged molecules can lead to an enormous wealth of knowledge about the nature of interatomic interactions. The interactions of such molecules with other, neutral atomic and molecular entities lead to a variety of processes that are important from the viewpoint of gaining a better, more quantitative insight into the structure of the ions themselves, and the mechanisms and kinetics involved in their reactions. For instance, experimental information on the stability of multiply charged molecules, and the energetics involved in their formation, plays an important role in assessing the effectiveness of contemporary quantum-chemical techniques of generating potential energy surfaces [8.3]. Consider the potential energy surfaces of multiply charged diatomic hydrides XH^{n+} ($X = C, N, O, S, \ldots$). Such surfaces are crucial ingredients for high-level calculations of electron capture cross sections for multiply charged atomic ions, X^{n+}, on atomic hydrogen. These cross sections [8.4, 5]

Springer Series in Chemical Physics, Vol. 54
Deepak Mathur (ed.): Physics of Ion Impact Phenomena
© Springer-Verlag Berlin Heidelberg 1991

can be extremely sensitive to the energy-level structure of the XH^{n+} molecules. The prediction of such energy-levels within single channel descriptions provided by the Born-Oppenheimer approximation is often inadequate because of the strong coupling between attractive electronic states $(X^{n+}$-H$)$ and one or more Coulomb repulsive states $(X^{(n-1)+}$-H$^+)$ into which charge transfer can occur. Charge transfer processes are of importance not only in hot astrophysical and terrestrial plasmas but also in low temperature situations. A large variety of molecular ions have been identified in cold astrochemical environments – the interstellar medium, cometary halos, and in planetary atmospheres. Electron capture by multiply charged molecular ions has been proposed [8.6] as a mechanism for generating ions in the Earth's upper atmosphere.

In this chapter we outline various recipes for carrying out experimental and theoretical studies on multiply charged molecular ions. Experimental studies usually attempt to obtain quantitatively reliable information on formation energetics; contemporary techniques include Auger spectroscopy and conventional electron impact and photoionization mass spectrometry as well as somewhat more modern methods involving ion-neutral collisions leading to charge stripping and double charge transfer reactions. Dissociation dynamics have also become amenable to investigation by ion collision methods of the type employed in collision-induced dissociation and Coulomb explosion experiments. Reactions involving multiply charged molecules, particularly those resulting in excitation to, and de-excitation from, electronically excited states as well as those leading to electron capture also come within the purview of ion impact experiments. Theoretical investigations primarily address the question of the stability (or metastability) of multiply charged molecules. The rationale behind the development of various quantum-chemical techniques described here is to attempt to answer some of the following questions:

1. Why are multiply charged molecular ions stable (or metastable)?
2. What is the equilibrium geometry for which they are stable?
3. What are the electronic configurations for which they are stable?
4. What are their dissociation pathways, and what potential barriers do they cross (or tunnel through) before dissociation?

8.1 Stability of Multiply Charged Molecular Ions

8.1.1 Potential Energy Functions

The stability of a molecular ion is strongly correlated with the nature of its potential energy surface. Consider a diatomic molecule AB which may dissociate into $A + B$. If the potential energy curve describing this species (Fig. 8.1) has a minimum at R_e such that $E(R_e) < E(A) + E(B)$, where $E(A)$ and $E(B)$ correspond to the energies of the atoms (or ions) A and B, respectively, then the molecule AB is considered to be *stable*. If, however, the potential energy curve does not possess any minimum, as is the case depicted in Fig. 8.1b, then the molecule AB is said to be *unstable*.

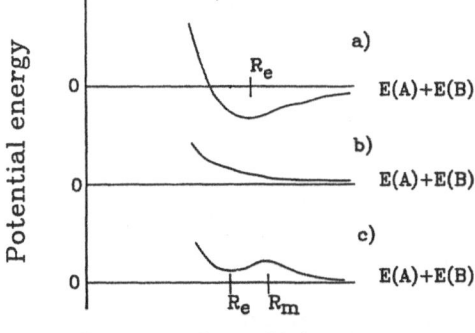

Fig. 8.1. Schematic potential energy curves of a hypothetical diatomic molecule, AB, in an electronic state which is a) stable, b) dissociative and c) metastable

Diatomic potential energy curves can also be obtained in which there exists both a minimum as well as a maximum. Figure 8.1c shows such a curve, in which a potential maximum is located at an internuclear distance R_m and a minimum occurs at a distance R_e such that $E(R_e) > E(A) + E(B)$. Molecules described by such a curve are considered to be *metastable*. Any molecular species trapped within the potential well shown in Fig. 8.1c will ultimately fragment due to the tunneling effect. The lifetime of such a metastable species would depend on a) the barrier height, defined as $E(R_m) - E(R_e)$ and b) the width of the barrier.

Potential energy curves describing multiply charged molecules will always be of the type depicted in either Fig. 8.1b or Fig. 8.1c. In order to cover the cases in which the molecular ion comprises more than two atomic nuclei, we state the generalization that if there exists a local minimum at some internuclear separation R_e and $E(R_e)$ is greater than $\sum_i E(A_i)$, then the molecular ion is expected to be metastable. If there is no local minimum to be found anywhere on the global potential energy surface describing the molecule, then it is expected to be totally unstable.

The origin of the potential barrier shown in Fig. 8.1c can be understood in a qualitative fashion by considering a hypothetical diatomic molecule AB^{2+} and taking into account avoided crossings between the two adiabatic potential energy curves corresponding to $A^{2+} + B$ and $A^+ + B^+$. In the former case, at large internuclear distances r_{AB}, the potential energy curve will have a weakly attractive form due to charge polarization effects

$$V_{A^2+B} \propto -r_{AB}^{-4} \, . \tag{8.1}$$

In the case of $A^+ - B^+$, however, the long-range part of the potential energy curve will be dominated by Coulomb repulsion, with a minor attractive (polarization) component

$$V_{A+B+} \propto +r_{AB}^{-1} - r_{AB}^{-4} \, . \tag{8.2}$$

227

Mutual perturbations between such potential functions can give rise to the formation of a potential well whose depth and width are sometimes sufficient to permit the existence of one or more quasibound states of diatomic AB^{2+} ions.

8.1.2 A Qualitative Molecular Orbital Picture of Stability

It is generally known that polarization of the electron density in the interatomic region of a molecule is responsible for bonding of the constituent atoms. The same concept holds, in a much stronger fashion, in descriptions of the stability, or otherwise, of multiply charged molecular ions.

Consider a molecular orbital for the diatomic species AB

$$\psi(r) = C_a\phi_a(r) + C_b\phi_b(r) , \tag{8.3}$$

where ϕ_a, ϕ_b are atomic orbitals centered at nuclei A and B, respectively. The electron density due to occupation of this molecular orbital is given as

$$|\psi(r)|^2 = C_a^2|\phi_a(r)|^2 + C_b^2|\phi_b(r)|^2 + 2C_aC_b\phi_a(r)\phi_b(r) . \tag{8.4}$$

We define the *total electron density shared by the two nuclei* as

$$\rho_{AB}(r) = \sum_{a\in A}^{occ}\sum_{b\in B}^{occ} C_aC_b\phi_a(r)\phi_b(r) . \tag{8.5}$$

This electron density, shared by the two nuclei, is fundamentally responsible for the stability of the molecular ion AB, particularly if AB happens to be multiply charged. If the effective atomic (ionic) charges are defined as

$$q_{AA} = Z_A - \sum_{a\in A}^{occ} \int C_a^2|\phi_a(r)|^2 \, dr \qquad \text{and} \tag{8.6}$$

$$q_{BB} = Z_B - \sum_{b\in B}^{occ} \int C_b^2|\phi_b(r)|^2 \, dr , \tag{8.7}$$

then, for a multiply charged molecular ion, q_{AA} and q_{BB} will have the same sign and would be responsible for the strong repulsion between A and B. It is only $\rho_{AB}(r)$ which is responsible for binding A and B together, temporarily, to yield a metastable state.

Based on this qualitative picture, the following three factors can generally be considered to be responsible for the stability of multiply charged molecular ions:

1. The atomic orbitals should be sufficiently extended so as to give maximum overlap of orbitals and, consequently, the shared electron density in the interatomic region;

2. The difference in the ionization energies of the constituent ions should be minimum, leading to covalent bonding; in the treatment given above, C_a, C_b should be comparable; and

3. There should be more electrons in bonding orbitals compared to those in anti-bonding orbitals so that $\int \rho_{AB}(r)\, dr$ is a positive quantity which is large enough to hold the interacting ions together.

8.2 Contemporary Ion-Impact Methods of Studying Multiply Charged Molecules

8.2.1 Translational Energy Spectrometry

The technique of translational energy spectrometry is based, in a phenomenological sense, upon the physics of collisional energy transfer processes. In this technique, which has found widespread utility in experimental studies of the spectroscopic and dynamical properties of multiply charged molecules, an ion beam possessing a few keV kinetic energy collides with an essentially static target atom or molecule. Conversion of the projectile ion's kinetic energy into internal (potential) energy in the target, or in the projectile itself, makes it possible for an entire gamut of reactions to occur. Amongst the ion-neutral processes of interest in studies of multiply charged molecules are those in which conversion of kinetic energy into potential energy leads to:

i) further ionization of singly charged projectile ions, leading to formation of doubly charged species which may be metastable to the extent of making possible their subsequent detection. This process is known as charge stripping;

ii) double ionization of the neutral target following capture of two electrons by the incident singly charged ion. This process is referred to as double electron capture;

iii) dissociation of metastable doubly charged projectile molecular ions; and

iv) collisional excitation of, and electron capture by, metastable doubly charged molecular ions.

Measurement of the fraction of kinetic energy that is converted into internal energy renders amenable experimental studies of all the above processes on a quantitative basis. The general methodology adopted in such studies entails ion transportation through electrostatic and magnetic fields in different instrumental configurations. It is therefore instructive to briefly review the equations which govern transmission of singly and doubly charged ions of given kinetic energy and momentum through such fields.

8.2.2 Transmission of Singly and Doubly Charged Ions Through Electrostatic and Magnetic Fields

Firstly, we compare the transmission of singly (m^+) and doubly (m^{2+}) charged ions through d.c. electrostatic fields. In the former case, equating the electrostatic force with that describing circular motion yields

$$eE = (m^+ v^2)/r \,, \tag{8.8}$$

where v is the velocity of an ion of mass m, E is the electric field strength and r is the radius of the mean path within the sector containing the electric field. As the kinetic energy of the ion is given by

$$(m^+v^2)/2 = eV ,$$ (8.9)

where V is the accelerating potential the ion experiences, substituting for the ion velocity, v,

$$eE = (m^+2eV)/rm^+$$ (8.10)

yields an expression relating the electric field strength to the accelerating potential

$$E = 2V/r .$$ (8.11)

In the case of doubly charged ions, the electrostatic force on m^{2+} is twice that on m^+

$$2eE = (m^{2+}v^2)/r .$$ (8.12)

Consequently, the ion energy becomes

$$(m^{2+}v^2)/2 = 2eV .$$ (8.13)

Substituting for ion velocity,

$$2eE = (m^{2+}4eV)rm^{2+}$$ (8.14)

yields

$$E = 2V/r .$$ (8.15)

As the final equations are identical in the two cases, it is clear that singly and doubly charged ions will be transmitted through electrostatic fields under the same conditions.

In the case of ion transmission through a magnetic field (of field strength B), the mass-to-charge ratio (m/q) also has to be considered

$$m/q = (B^2r^2)/2V .$$ (8.16)

Thus a doubly charged ion will be detected at half the ratio of the corresponding singly charged ion.

The practical implications of these equations manifest themselves in the following sections which describe different types of contemporary ion translational energy spectrometry experiments which yield information on various properties of multiply charged molecules.

8.2.3 Charge-Stripping Studies

Much of the contemporary experimental work on multiply charged molecules has been carried out using commercial mass spectrometers (or custom-made ion scattering apparatus) in which the reversed-geometry configuration is employed. Here, a mass filtering device (usually a magnetic sector) preceeds the electrostatic energy analyser, and a collision gas cell is placed in the field-free region between the two sectors. Ions formed in a conventional electron impact ion source are accelerated to a few keV energy and are then selected on the basis of mass/charge ratio. These mass-selected ions undergo single collisions with essentially static neutral atomic or molecular species and the fast moving products are then energy analysed in the post-collision region and detected by a particle multiplier. All the major types of collision experiments mentioned above can be carried out using such an experimental arrangement.

In charge stripping measurements, the experimental procedure involves measurement of the translational energy spectrum of doubly charged product ions and determination of the endoergic energy defect, ΔE, for the single electron loss process

$$A^+ + B \rightarrow A^{2+} + B + e - \Delta E , \tag{8.17}$$

where ΔE is a measure of the projectile ion's translational (kinetic) energy which is converted into internal (potential) energy required for ionization of a given electronic state of the singly charged ion. If the projectile singly charged ion beam consists of a mixed fraction of ground state and metastable excited state components, distinct peaks will be observed in the doubly charged product ion translational energy spectrum, each peak being associated with a different value of ΔE.

By way of illustration Fig. 8.2 shows a schematic diagram of a laboratory-made ion translational energy spectrometer which has been used rather extensively in recent years for various studies of multiply charged molecular ions [8.7–9]. Ions formed when gas is introduced into a low-voltage arc type of electron impact ion

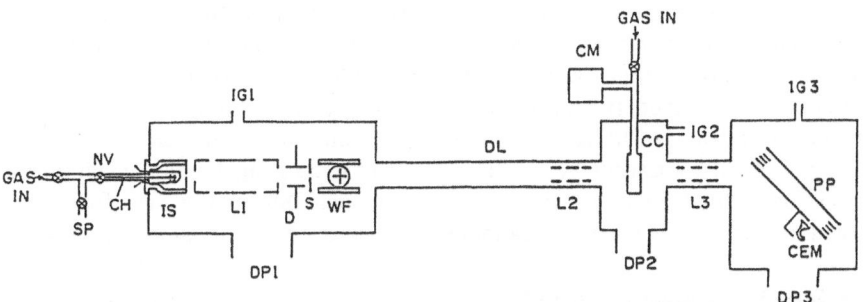

Fig. 8.2. Ion translational energy spectrometer used at TIFR. IS: low-voltage arc ion source, L1,2,3: electrostatic cylindrical lenses, S: slit, WF: Wien velocity (mass) filter, DL: drift length, CC: collision cell, PP: parallel plate electrostatic energy analyzer, CEM: channel electron multiplier, CM: capacitance manometer, IG1,2,3: ionization gauges, DP1,2,3: diffusion pumps, CH: solid charge holder, SP: sorption pump, NV: needle valve

source are extracted at an accelerating potential in the range 2–5 kV. These ions are focused by a three-element electrostatic lens on to the entrance plane of a Wien filter (a region of crossed electric and magnetic fields) where separation occurs on the basis of ion velocities. As all ions emerging from the ion source are accelerated through the same potential, an ion velocity spectrum corresponds to a spectrum of mass-to-charge ratios. Mass-selected ions are passed through a differentially pumped static gas collision cell which is generally maintained at room temperature. In the post-collision region, the scattered ions are energy analyzed by a large electrostatic parallel-plate energy analyzer whose overall energy resolution is generally of the order of 0.08 %. Ion detection is by means of a channel electron multiplier (operating in the particle counting mode) coupled to conventional pulse counting electronics which is interfaced to a laboratory microcomputer which carries out data acquisition in an on-line mode.

In such an apparatus, typical gas pressures are generally of the order of 10^{-1} Torr within the ion source, 5×10^{-7} Torr immediately outside the ion source and in the mass analysis region, and 1×10^{-7} Torr in the energy analysis and detection region. Experimental charge stripping measurements are carried out in the following fashion. If V is the energy analyzer voltage at which the elastically scattered singly charged A^+ peak is transmitted, then $0.5V - \Delta V$ will be the voltage at which the analyzer transmits the doubly charged product ions. The *minimum* value of ΔV, which is related to the minimum energy defect, $\Delta E_{\min}$, of the charge stripping reaction (8.17) by the energy analyzer's geometrical factor, is the measured quantity which enables deduction to be made of the ionization energy of the singly charged ion. Charge stripping peaks usually exhibit relatively sharp and well-defined onsets, which enables determinations to be made of $\Delta V_{\min}$ with greater accuracy than is generally possible in the analogous threshold extrapolation procedure adopted in conventional electron impact ionization efficiency curve measurements for doubly charged ions. In actual practice, the value of $\Delta E_{\min}$ is obtained from the high energy side of the charge stripping peak (by extrapolation to the baseline); measurement of the peak onset in this way eliminates additional translational energy loss from any other inelastic process which may interfere [8.10].

The charge stripping technique yields only differences between the single and double ionization energies and, consequently, accepted values of the single ionization energy of a molecular species must be available in order to deduce a value for the double ionization energy. Furthermore, the singly ionised projectile must be assumed to be, or ensured to be, in its ground electronic state. It has been discovered that one of the major advantages of this method of measuring double ionization energies is the ability of charge stripping reactions to yield information even for doubly charged molecular species which cannot be formed in conventional ion sources.

8.2.4 Double Electron Capture

Translational energy spectrometry has also been applied to studies of double electron capture reactions of the type

$$A^+ + XY \rightarrow A^- + XY^{2+} - \Delta E . \tag{8.18}$$

Determination of the endoergic energy defect, ΔE, for such a process yields the double ionization energy, $IE_2(XY)$, of the target molecule XY

$$\Delta E = IE_2(XY) - [IE_1(A) + EA(A)] \qquad (8.19)$$

as long as the single ionization energy, IE_1, and the electron affinity, EA, of the projectile species is known. Such experiments are usually conducted with H^+ projectiles (for which values of IE_1 and EA are well established) [8.11] although molecular projectiles have also been used recently. Results of experiments on H_2, yielding H_2^{2+} ions, and on the rare gases, indicate that the Franck-Condon principle is obeyed and that the Wigner spin-conservation rule is also respected.

The double electron capture technique, which is the subject of fuller discussion in Chap. 7, yields information which is complementary to that obtainable from charge stripping measurements. The latter technique yields double ionization energy data for those species which are metastable for at least the time required for the product ion to traverse the distance between the region where it is formed and the detector (the typical time period being a few μs). On the other hand, there is no such lifetime criterion imposed on the doubly charged species formed in double electron capture experiments where it is the fast negatively charged ion that is detected. The only constraint in these measurements, other than that imposed by the spin-conservation rule, is that the projectile species must possess a positive value of electron affinity so that long-lived negative ions may be formed.

8.2.5 Dissociation Studies

Collision induced dissociation of multiply charged molecules is an important subset of the overall class of energy transfer reactions that are readily amenable to investigation by energy spectrometric methods in which the momentum or energy distributions of fragment ions can be monitored. By conducting experiments at collision energies of a few keV, kinetic energy releases of a few eV in the centre-of-mass frame become amplified in the laboratory frame. If E is the kinetic energy of a projectile homonuclear diatomic molecule and ε_{cm} is the kinetic energy released (in the centre-of-mass) upon dissociation, the measured energy in the laboratory frame, ε_{lab}, can be deduced simply by addition of collision velocities

$$\varepsilon_{lab} = E/2 \pm 2(\varepsilon_{cm}E)^{1/2} + \varepsilon_{cm} . \qquad (8.20)$$

The amplification of ε_{cm} in the laboratory frame can be demonstrated most easily by considering a typical kinetic energy release of 5 eV following dissociation of a homonuclear diatomic doubly charged molecule. For an incident molecular ion energy of 3 keV, the fragment ion possessing this value of kinetic energy release will appear in the laboratory frame at 1505 ± 245 eV; in practice two peaks will be seen in the translational energy spectrum around 1.5 keV, each separated from the other by 490 eV (corresponding to fragment ions which are forward and backward scattered). This energy 'amplification' makes it possible to study dissociation processes even in instruments possessing relatively modest energy resolution capabilities.

Such energy amplification also offers the possiblity of obtaining sub-Doppler resolution which occurs due to the narrowing down of the velocity distribution of an ion beam when it is accelerated, a notion known as kinematic compression. If an ion beam possessing an energy spread given by ΔE is extracted from an ion source and accelerated to an energy E, the intrinsic energy spread is maintained at ΔE no matter what the value of E. However, since $E = mv^2/2$, the velocity spread of the accelerated beam is substantially reduced from its value prior to acceleration. For example, if an ion beam with an energy spread of 1 eV is accelerated to 3 keV laboratory energy, the velocity spread in the fast beam corresponds to a ion temperature of less than 1 K!

Multiply charged molecules may also dissociate into charged fragments in a collisionless (unimolecular) fashion. Consider the spontaneous process

$$M_1^{x+} \rightarrow M_2^{y+} + M_3^{z+} , \qquad (8.21)$$

where x, y and z represent the charge states on M_1, M_2 and M_3, respectively. The kinetic energy, ε, that is spontaneously released in such a process can be deduced from the translational energy spectrum of M_2^{y+} and M_3^{z+}

$$\varepsilon = \frac{y^2 z^2 M_1^2 e V_{\text{accl}} (\Delta V_{\text{anal}})^2}{16 x M_2 M_3 V_{\text{el}}^2} , \qquad (8.22)$$

where V_{accl} is the voltage with which the incident ion beam was accelerated, ΔV_{anal} is the width of the fragment ion peak, measured in terms of analyser voltage, that is observed in the translational energy spectrum and V_{el} is the analyser voltage at which elastically scattered stable ions are transmitted. As the acceleration voltage increases, so does the precision with which ΔV_{anal} can be measured as $(\Delta V_{\text{anal}})^2$ is proportional to $V_{\text{el}}^2 / V_{\text{accl}}$. The experimental limitation is usually imposed by the maximum translational energy that projectile ions can be accelerated to in any given apparatus; in most commercial mass spectrometers this limit is usually in the region of 12–15 keV.

Dissociation of doubly charged molecules can also occur by collision processes of the type

$$XY^{2+} + M \rightarrow XY^{+*} + M^+ \rightarrow X^+ + Y + M^+ , \qquad (8.23)$$

in which electron capture in a highly excited (dissociative) electronic state of the singly charged species, XY^{+*}, preceeds fragmentation. In the case of homonuclear diatomic projectiles such a process gives rise to translational energy peaks which coincide with the elastically scattered XY^{2+} peak; additional peaks with larger kinetic energy releases may also be obtained as symmetrical structure on both the low energy as well as the high energy side.

8.2.6 Excitation (De-excitation) and Electron Capture Reactions

As described in Chap. 6, translational energy spectrometry has also been used to study collisional excitation and de-excitation processes in multiply charged molecular ions. Consider the following ion-neutral collisions:

$$XY^{2+} + M \rightarrow XY^{2+*} + M - \Delta E , \tag{8.24}$$

$$XY^{2+*} + M \rightarrow XY^{2+} + M + \Delta E . \tag{8.25}$$

The first collision is an inelastic process in which the projectile ion's translational energy is converted into internal excitation energy of XY^{2+}, giving rise to an overall endoergicity (which results in a negative value of energy defect). The second collision, on the other hand, is a superelastic one in which the internal excitation energy of the projectile ion, XY^{2+*}, is converted to translational energy, giving rise to an overall exoergicity (which results in a positive value of energy defect). It is assumed here that the target M remains a spectator; in practice, use of rare gas targets is an extremely good approximation to this idealized situation because, for such target species, collisional excitation and ionization are precluded by cross sections which are considerably lower than those for processes involving other species.

Measurement of the overall energy defects for the above inelastic collisions enables spectroscopic information to be obtained for ionic species which are not readily amenable to conventional optical spectroscopic techniques. However, in order to carry out excitation studies, the experimental constraints are severe in that an extremely high energy resolution is demanded; the use of commercial mass spectrometric apparatus is generally not possible. Details of the instrumentation employed in such experiments are given in Chaps. 3 and 6.

An important category of inelastic ion-neutral collisions is that which leads to electron capture

$$XY^{2+} + M \rightarrow XY^+ + M^+ \pm \Delta E , \tag{8.26}$$

where ΔE represents the overall energy defect for the process. The overall energy budget for such processes can be expressed as

$$\begin{aligned}
\Delta E = &[IE_2(XY) + E_{\text{exc}}(XY^{2+})] \\
&- [IE_1(XY) + E_{\text{exc}}(XY^+) + IE_1(M) + E_{\text{exc}}(M^+)] ,
\end{aligned} \tag{8.27}$$

where $IE_{1,2}$ represent the single and double ionization energies and E_{exc} is the excitation energy pertaining to a specific electronic state.

The methodology adopted for measuring ΔE is somewhat similar to that adopted in charge stripping experiments. If V is the energy analyser voltage at which the elastically scattered XY^{2+} projectiles are transmitted, then $(2V - \Delta V)$ will be the voltage at which the singly charged electron capture product, XY^+, will be detected. As before, ΔV is related to the energy defect ΔE simply by the energy analyser's geometrical factor. Experimental determination of ΔE also makes it possible to deduce the internal excitation energies of the projectile and product ions, thus facilitating quantum-state-diagnosed studies of the electron capture process. Examples of some high resolution translational energy spectrometric studies of such state-diagnosed electron capture and collisional excitation processes have been described in Chap. 6.

8.2.7 Studies Using Forward-Geometry Mass Spectrometers

The dissociation of a doubly charged molecular ion into two singly charged fragments,

$$m_1^{2+} \rightarrow m_2^+ + m_3^+ \,, \tag{8.28}$$

can also be studied in conventional, commercially available mass spectrometers comprising an electrostatic energy analyser followed by a magnetic sector (known as forward-geometry, or E–B, instruments). If we assume that the ion is accelerated out of the ion source as a doubly charged species but dissociates into singly charged ions in the electrostatic analyser, the electric field strength, E_2, required to transmit the fragment ions in such a case will be

$$E_2 = (2m_2 2V)/(m_1 r) \,, \tag{8.29}$$

which is clearly twice the corresponding field strength (E_1) associated with dissociation of a singly charged molecular ion

$$E_2 = 2E_1(m_2/m_1) \,. \tag{8.30}$$

If m_1^{2+} is a polyatomic species, the electrostatic analyser must be scanned from $2E_1$ downwards to allow transmission of all the possible singly charged fragments. Similarly, for transmission of fragment ions through the magnetic sector, it can be shown that

$$B_2 = 2B_1(m_2/m_1) \,. \tag{8.31}$$

The results of these considerations show that the mass spectrometer has to be operated in a linked-scan mode whereby, after setting up the instrument with the parent doubly charged molecular ion at B_1 and E_1, a scan has to be carried out with the ratio B/E held constant from $2B_1$ and $2E_1$ downwards. The resulting spectrum will yield peaks corresponding to fragmentations of doubly charged molecules into singly charged ions in the range from $2B_1$ to B_1; from B_1 downwards, the spectrum will also yield additional peaks resulting from dissociations of doubly charged molecules into doubly charged fragments. This linked-scan technique has been recently applied to gain hitherto unavailable information on doubly charged peptides [8.12]

Experimental measurements of electron impact double ionization energies using conventional E–B instruments are hampered by the existence of low mass $(m/2)^+$ ions which appear at the same value of m/q as even mass m^{2+} species. An electron capture technique has been utilised [8.13] which separates all interfering singly charged ions. In this method, all singly and doubly charged ions which are produced in the ion source are accelerated by a few keV energy into a collision cell located in the field-free region preceding the electrostatic analyser. Here, the doubly charged ions undergo single electron capture reactions with the collision chamber gas

$$m^{2+} + n \rightarrow m^+ + n^+ . \tag{8.32}$$

The fast-moving m^+ product ions from such reactions possess essentially the same velocity as the doubly charged m^{2+} reactants. However, because they possess only half the amount of charge, they can only be transmitted through the electrostatic sector when it is operated at twice the voltage required to transmit the elastically scattered m^{2+} ions. Fast m^+ product ions which exit the electrostatic field are focused into the magnetic sector (for momentum analysis) and thence to a particle multiplier. Operation of the electrostatic analyser in this mode serves to filter out background contributions from singly and doubly charged species generated in the ion source, particularly in the case of polyatomic molecules. The enhanced signal-to-noise ratios obtained using this technique facilitate measurement of doubly charged ion appearance energies; such measurements are carried out by varying the electron energy within the ion source and monitoring the signal from fast singly charged electron capture product ions.

8.2.8 Studies Using High-Energy Accelerators

High-energy ion accelerators have also found utility in studies of metastable and dissociating states of multiply charged molecules. In the earliest accelerator-based observations of metastable doubly charged diatomic ions [8.14], a sputter ion source in a tandem accelerator was used to produce a beam of vibrationally hot ($T_{vib} = 5000$ K) XY^- molecules which were accelerated to energies in the MeV range. High energy multiple charge stripping reactions with an inert gas target resulted in beams of doubly charged molecular ions, some of which were formed in metastable states whose lifetime was long enough (several μs) to enable their transmission through a pair of magnetic sector analyzers and subsequent detection. The identification of these ions was by observing the fragmentation which occurred after passing through a carbon foil placed in front of a conventional silicon surface barrier detector. MeV beams of diatomic species such as B_2^{2+}, $^{10}B^{11}B^{2+}$, BO^{2+}, CN^{2+} and triatomic ions such as CH_2^{2+}, C_2H^{2+} were observed [8.14] with fluxes as high as 10^8 molecules s^{-1}.

More recently, high energy (> 1 MeV) stripping of singly charged molecular species has been utilised to conduct Coulomb explosion experiments on metastable doubly charged molecules which have yielded experimental information on the equilibrium bond length in the doubly charged species. Figure 8.3 shows a schematic diagram of the experimental arrangement which has been used [8.15] for Coulomb explosion studies on He_2^{2+} ions. 2 MeV beams of singly charged $^3He^4He^+$ ions are made to undergo charge stripping collisions with N_2 gas. The resulting $^3He^4He^{2+}$ ions are mass selected and transported through a thin Formvar film in which the remaining electrons in the projectile ion beam are stripped, resulting in a rapid, violent Coulomb explosion that separates the two helium nuclei. The time period over

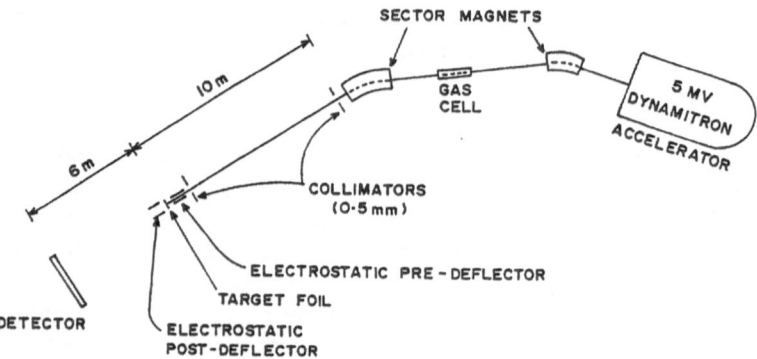

Fig. 8.3. Apparatus for studying Coulomb explosion of doubly charged molecular ions

which this explosion occurs is very much smaller than the characteristic vibrational and rotational time periods so that, for each molecular ion in the incoming beam, the final velocities of the dissociating fragments are essentially determined by the spatial geometry of the nuclei at the instant that the molecule enters the foil. After emerging from the stripping foil, the fragments are charge and mass separated by an electrostatic field and then detected by means of a position sensitive multiwire proportional counter which monitors the two-dimensional positions and relative arrival times for both nuclei in each molecular ion. The size of the spatial distributions of the explosion fragments reflects the differences in the mean bond length of the doubly charged molecular ion. This technique is clearly applicable to those doubly charged species whose lifetime is in excess of a few μs, which is the transit time to traverse the distance within the apparatus between the initial charge stripping cell and the Coulomb explosion foil.

In contrast, accelerator-based experiments have also produced beams of metastable multiply charged molecules possessing extremely small amounts (few tens or hundreds of meV) of translational energy. It is well established that when high energy, multiply charged heavy ion projectiles undergo collisions with neutral gas targets, a large number of target electrons can be ejected without imparting much kinetic energy to the target nucleus. Such collisions between MeV ions and neutral targets can give rise to the formation of highly charged, low energy ions which can be used for subsequent collisional studies; Chap. 3 discusses details of such experiments.

An example of an experimental arrangement used recently for production of ultra-low-energy doubly charged CO ions is shown in schematic form in Fig. 8.4. [8.16]. A crossed-beam geometry has been utilized to study collisions between a molecular beam of CO with either fast electrons or a beam of highly charged heavy ions from the TIFR 14 MV tandem Pelletron accelerator. In the latter case, collisions between the fast (tens of MeV) ions and CO result in production of ions which are extracted by an electrostatic field into a retarding-potential energy analyzer which then focuses the energy-selected ion beam on to the entrance plane of a quadrupole

238

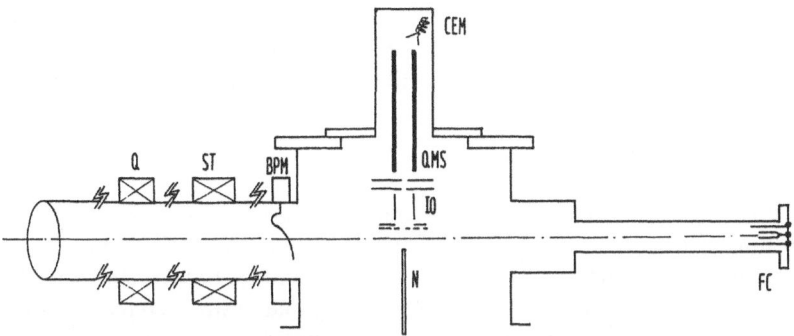

Fig. 8.4. Apparatus for production of ultra-low-energy multiply charged recoil ions. Q, ST and BPM are, respectively, a quardupole magnet, beam steerer and beam profile monitor in the beam line of a high-energy accelerator. N is the nozzle producing a molecular beam, IO is the ion optics which extracts slow recoil ions into a quadrupole mass spectrometer (QMS) and a channel electron multiplier (CEM) detector. FC is the Faraday cup which monitors the fast ion beam current

mass spectrometer. Ions which are analyzed on the basis of their kinetic energy and their mass/charge ratio are either detected by an off-axis channel electron multiplier (as in the experimental arrangement shown in Fig. 8.4) or are utilized for further collisional or spectroscopic experiments, as discussed in Chap. 1; the most probable transverse kinetic energy of metastable CO^{2+} ions produced in such an experiment is measured to be only a few tens of meV.

The method of producing low energy recoil ions using MeV projectile beams can also be utilized to study multiply charged molecular ions which are formed in dissociative states [8.17]. In such cases, the kinetic energies of multiply charged atomic fragment ions, arising from dissociation of multiply charged molecular species, can be measured by means of magnetic sectors or by time-of-flight spectrometry. Atomic fragment ions are found to be energy shifted by as much as 200–300 eV compared with the energy of corresponding ions produced with atomic targets [8.18].

Availability of pulsed beams of fast projectile ions facilitates more detailed studies of molecular dissociation processes; by measuring simultaneously the energy and the time-of-flight spectra of fragment ions it becomes possible to deduce information on the dissociation dynamics of specific charge states of multiply charged molecules [8.19, 20]. This is in contrast to conventional electron impact dissociative ionization experiments in which it is not directly possible to predict whether the precursor of a particular fragment ion is a singly charged or multiply charged molecular state. A further refinement of this energy-time spectroscopy technique has been the measurement of angular correlations of N^+ ions produced in the dissociative double ionization of N_2 induced by 1 MeV H^+ and He^+ collisions [8.21]. In such collisions if N_2^{2+} ions were formed with zero recoil energy, the pair of N^+ fragments produced upon dissociation of the molecular ion would possess equal and opposite velocity vectors in the laboratory frame of reference. By measuring the coincidence yield of N^+ pairs as a function of the angle between their velocity vectors it becomes possible to estimate the component of momentum which is imparted to the N_2^{2+} ion in the beam direction. Such experiments also make possible the measurement of cross sections for dissociative double ionization as a function of the orientation of the target diatomic molecules relative to the fast projectile beam's axis [8.21].

8.2.9 Energy Calibration

The reliability of quantitative information which is deduced from translational energy spectrometry measurements is, in the final analysis, determined essentially by the reliability with which the energy scale has been determined. *Langford* et al. [8.22] have carried out an evaluation of some of the generally underestimated problems associated with calibration of energy scales. As an illustration we outline here typical energy calibration procedures that can be applied to most translational energy spectrometry experiments. For sake of clarity, we focus only on one particular type of ion translational energy measurement – that involving determination of the minimum energy loss in charge stripping collisions:

$$XY^+ + M \rightarrow XY^{2+} + M + e - \Delta E .\tag{8.33}$$

The treatment that follows is also applicable to all other types of energy spectrometry experiments.

In most charge stripping measurements, the minimum energy defect, ΔE_{min}, is measured after the energy loss scale has been calibrated with reference to the value of ΔE_{min} obtained from a known charge stripping reaction. There are two distinct methods which have been used for carrying out such calibration procedures, which have also been applied to other experimental studies, such as single or double electron capture.

In the 'linear offset' method, the value of ΔE_{min} for a known charge stripping process is found by measuring the energy difference, ΔE_{cal}, between the high energy onsets of both the elastic scattering and the inelastic charge stripping peaks. This result is then subtracted from the known energy defect of the calibrant ion, Q_{cal}, to yield a calibrating offset, θ

$$\theta = Q_{\mathrm{cal}} - \Delta E_{\mathrm{cal}} .\tag{8.34}$$

When an ion of unknown energy defect is charge stripped, this offset is added to the measured translational energy loss, ΔE_{exp}, in order to obtain a corrected value of minimum energy loss, ΔE_{min}.

Repeated measurements involving both the calibrant ion and the ion under study show that ΔE_{cal} and ΔE_{exp} are only accurate within the bounds of random errors δE_{cal} and δE_{exp}, respectively, so that

$$\theta \pm \delta\theta = Q_{\mathrm{cal}} - [\Delta E_{\mathrm{cal}} \pm \delta E_{\mathrm{cal}}] ,\tag{8.35}$$

and the value of minimum energy loss for the ion under study can be expressed as

$$\Delta E_{\mathrm{min}} = \Delta E_{\mathrm{exp}} + Q_{\mathrm{cal}} - \Delta E_{\mathrm{cal}} \pm \delta E_{\mathrm{exp}} \pm \delta E_{\mathrm{cal}} .\tag{8.36}$$

Thus the maximum error in ΔE_{min} is $\pm (\delta E_{exp} + \delta E_{cal})$ and the most probable error is $\pm (\delta E_{exp}^2 + \delta E_{cal}^2)^{1/2}$. In practice, the errors δE_{exp} and δE_{cal} are usually equal in magnitude and thus the most probable error in ΔE_{min} due to the uncertainties in measuring ΔE_{exp} and ΔE_{cal} is $\pm 2^{1/2}\delta E_{exp}$.

It follows that, in the linear offset method, random errors are propagated as a constant, being independent of the actual value of the double ionization energy of either the ion under study or the calibrant ion.

A few workers have adopted a somewhat different approach, referred to as the 'multiplicative method'. Here, a charge stripping measurement is made for a known (calibrant) collision system and a parameter ϕ is evaluated where

$$\phi = \frac{Q_{cal}}{\Delta E_{cal}} . \tag{8.37}$$

For charge stripping of the unknown ion, the resulting value of minimum energy loss, ΔE_{exp}, is multiplied by the factor ϕ in order to yield the final result

$$\Delta E_{min} = \phi \Delta E_{exp} . \tag{8.38}$$

If the inherent random errors are once again taken into consideration, it can be shown [8.22] that the maximum random error in ΔE_{min} in this method becomes $\pm \delta E_{exp}(Q_{cal} + \Delta E_{min})/\Delta E_{cal}$ and the most probable error is $\pm \phi \delta E_{exp}(1 + \Delta E_{min}^2/Q_{cal}^2)^{1/2}$.

Thus the error using the multiplicative method depends not only on δE_{exp} but also on the magnitude of the double ionization energy of the ion under study as well as the calibrant ion. This is clearly an undesirable feature; experimental evaluation of the two methods by conducting charge stripping studies on singly charged ions of Ne, Ar and Kr has led *Langford* et al. [8.22] to recommend that the linear offset method rather than the multiplicative method should be used for energy calibration purposes.

8.3 Other Experimental Techniques

Although the subject matter of this chapter is confined mainly to studies of multiply charged molecules using ion collision techniques, it is, nevertheless, obligatory to briefly outline some of the other methods that have been utilised in several contemporary investigations.

8.3.1 Auger Spectroscopy

Some of the earliest experimental information on electronic states of doubly charged molecules came from Auger spectroscopy experiments on diatomic molecules [8.23, 24]. In such measurements, a fast beam of electrons, or photons from a synchrotron radiation source, collide with the target molecule, XY, so as to eject an electron from one of the inner shells. The subsequent Auger rearrangement process may result in formation of a doubly charged molecule XY^{2+}

$$\frac{X-\text{ray}}{\text{keV electron}} + XY \rightarrow X^{K+}Y \rightarrow XY^{2+} + e. \tag{8.39}$$

Here, $X^{K+}Y$ designates an extremely short-lived molecular state in which there exists a vacancy in the X-atom's K-shell. The experimental measurement is of the kinetic energy of the fast Auger-ejected electron; the doubly charged species is not directly observed and, consequently, no information can be deduced on the nature (bound or repulsive) of the XY^{2+} electronic state that is accessed in such an experiment.

The question of whether the Auger technique yields adiabatic or vertical double ionization energies has to be determined in the light of the applicability of the Franck-Condon principle for the overall process on a case by case basis. If the Auger emission timescale is shorter than the characteristic vibrational period of the molecule in question, the final states of the doubly charged species will lie within the Franck-Condon region of the neutral precursor. If the converse situation holds, it is the Franck-Condon region defined by the K-hole intermediate state, $X^{K+}Y$, that would determine the internuclear distance in the XY^{2+} electronic state.

8.3.2 Photoionization Methods

Conventional photoionization mass spectrometry entails measurement of the yield of multiply charged product ions (usually identified by the time-of-flight technique) as a function of the wavelength of the incident photons (usually from a synchrotron radiation source) [8.25]. As with charge stripping and conventional electron impact experiments, this technique also requires the multiply charged ion to possess a lifetime of at least a few μs in order to be detected.

In recent years more sophisticated experimental methods, such as the photoion-photoion coincidence technique (PIPICO), have yielded much new information on electronic states of multiply charged diatomic and triatomic species [8.26, 27]. In these experiments, the two ionic fragments X^+ and Y^+ which are produced by dissociative photoionization of the molecule XY, are detected by a single particle multiplier. The difference in flight times of the respective fragment ions is measured in delayed coincidence. The average difference in time-of-flight for each of the species X^+ and Y^+ is proportional to the difference in the square roots of the masses of X and Y whereas the distribution of flight times about such an average yields information on the initial kinetic energy which is released upon dissociation of the XY^{2+} precursor. From determination of these two parameters the energies of doubly charged electronic states can be estimated. The photoion-photoion coincidence technique complements the conventional photoionization method in that the former detects only rapidly dissociating electronic states of multiply charged ions whereas the latter accesses only those states which are relatively long-lived.

Recently, *Eland* et al. [8.28] have developed a triple coincidence technique (PEPIPICO – photoelectron-photoion-photoion coincidence) which has been applied to studies of the dynamics of fragmentation in doubly charged triatomic molecules. The method entails detection in coincidence of an electron and two ions from a dissociative double photoionization process of the type

$$h\nu + XYZ \rightarrow XY^+ + Z^+ + 2e \ . \tag{8.40}$$

Detection of the electron defines the starting point in time from which times of arrival of all subsequent fragment ions can be determined in an absolute manner. These arrival times correspond to the flight times in a time-of-flight spectrometer, yielding the masses of the different fragments. This technique enjoys a number of advantages over the related PIPICO method, chief among them being the fact the absolute measurements of flight times (rather than relative time differences) facilitates unambiguous identification of fragment ion pairs and facilitates a direct characterisation of the various decay routes followed in the fragmentation process. In addition, the technique can also yield information on angular correlations between the ejected electrons and ion fragments [8.28].

8.3.3 Optical Spectroscopy

The first, and until very recently the only, optical spectroscopic study of a multiply charged molecule, N_2^{2+}, was reported by *Carrol* [8.29] more than thirty years ago. These pioneering measurements have been confirmed, and considerably extended, in more recent studies conducted by means of laser photofragment spectroscopy [8.30] as well as with conventional emission spectroscopy in the 1550–1650 Å region using a large 10-m vacuum ultraviolet spectrograph [8.31]. Optical emissions have also been recently discovered in two other diatomic species, NO^{2+} [8.32] and O_2^{2+} [8.33].

8.3.4 Electron Impact Experiments

The earliest experimental evidence for the existence of metastable doubly charged molecular ions came from electron impact studies carried out in a conventional, single-focusing mass spectrometer nearly sixty years ago [8.2]. In these remarkable experiments the electron impact double ionization energies of CO, CO_2, NO and NO_2 were measured and products resulting from unimolecular (spontaneous) and collision induced dissociation were determined. The earliest survey of multiply charged molecular positive ions was published nearly thirty years ago, and listed the results of electron impact mass spectrometric determinations of the appearance energies of more than fifty doubly charged diatomic and polyatomic ions [8.34]. More recent compilations of pertinent references [8.7, 35] indicate that although the electron impact mass spectra of diatomic and small polyatomic species yield evidence for less than 1 % abundance of multiply charged molecular ions, corresponding abundances of up to 20 % have been observed in the case of large aromatic hydrocarbons and organometallic complexes.

Apart from the sheer volume of data that has been acquired using electron impact techniques (such as, ionization efficiency curve measurements using conventional mass spectrometers), and its importance from a historical viewpoint, the electron beam method continues to hold much promise for future investigations, particularly when used in conjunction with contemporary coincidence techniques. We present, in outline form, some of the exciting possiblities which are likely to be, or are already in the process of being, implemented. A cogent and detailed account of the various coincidence measurements that can be carried out in multiple photoionization studies has been presented by *Eland* et al. [8.36].

In an electron-molecule collision leading to double ionization

$$e + XY \rightarrow [XY^{2+} + 3e] \rightarrow X^+ + Y^+ + e + e_1 + e_2 , \qquad (8.41)$$

several coincidence measurements are possible which can yield information on the dynamics of the double ionization process in molecules.

i) XY^{2+} in coincidence with e. Such measurements yield quantitative information on electron-electron correlations and have already been implemented in photon-impact experiments in the form of PEPICO, photoelectron-photoion coincidence [8.36].

ii) X^+ (or Y^+) in coincidence with e. Dissociation of a doubly charged molecule results in two singly charged fragments which possess high kinetic energies; coincidence measurements between one of the fragments and the incident electron affords the possibility of discriminating against dissociative single ionization. Coincidence measurements which are carried out in conjunction with electron energy analysis give access to specific electronic states of the doubly charged molecule.

iii) X^+ in coincidence with Y^+. Such experiments have been carried out extensively using photons (the PIPICO method, already referred to above) [8.36]. There is also one report of such measurements using an electron beam [8.37]. These experiments measure the kinetic energy released upon dissociation and, in the case of multiply charged triatomic and polyatomic ions, can also yield information on branching ratios for various dissociation channels.

iv) e_1 in coincidence with e_2. Such experiments, which are yet to be undertaken, hold much promise of enabling insight to be obtained into the dynamics of the double ionization process. In particular, an important goal of such experiments would be to determine the manner in which excess energy above the double ionization threshold is distributed between the ejected electrons.

v) triple coincidence between e, X^+ and Y^+. The electron signal is used to initiate time-of-flight measurements of X^+ and Y^+, yielding flight times t_X and t_Y, respectively. The difference between these two times gives information which is analogous to that obtained in double coincidence measurements (iii). However, the flight time measurements can also yield additional information. If t_X^0 and t_Y^0 represent, respectively, the flight times of thermal energy X^+ and Y^+ ions, then determination of $t_X - t_X^0$ and $t_Y - t_Y^0$ enables distinction to be made between unimolecular (spontaneous) dissociation due to Coulomb explosion and sequential dissociation brought about by avoided crossings of potential energy curves. Pioneering experiments along these lines, using photons to doubly ionise triatomic molecules, have recently been reported by *Eland* et al. [8.28].

vi) triple coincidence betwen e_1, e_2 *and* XY^{2+}. Such experiments have not been attempted yet but can be expected to yield information on the reactivity of specific electronic states of the doubly charged molecule.

8.4 Theoretical Description of Molecular Ions

8.4.1 The Schrödinger Equation and the Born-Oppenheimer Approximation

A non-empirical calculation in molecular quantum mechanics is normally meant to be the solution of a time-independent Schrödinger equation

$$\mathcal{H}_{\text{total}} \, \Psi_{\text{total}} = E_{\text{total}} \, \Psi_{\text{total}} \, , \tag{8.42}$$

where Ψ_{total} is a total wavefunction for all particles in the molecule, and E_{total} is the total energy of the system.

Consider a molecule containing N nuclei and n electrons. The many particle Hamiltonian operator $\mathcal{H}_{\text{total}}$ is given, in atomic units, as

$$\hat{\mathcal{H}}_{\text{total}} = -\sum_{k=1}^{N} \frac{1}{2M_K} \nabla_K^2 + \sum_{K>K'}^{N} \frac{Z_K Z_{K'}}{|R_K - R_{K'}|} - \sum_{i=1}^{n} \frac{1}{2} \nabla_i^2$$
$$-\sum_{k=1}^{N} \sum_{i=1}^{n} \frac{Z_K}{|R_K - r_i|} + \sum_{i>j}^{n} \frac{1}{|r_i - r_j|} \, , \tag{8.43}$$

where M_K, Z_K and R_K are the mass, charge and the position vector, respectively, of nucleus K, and r_i is the position vector of electron i. Summations involving indices K and K' are over atomic nuclei and those involving i and j are over electrons.

As the masses of the nuclei are several thousand times larger than the electronic mass, they move much more slowly compared to electrons. Therefore, it is reasonable to assume that the electrons adjust themselves to new nuclear positions so rapidly that, at any instant, the electron motion is just as it would be if the nuclei were at rest at the position they occupy at that instant. This separation of electronic and nuclear motions is known as the Born-Oppenheimer approximation. Using this, we need consider only that part of the Hamiltonian which depends on the position but not on the momenta of the nuclei. This is the electronic Hamiltonian operator, $\hat{\mathcal{H}}_{\text{el}}$, given as

$$\hat{\mathcal{H}}_{\text{el}} = \sum_{i=1}^{n} \hat{h}_i + \sum_{i>j}^{n} \frac{1}{|r_i - r_j|} \, , \tag{8.44}$$

where

$$\hat{h}_i = \frac{1}{2} \nabla_i^2 - \sum_{k=1}^{N} \frac{Z_K}{|R_K - r_i|} \, . \tag{8.45}$$

245

The modified Schrödinger equation is then

$$\hat{\mathcal{H}}_{el}\Psi_{el} = E_{el}\Psi_{el} \,, \tag{8.46}$$

where E_{el} is the electronic energy and Ψ_{el} is a wavefunction for all electrons in the field of fixed nuclei. The total 'potential' energy of the system, for a given set of nuclear positions, is then

$$E_{pot}(R_1, R_2, \ldots, R_N) = E_{el} + \sum_{K>K'}^{N} \frac{Z_K Z_{K'}}{|R_K - R_{K'}|} \,. \tag{8.47}$$

Heretofore, we shall denote E_{el} and Ψ_{el} simply by E and Ψ, respectively.

For an n-electron system, it is simple to associate the n electrons with an orthonormal set of one-electron functions $\tilde{\psi}_1, \tilde{\psi}_2, \ldots, \tilde{\psi}_n$, called spin orbitals. Such a spin orbital is usually assumed to have the form

$$\tilde{\psi}_i(r, \sigma) = \psi_i(r)\chi_i(\sigma) \,. \tag{8.48}$$

in which the orbital $\psi_i(r)$ describes the motion in space of the electron and the spin function $\chi_i(\sigma)$ describes its spin state.

The total n-electron wavefunction is now built up as an antisymmetrized product of molecular spin orbitals

$$\Phi = (n!)^{-1/2} \begin{vmatrix} \tilde{\psi}_1(1) & \tilde{\psi}_2(1) & \ldots & \tilde{\psi}_n(1) \\ \tilde{\psi}_1(2) & \tilde{\psi}_2(2) & \ldots & \tilde{\psi}_n(2) \\ \vdots & \vdots & \ddots & \vdots \\ \tilde{\psi}_1(n) & \tilde{\psi}_2(n) & \ldots & \tilde{\psi}_n(n) \end{vmatrix} \,. \tag{8.49}$$

Such a determinant of spin orbitals, known as a Slater determinant, is the simplest normalized wavefunction which satisfies the antisymmetry principle.

8.4.2 Hartree-Fock Theory

Substituting the above wavefunction into the expression for energy, we get

$$E = \frac{\int \Phi^* \hat{\mathcal{H}} \Phi \, d\tau}{\int \Phi^* \Phi \, d\tau} = \sum_{i=1}^{n} h_i + \frac{1}{2}\sum_{i,j}^{n} J_{i,j} - \frac{1}{2}\sum_{i,j}^{n} K_{i,j} \,, \tag{8.50}$$

where the orbital energies h_i, the Coulomb integrals $J_{i,j}$ and the exchange integrals $K_{i,j}$ are defined by

$$h_i = \int \psi_i^*(r)\hat{h}\psi_i(r) \, dr \,. \tag{8.51}$$

$$J_{i,j} = \int \frac{\psi_i^*(r_1)\psi_i(r_1)\psi_j^*(r_2)\psi_j(r_2)}{|r_1 - r_2|} \, dr_1 \, dr_2 \,, \quad \text{and} \tag{8.52}$$

$$K_{i,j} = \delta(m_{si}, m_{sj}) \int \frac{\psi_i^*(\mathbf{r}_1)\psi_j(\mathbf{r}_1)\psi_j^*(\mathbf{r}_2)\psi_i(\mathbf{r}_2)}{|\mathbf{r}_1 - \mathbf{r}_2|} \, d\mathbf{r}_1 \, d\mathbf{r}_2 \; . \tag{8.53}$$

According to the variational principle, the best molecular orbitals are obtained by varying the one-electron functions $\tilde{\psi}_1, \tilde{\psi}_2, \ldots, \tilde{\psi}_n$ until the energy achieves its minimum value. The condition for the minimum of energy leads to the set of differential equations

$$\mathcal{H}_i^{HF} \psi_i = \varepsilon_i \psi_i \; , \tag{8.54}$$

where ε_i are the orbital energies and the $\mathcal{H}_i^{HF}$ operator is given by

$$\mathcal{H}_i^{HF} = \hat{h}_i + \sum_j (\hat{J}_j - \hat{K}_j) \; , \tag{8.55}$$

where the Coulomb operator $\hat{J}_j$ and the exchange operator $\hat{K}_j$ are defined as

$$\hat{J}_j(\mathbf{r}_1) = \int \frac{\psi_j^*(\mathbf{r}_2)\psi_j(\mathbf{r}_2)}{|\mathbf{r}_1 - \mathbf{r}_2|} \, d\mathbf{r}_2 \tag{8.56}$$

and

$$\hat{K}_j(\mathbf{r}_1)\psi_i(\mathbf{r}_1) = \delta(m_{si}, m_{sj})\psi_j(\mathbf{r}_1) \int \frac{\psi_j^*(\mathbf{r}_2)\psi_i(\mathbf{r}_2)}{|\mathbf{r}_1 - \mathbf{r}_2|} \, d\mathbf{r}_2 \; . \tag{8.57}$$

a) Matrix Formulation of Hartree-Fock Theory (LCAO Method). For a molecular system of any size, direct solution of the above equations is impossible. It is convenient to approximate the HF orbitals by a linear combination of atomic orbitals (LCAO). In this approach, each molecular orbital is of the form

$$\psi_i(\mathbf{r}) = \sum_{\mu=1}^{M} C_{\mu i} \phi_\mu(\mathbf{r}) \; , \tag{8.58}$$

where the ϕ_μ are atomic orbitals. Since the orbitals ψ_i form an orthonormal basis set, we require that

$$\int \psi_i^*(\mathbf{r})\psi_j(\mathbf{r}) \, d\mathbf{r} = \delta_{ij} = \sum_{\mu\nu} C_{\mu i}^* C_{\nu j} S_{\mu\nu} \; , \tag{8.59}$$

where $S_{\mu\nu}$ is the overlap integral between atomic orbitals ϕ_μ and ϕ_ν.

In the LCAO approximation, the electron density at a point $\mathbf{r}_0$ is given as

$$\rho(\mathbf{r}_0) = \sum_i n_i \psi_i^*(\mathbf{r}_0)\psi_i(\mathbf{r}_0) = \sum_{\mu\nu} P_{\mu\nu} \phi_\mu^*(\mathbf{r}_0)\phi_\nu(\mathbf{r}_0) \; , \tag{8.60}$$

where

$$P_{\mu\nu} = \sum_{i=1}^{M} n_i C_{\mu i}^* C_{\nu i} \; . \tag{8.61}$$

The matrix $P_{\mu\nu}$ is known as the bond order or density matrix, and n_i is the occupation number of the ith molecular orbital.

b) Restricted Hartree-Fock (RHF) Model. Given the the form of the spin orbitals

$$\tilde{\psi}_n(\boldsymbol{r}, \sigma) = \psi_n(\boldsymbol{r})\chi_n(\sigma) , \tag{8.62}$$

one of the most frequently made assumptions is that the spin orbitals occur in pairs $\psi_n(r)\alpha(\sigma)$ and $\psi_n(r)\beta(\sigma)$ with a common orbital but with different spin factors. Calculations performed using such approximations are referred to as Restricted Hartree-Fock (RHF) calculations. Substituting the molecular orbitals ψ_i in the molecular integrals h_i, J_{ij} and K_{ij}, the total electronic energy of the system is given by

$$E_{\text{el}} = \sum_{\mu\nu} P_{\mu\nu} h_{\mu\nu} + \frac{1}{2} \sum_{\mu\nu} \sum_{\lambda\sigma} P_{\mu\nu} P_{\lambda\sigma} \left[(\mu\nu|\lambda\sigma) - \frac{1}{2}(\mu\lambda|\nu\sigma) \right] , \tag{8.63}$$

where

$$h_{\mu\nu} = \int \phi_\mu^*(r)\hat{h}\phi_\nu(r)\, dr \tag{8.64}$$

and

$$(\mu\nu|\lambda\sigma) = \int \int \frac{\phi_\mu^*(r_1)\phi_\nu(r_1)\phi_\lambda^*(r_2)\phi_\sigma(r_2)}{|r_1 - r_2|}\, dr_1\, dr_2 . \tag{8.65}$$

The coefficients $C_{\mu i}^*$ and $C_{\nu i}$ which enter $P_{\mu\nu}$ are now optimized so as to give minimum energy. This leads to the condition

$$\sum_\nu (F_{\mu\nu} - \varepsilon_i S_{\mu\nu})C_{\nu i} = 0 , \tag{8.66}$$

with the Fock matrix $F_{\mu\nu}$ given by

$$F_{\mu\nu} = h_{\mu\nu} + \sum_{\lambda\sigma} P_{\lambda\sigma} \left[(\mu\nu|\lambda\sigma) - \frac{1}{2}(\mu\lambda|\nu\sigma) \right] . \tag{8.67}$$

These are the Roothaan equations; they differ from the Hartree-Fock equations in that they are algebraic rather than differential equations.

Since the matrix elements of the Fock matrix, $F_{\mu\nu}$, are dependent on the molecular orbitals through the elements $P_{\mu\nu}$, the Roothaan equations are solved by a series of steps:

1) An initial guess of the LCAO coefficients $C_{\nu i}$ is made;

2) Electrons are assigned in pairs to the molecular orbitals ψ_i with lowest energies;

3) The density matrix $P_{\mu\nu}$ is calculated and then used to form a Fock matrix $F_{\mu\nu}$;

4) A diagonalization procedure is effected and a new set of coefficients $C_{\mu\nu}$ is obtained;

5) Steps 2, 3 and 4 are repeated until the coefficients $C_{\nu i}$ no longer change within a given tolerance on repeated iteration.

c) Unrestricted Hartree-Fock Model (UHF). The pairing of spin orbitals, which is associated with the concept of double occupancy of orbitals, is not generally exhibited by the spin orbitals obtained by an unconstrained solution of the Hartree-Fock equation. There are at least two properties for which the RHF model can yield a quite incorrect description of the system. One is the Fermi contact interaction, or spin density at a nucleus, which is an important contribution to the hyperfine structure of atoms and molecules in open shell states. Only the s-orbitals make a non-zero contribution to the charge density at the nucleus and the double occupancy of these orbitals means that there is no net electron spin at the nucleus. The RHF wavefunction, therefore, gives a zero Fermi contact term whereas the experimental value is certainly non-zero. This deficiency can be overcome by relaxing the double occupancy constraint, as is done in the unrestricted Hartree-Fock (UHF) model.

The second property for which the RHF model yields an incorrect description is the dissociation energy and the nature of the dissociation products. RHF dissociation energies are generally too small and, in some cases, negative values are obtained, indicating that the RHF molecule is unstable towards dissociation.

Because of the unequal numbers of α and β spins, the orbitals ψ_n^α associated with α spins satisfy a different eigenvalue equation from those, ψ_n^β, associated with β spin. In the LCAO approximation, the two sets of molecular orbitals are written as

$$\psi_i^\alpha = \sum_\mu C_{\mu i}^\alpha \phi_\mu \tag{8.68}$$

and

$$\psi_i^\beta = \sum_\mu C_{\mu i}^\beta \phi_\mu . \tag{8.69}$$

A separate density matrix may be obtained for α and β electrons

$$P_{\mu\nu}^\alpha = \sum_{i=1}^{n_\alpha} n_i^\alpha C_{\mu i}^{*\alpha} C_{\nu i}^\alpha \tag{8.70}$$

and

$$P_{\mu\nu}^\beta = \sum_{i=1}^{n_\beta} n_i^\beta C_{\mu i}^{*\beta} C_{\nu i}^\beta , \tag{8.71}$$

where n_i^α and n_i^β are the occupation numbers of ψ_i^α and ψ_i^β, respectively. The full density matrix is the sum of these

$$P_{\mu\nu} = P_{\mu\nu}^\alpha + P_{\mu\nu}^\beta \tag{8.72}$$

and the spin density matrix

$$P_{\mu\nu}^s = P_{\mu\nu}^\alpha - P_{\mu\nu}^\beta . \tag{8.73}$$

When these LCAO molecular orbitals are used, the electronic energy of the system can be written as

$$E = \sum_{\mu\nu} P_{\mu\nu} H_{\mu\nu}$$
$$+ \frac{1}{2}\sum_{\mu\nu}\sum_{\lambda\sigma}\left[P_{\mu\nu}P_{\lambda\sigma}(\mu\nu|\lambda\sigma) - (P_{\mu\nu}^{\alpha}P_{\lambda\sigma}^{\alpha} + P_{\mu\nu}^{\beta}P_{\lambda\sigma}^{\beta})(\mu\lambda|\nu\sigma)\right] . \tag{8.74}$$

The coefficients $C_{\mu i}^{\alpha}$, $C_{\nu i}^{\alpha}$ and $C_{\mu i}^{\beta}$, $C_{\nu i}^{\beta}$ which enter $P_{\mu\nu}^{\alpha}$ and $P_{\mu\nu}^{\beta}$ are now independently optimized to give minimum energy. This leads to the conditions

$$\sum_{\nu}(F_{\mu\nu}^{\alpha} - \varepsilon_i^{\alpha} S_{\mu\nu})C_{\nu i}^{\alpha} = 0 \qquad \text{and} \tag{8.75}$$

$$\sum_{\nu}(F_{\mu\nu}^{\beta} - \varepsilon_i^{\beta} S_{\mu\nu})C_{\nu i}^{\beta} = 0 , \tag{8.76}$$

where the two Fock matrices are given by

$$F_{\mu\nu}^{\alpha} = h_{\mu\nu} + \sum_{\lambda\sigma}[P_{\lambda\sigma}[(\mu\nu|\lambda\sigma) - P_{\lambda\sigma}^{\alpha}[(\mu\lambda|\nu\sigma)] \qquad \text{and} \tag{8.77}$$

$$F_{\mu\nu}^{\beta} = h_{\mu\nu} + \sum_{\lambda\sigma}[P_{\lambda\sigma}[(\mu\nu|\lambda\sigma) - P_{\lambda\sigma}^{\beta}[(\mu\lambda|\nu\sigma)] . \tag{8.78}$$

These equations represent a generalization of Roothaan's equations and are solved by an iterative procedure similar to that described for the closed shell case.

An inherent defect in the UHF procedure is that it produces molecular wavefunctions which are eigenfunctions of S_z but not of the spin-squared operator S^2. Instead, Φ^{UHF} contains components of several spin states [8.38]:

$$\Phi^{\text{UHF}} = \sum_{m=0}^{\infty}[C_{S'+m}\Phi_{S'+m}] , \tag{8.79}$$

where S' is the M_s value for which the UHF calculations are performed and $\Phi_{S'+m}$ is an eigenfunction of S^2 with multiplicity $2(S'+m)+1$. This 'spin contamination' partially offsets the advantage gained by the UHF procedure and is especially inconvenient for discussing the spectroscopic and spin properties of the system under consideration. It is possible, in principle, to select wavefunctions corresponding to pure spin states from Φ^{UHF} by application of an appropriate spin projection operator O_s [8.39, 40] giving

$$\Phi^{\text{PUHF}} = O_s \Phi^{\text{UHF}} , \tag{8.80}$$

where O_s is defined by

$$O_s = \prod_{k \neq s} \frac{s^2 - k(k+1)}{s(s+1) - k(k+1)} . \tag{8.81}$$

However, the PUHF wavefunction is not strictly variational in the sense that the energy minimization is carried out for the UHF wavefunction, and not the PUHF wavefunction.

An alternative procedure is to form a symmetry adapted wavefunction in the form of a projected Slater determinant prior to the calculation of orbitals. This extended Hartree-Fock (EHF) model produces orbitals which are consistent with the symmetry properties of the state under consideration [8.41].

d) Representation of the orbitals. As has already been indicated, the most generally applicable and useful representation of a molecular orbital (MO) is a linear combination of atomic orbitals (LCAO). In current MO calculations, an 'atomic orbital' is understood to be any convenient one-centre basis function which is normally, but not invariably, centred on a nucleus of the molecule. Two types of basis, each with its own advantages and disadvantages, are used for the construction of molecular orbitals: Slater type orbitals (STO) and Gaussian type orbitals (GTO).

The atomic orbitals (AO) are most conveniently represented as linear combinations of STOs which have the form

$$\chi_{nlm}(r, \zeta) = R_{nl}(r, \zeta) Y_{lm}(\theta, \phi) , \tag{8.82}$$

where, for $n > l$,

$$R_{nl}(r, \zeta) = r^{n-1} e^{-\zeta r} . \tag{8.83}$$

The angular function $Y_{lm}(\theta, \phi)$ is a normalized spherical harmonic which determines the spatial properties of the orbital. The exponent ζ is a parameter which can be calculated variationally.

An atomic orbital with angular momentum quantum numbers l and m is then a linear combination of the basis functions

$$\psi(r, \theta, \phi) = R(r) Y_{lm}(\theta, \phi) , \tag{8.84}$$

where

$$R(r) = \sum_{n} \sum_{i} C_{ni} R_{nl}(r, \zeta_{ni}) . \tag{8.85}$$

Given a set of STOs with specified values of ζs the expansion coefficients are calculated as solutions of the SCF equations. It is found quite generally that a basis of three to five STOs per AO is required to produce highly accurate SCF wavefunctions and energies. Accuracies which are adequate for many purposes can, however, be obtained from a basis of two STOs per AO. This is called a double-zeta basis.

Although STOs form a satisfactory basis for the representation of molecular orbitals, accurate calculations in terms of these have been performed only for diatomic and some polyatomic molecules. Unfortunately, a great difficulty is associated with the evaluation of the electronic interaction integrals $(\mu\nu|\lambda\sigma)$ encountered in the evaluation of the Fock matrix in RHF and UHF procedures. The difficulties are most severe for three- and four-centre integrals when STOs are centred on three or four non-collinear atoms in a molecule. To bypass the molecular integral problem, *Boys* [8.42] suggested the use of functions of the form

$$R(r, \alpha) = x^l y^m z^n e^{-\alpha r^2} .\tag{8.86}$$

These are referred to as Gaussian type orbitals (GTOs).

Since the product of two Gaussians on different centres is equivalent to a single Gaussian on a new centre, these functions have a very clear advantage over STOs with regard to the evaluation of electronic interaction integrals.

The main disadvantage in using a Gaussian basis is that in order to produce a given accuracy, the GTOs required are two to five times the number of STOs. A basis of ten 1s-type and six sets of 2p-type GTOs is required to produce double zeta accuracy for the first row atoms. Because the number of electron interaction integrals is proportional to the fourth power of the basis size, a calculation for a molecule of quite modest dimensions can involve several hundred GTOs and 10^8 to 10^{10} integrals.

One way of reducing this considerable computational problem to managable proportions is by the use of 'contracted' Gaussian functions. A contracted Gaussian is a linear combination of primitive Gaussians with fixed coefficients. For example, the ten s-type GTOs for the first row atoms can be contracted into four linear combinations which simulate the double zeta representation of the 1s and 2s orbitals in these atoms. Similarly, the six sets of p-type GTOs can be contracted into two sets of linear combinations to simulate the double zeta representation of the 2p atomic orbitals.

Although the atomic basis sets give accurate description of the atoms, in order to obtain high accuracy in molecular calculations these sets must be augmented with polarization functions, for example, 2p for hydrogen, 3d for first row and second row atoms, and so on. Polarization functions are necessary to describe the distortions in the atomic orbitals resulting from the molecule formation process.

One alternative to the use of higher angular momentum functions on the nuclei in order to describe polarization effects is the use of bond functions centred between the nuclei [8.43, 44]. The bond functions (1s,2p) account for almost 90 % of the energy lowering obtained using 3d polarization functions at a considerable saving in computation time. Bond functions, however, appear to be less useful in calculations of potential energy surfaces for polyatomic molecules where ambiguity may arise in the location of the functions.

8.4.3 Electron Correlation

In the Hartree-Fock (HF) approximation the motion of each electron in an atom or molecule with n electrons is considered in the presence of the average potential created by the remaining $(n - 1)$ electrons. As such the HF approximation neglects the instantaneous repulsions between pairs of electrons. The contribution of such instantaneous repulsions to the total energy is called the correlation energy. The method of dealing with correlation energy is known as Configuration Interaction (CI).

For any atom or molecule, there are an infinite number of orbitals in addition to the Hartree-Fock orbitals. These orbitals can be used to construct other possible configurations in the form of Slater determinants. The CI wavefunction for an N-electron system is then represented as a linear combination of a complete orthonormal set of Slater determinants

$$\Psi_m = \sum_a A_{am} \Phi_a .$$

(8.87)

The coefficients A_{am} are determined to minimize the electronic energy

$$E = \int \Psi_m \mathcal{H} \Psi_m \, d\tau .$$

(8.88)

Application of the variation principle leads to the matrix equations

$$H A = A E ,$$

(8.89)

where H is composed of matrix elements between the configurations

$$H_{ab} = \int \Phi_a \mathcal{H} \Phi_b \, d\tau .$$

(8.90)

Hamiltonian matrix elements H_{ab} between configuration Φ_a and Φ_b of different symmetries are zero. Therefore, the secular equation is greatly simplified by only considering configurations which have the total symmetry of the particular electronic state being investigated. The diagonal elements of the Hamiltonian matrix are given by the total energy of the configuration

$$H_{aa} = \sum_i n_{ai} \left[h_i + \frac{1}{2} \sum_j n_{aj} \left(\langle ij | \frac{1}{r_{12}} | ij \rangle - \langle ij | \frac{1}{r_{12}} | ji \rangle \right) \right] ,$$

(8.91)

where n_{ai} ($=0$ or 1) is the occupation of orbital $\tilde{\psi}_i$ in Φ_a and the integrals $\langle ij | \frac{1}{r_{12}} | kl \rangle$ are defined as

$$\langle ij | \frac{1}{r_{12}} | kl \rangle = \int \int \frac{\tilde{\psi}_i^*(1) \tilde{\psi}_j^*(2) \tilde{\psi}_k(1) \tilde{\psi}_l(2)}{|\boldsymbol{r}_1 - \boldsymbol{r}_2|} \, d\boldsymbol{r}_1 \, d\boldsymbol{r}_2 \, d\sigma_1 \, d\sigma_2 .$$

(8.92)

Here $\tilde{\psi}_i$s are the spin orbitals and integration is over both space and spin coordinates for electrons 1 and 2.

Off-diagonal matrix element H_{ab} will be zero if two configurations Φ_a and Φ_b differ by more than two molecular spin orbitals. If Φ_a and Φ_b differ because $\tilde{\psi}_i$ is replaced by $\tilde{\psi}_k$ then

$$H_{ab} = h_{ik} + \sum_{j \neq i} n_{aj} \left(\langle ij | \frac{1}{r_{12}} | kj \rangle - \langle ij | \frac{1}{r_{12}} | jk \rangle \right) .$$

(8.93)

If Φ_a and Φ_b differ because $\tilde{\psi}_i$ is replaced by $\tilde{\psi}_k$ and $\tilde{\psi}_j$ by $\tilde{\psi}_l$, then

$$H_{ab} = \langle ij | \frac{1}{r_{12}} | kl \rangle - \langle ij | \frac{1}{r_{12}} | lk \rangle .$$

(8.94)

If two determinants differ by three or more $\tilde{\psi}_i$s, $H_{ab} = 0$.

The full CI method represents the most complete treatment possible within the limitations imposed by the basis set. As the basis set, M, becomes more complete, that is, $M \to \infty$, the result of a full CI treatment will approach the exact solution of the electronic part of the Schrödinger equation. The full CI method is well-defined, size-consistent and variational. It is, however, not practical except for very small systems because of the very large number of substituted determinants, the total number of which is $2M!/[n!(2M-n)!]$. CI calculations can be greatly simplified if they are carried out in terms of configurations, rather than Slater determinants. For example, in a calculation on Ne by *Viers* et al. [8.45], 8392 distinct Slater determinants contributed to the ground state with ^{1}S symmetry. However, only 434 configurations of ^{1}S symmetry can be constructed as linear combinations of the 8392 determinants and CI calculations for these 434 configurations was relatively straightforward. Two methods exist for the construction of symmetry adapted configurations. These involve (a) the use of projection operators [8.39, 40] and (b) the direct diagonalization of the appropriate symmetry operators (e.g. S^2) [8.46].

a) Limited configuration interaction conditions. The most straightforward way of limiting CI calculations is to follow the procedure:

(a) solve the Hartree-Fock equations to obtain an orthonormal set of HF and other (virtual) orbitals,

(b) generate the different configurations by substitution of various HF orbitals by appropriate virtual orbitals and

(c) to truncate the series of the CI expansion at a given level of orbital substitution.

Inclusion of single substitution only, termed as CIS, is given by

$$\Psi_{\text{CIS}} = A_0 \Phi_0 + \sum_{ia} A_i^a \Phi_i^a , \tag{8.95}$$

where Φ_i^a denotes a Slater determinant obtained by substituting occupied $\tilde{\psi}_i$ with virtual $\tilde{\psi}_a$ in the HF solution Φ_0. Such a truncation would lead to no improvement relative to the HF wavefunction or energy since, by Brillouin's theorem, all the matrix elements of the Hamiltonian between Φ_0 and singly-substituted configuration Φ_i^a would identically vanish. In general, the simplest procedure to have any effect on the calculated energy is limited to double substitution only and is termed CID

$$\Psi_{\text{CID}} = A_0 \phi_0 + \frac{1}{4} \sum_{ijab} A_{ij}^{ab} \Phi_{ij}^{ab} . \tag{8.96}$$

This is an important practical procedure in which two major computational tasks are involved. First is a transformation of two electron integrals $\langle \alpha\beta|\gamma\delta \rangle$ into corresponding integrals $\langle ij|\frac{1}{r_{12}}|kl \rangle$ with HF spin orbitals $\tilde{\psi}_i$s replacing the basis functions ϕ_αs. The second is the determination of the lowest (or lowest few) eigenenergies and the associated wavefunction coefficients. Both tasks are computationally significant, and considerable effort has been expended towards the development of efficient algorithms [8.47, 48].

At a slightly higher level of theory, both single and double substitutions can be included in the configuration interaction treatment. The model is termed CISD. The trial wavefunction is given by

$$\Psi_{\text{CISD}} = A_0 \Phi_0 + \sum_{ia} A_i^a \Phi_i^a + \frac{1}{4} \sum_{ijab} A_{ij}^{ab} \Phi_{ij}^{ab} . \tag{8.97}$$

Here all the coefficients are varied to minimize the expectation value of the energy. Although the single substitutions do not contribute by themselves (in CIS), they do contribute in CISD since there are non-zero matrix elements of the Hamiltonian between singly and doubly substituted determinants. However, since the participation is indirect, the energy lowering due to inclusion of single substitution is considerably less than due to double substitutions.

The most serious deficiency of the CID and CISD limited configuration interaction methods is that they fail to satisfy the size consistency condition. When a method is applied to an assembly of isolated molecules, the results should be the sum of the energies calculated by applying the same method to the molecules individually. This is known as the size consistency condition. If this condition is not satisfied, the theoretical method is unlikely to provide an adequate description of the relative energies of molecules of different sizes.

Various attempts have been made to obtain corrections to these methods. The most commonly used of these is due to *Langhoff* and *Davidson* [8.49] and *Pople* et al. [8.50], who proposed the approximate formula

$$\Delta E_{\text{correction}} = (1 - A_0^2) \Delta E_{\text{CISD}} . \tag{8.98}$$

where ΔE_{CISD} is the correlation energy at the CISD level and A_0 is the coefficient of the Hartree-Fock function in the CISD expression. This corrects a major part of the discrepancy. However, the total energy is still not precisely size consistent.

b) Coupled-cluster methods. To ensure that the size consistency condition is satisfied, CI techniques such as coupled cluster methods have been developed which are based upon an exponential wavefunction

$$\Psi_{\text{cc}} = \exp(T) \Phi_0 , \tag{8.99}$$

where T is defined to be an operator that creates configurations with various substitutions of HF orbitals

$$T = \sum_i T_i = T_1 + T_2 + T_3 + \dots , \tag{8.100}$$

where

$$T_1 \Phi_0 = \sum_{ia} A_i^a \Phi_i^a , \tag{8.101}$$

$$T_2 \Phi_0 = \frac{1}{4} \sum_{ijab} A_{ij}^{ab} \Phi_{ij}^{ab} . \tag{8.102}$$

In general,

$$T_n \Phi_0 = \frac{1}{(n!)^2} \sum_{ijk...abc...} A_{ijk...}^{abc...} \Phi_{ijk...}^{abc...} .$$ (8.103)

The coupled cluster method with double substitutions (CCD) uses a wavefunction that allows for all double substitutions, including the simultaneous substitutions that are omitted in CID theory. This method has also been referred to as the coupled pair many-electron theory (CPMET) [8.51, 52]. It satisfies size consistency conditions, but it does not have the variational property.

The CCD theory uses a trial function Ψ_{CCD} defined by

$$\Psi_{CCD} = \left[1 + T_2 + \frac{1}{2} T_2^2 \right] \Phi_0$$

$$= \Phi_0 + \frac{1}{4} \sum_{ijab} A_{ij}^{ab} \Phi_{ij}^{ab} + \frac{1}{32} \sum_{ijab} \sum_{klcd} A_{ij}^{ab} A_{kl}^{cd} \Phi_{ijkl}^{abcd} .$$ (8.104)

Explicit equations for the A coefficients are obtained by taking the function

$$(H - E)\Psi_{CCD} \equiv 0$$ (8.105)

and requiring that its projection on Φ_0 and all Φ_{ij}^{ab} is zero. Thus

$$\langle \Phi_0 | (H - E) | \Psi_{CCD} \rangle = 0$$ (8.106)

and

$$\langle \Phi_{ij}^{ab} | (H - E) | \Psi_{CCD} \rangle = 0 .$$ (8.107)

These equations suffice to determine the energy E and the unknown coefficients A_{ij}^{ab}

In an augmented coupled cluster method [8.53] known as ST4CCD, the advantages of both coupled cluster and perturbation methods are combined. The method is carried out in two stages. First, a complete CCD calculation is carried out to convergence. The resulting CCD wavefunction is then used in the evaluation of the contribution of single and triple substitutions in a manner analogous to fourth order perturbation theory described in the following section.

Pople et al. [8.54] have proposed a method of CI calculations using a truncated configuration space of single and double substitutions (QCISD). The method uses a trial wavefunction Ψ_{QCISD} which is defined as

$$\Psi_{QCISD} = \left[1 + T_1 + T_2 + T_1 T_2 + \frac{1}{2} T_2^2 \right] \Phi_0 .$$ (8.108)

As in the CCD procedure the explicit equations for the A coefficients are obtained by taking the function

$$(H - E)\Psi_{QCISD} = 0$$ (8.109)

and requiring that its projection on Φ_0 and on all the values of Φ_i^a and Φ_{ij}^{ab} is zero. In the QCISD(T) version of this method, the resulting QCISD wavefunction is used in the evaluation of the contribution from triple substitutions in a procedure analogous to the ST4CCD method.

The computational task in CID, CISD, CCD, QCISD, or any of the other CI methods that follow, can be reduced considerably by limiting the set of spin orbitals that are involved in the single or double substitutions. This is most conveniently done in terms of a window in which only a set of high energy occupied and low energy virtual spin orbitals is used.

c) Möller-Plesset perturbation theory. The theory of Möller and Plesset [8.55], closely related to many-body perturbation theory, is an alternative approach to the correlation problem. Möller-Plesset models are formulated by first introducing a generalized electronic Hamiltonian, $\tilde{H}_\lambda$

$$\tilde{H}_\lambda = \tilde{H}_0 + \lambda V . \tag{8.110}$$

Here H_0 is the Hartree-Fock (HF) Hamiltonian, given as a sum of one-electron HF operators, H_i^{HF}.

$$\tilde{H}_0 = \sum H_i^{\mathrm{HF}} , \tag{8.111}$$

V is the difference between the correct Hamiltonian and the HF Hamiltonian and λ is a dimensionless parameter. Ψ_λ and E_λ may now be expanded in powers of λ according to Rayleigh-Schrödinger perturbation theory

$$\Psi_\lambda = \Psi^{(0)} + \lambda \Psi^{(1)} + \lambda^2 \Psi^{(2)} + \dots , \tag{8.112}$$

$$E_\lambda = E^{(0)} + \lambda E^{(1)} + \lambda^2 E^{(2)} + \dots . \tag{8.113}$$

In order that the Schrödinger equation be satisfied for any arbitrary value of λ, it is necessary that the coefficient of each power of λ be separably zero. The solution of the perturbation equations gives

$$E^{(1)} = \int \Psi^{(0)} V \Psi^{(0)} \, d\tau , \tag{8.114}$$

$$E^{(2)} = \int \Psi^{(0)} V \Psi^{(1)} \, d\tau , \tag{8.115}$$

$$E^{(3)} = \int \Psi^{(1)} (V - E^{(1)}) \Psi^{(0)} \, d\tau , \tag{8.116}$$

which shows that the first order wavefunction determines the energy to third order. We have

$$\Psi_0 = \Phi_0 \qquad \text{and} \tag{8.117}$$

$$E_0 = \sum_i^{\mathrm{occ.}} \varepsilon_i , \tag{8.118}$$

where Φ_0 is the Hartree-Fock wavefunction and ε_is are the orbital energies. The Möller-Plesset energy to first order is the Hartree-Fock energy; thus

$$E^{(1)} = -\frac{1}{2} \sum_{ij} \left[\langle ij | \frac{1}{r_{12}} | ij \rangle - \langle ij | \frac{1}{r_{12}} | ji \rangle \right] . \tag{8.119}$$

The first order contribution to the wavefunction is

$$\Psi^{(1)} = -\frac{1}{4} \sum_{ijab} \frac{\langle \Phi_{ij}^{ab} | V | \Phi_0 \rangle}{\varepsilon_a + \varepsilon_b - \varepsilon_i - \varepsilon_j} \Phi_{ij}^{ab} , \tag{8.120}$$

where the summation is carried out over double substitution only. The second and third order contributions to the Möller-Plesset energy are given as

$$E^{(2)} = -\frac{1}{4} \sum_{ijab} \frac{|\langle \Phi_{ij}^{ab} | V | \Phi_0 \rangle|^2}{\varepsilon_a + \varepsilon_b - \varepsilon_i - \varepsilon_j} , \tag{8.121}$$

$$E^{(3)} = \frac{1}{16} \sum_{ijab} \sum_{klcd} \frac{\langle \Phi_0 | V | \Phi_{ij}^{ab} \rangle \langle \Phi_{ij}^{ab} | V - E^{(1)} | \Phi_{kl}^{cd} \rangle \langle \Phi_{kl}^{cd} | V | \Phi_0 \rangle}{(\varepsilon_a + \varepsilon_b - \varepsilon_i - \varepsilon_j)(\varepsilon_c + \varepsilon_d - \varepsilon_k - \varepsilon_l)} . \tag{8.122}$$

In both $E^{(2)}$ and $E^{(3)}$ the summations are carried out over double substitutions only. We refer to the methods by the highest order energy term allowed, that is, truncation after second order as MP2, after third order as MP3, after fourth order as MP4, and so on. At the MP4 level of theory, single, triple and quadruple substitutions also contribute, since they have non-zero Hamiltonian matrix elements with the double substitutions.

MP2, MP3 and MP4 are well defined and satisfy the size consistency requirement, as do Möller-Plesset energy expansions terminated at any order. This follows since full CI is size consistent with the Hamiltonian H_λ for any value of λ.

8.4.4 Multiconfiguration SCF Method (MCSCF)

Solution of the eigenvalue problem will yield the optimum values of the coefficients A_{am} so as to give minimum energy. However, the wavefunctions are not determined in a completely variational manner unless the forms of the molecular orbitals ψ_i have also been varied to minimize the total energy. For a specified set of configurations, the multiconfiguration HF wavefunction is the best wavefunction that can be obtained by simultaneously varying both the orbitals ψ_i and the CI coefficients A_{am}s.

In principle all CI calculations should be of the MCSCF variety. Unfortunately, solution of the MCSCF equations is very difficult and the calculations have to be limited to a small number (10–100) of configurations. Hence, an effective guide for the construction of higher order terms in building up an MCSCF wavefunction is necessary. In doing so, it is very important to distinguish between two types of excitations, the 'intra-atomic' and the 'inter-atomic' type. Those excitations which improve the wavefunctions of isolated atoms beyond the corresponding HF functions are interatomic and those which lead to an improved description of the total wavefunction when the atoms approach each other are the intra-atomic type. MCSCF calculations which only consider the inter-atomic type of excitations lead to what are known as 'optimized valence configuration' (OVC) wavefunctions. The OVC method usually consists of a small number of configurations and is found to predict accurate interaction curves for diatomic and polyatomic molecules [8.56].

a) **The density matrix and natural orbitals.** The concept of natural orbitals, introduced by *Löwdin* [8.57], provides a practical approach to the calculation of CI wavefunctions with optimized orbitals including a large number of configurations. For any electronic wavefunction Ψ the one-electron density function, ρ, is given by

$$\rho(r_1, \sigma_1) = n \int |\Psi(r_1, r_2, \ldots, r_n, \sigma_1, \sigma_2, \ldots, \sigma_n)|^2 \, d\tau_2 \, d\tau_3 \ldots d\tau_n , \quad (8.123)$$

in which the integration is over the spin and space coordinates of all but the first electron. For a CI wavefunction of the form $\Psi = \sum_a A_a \Phi_a$, constructed from orbitals ϕ, ρ is of the form

$$\rho(r, \sigma) = \sum_i \sum_j D_{ij} \tilde{\psi}_i^*(r, \sigma) \tilde{\psi}_j(r, \sigma) , \quad (8.124)$$

where D_{ij} form the density matrix.

$$D_{ii} = \sum_a A_a^2 n_{ai} , \quad (8.125)$$

and D_{ij}, $i \neq j$, arises from the off-diagonal term between Φ_a and Φ_b when they differ from each other only by a single excitation, $i \rightarrow j$. If ψ_is are chosen in such a way that the density matrix D is diagonal, the electron density reduces to the form

$$\rho(r, \sigma) = \sum_i b_i |\tilde{\psi}_i(r, \sigma)|^2 . \quad (8.126)$$

This choice of molecular orbitals are called natural orbitals (NO) and can be obtained by a similarity transformation diagonalizing the matrix D. The coefficients b_i are called occupation numbers and indicate the degree of importance of each spin orbital.

One reason for the importance of natural orbitals is that they give the most rapidly converging CI expansion as only those configurations built up from NOs with large occupation numbers are considered significant.

b) **The iterative natural orbital method.** A powerful method for carrying out large scale CI calculations with optimized orbitals is the iterative natural orbital (INO) method [8.58]. In its simplest form, a set of configurations is chosen, a CI calculation is carried out and the density matrix is diagonalized to give the natural orbitals for the wavefunction. Using the same set of configurations, the calculation is repeated in terms of the natural orbitals found from the first calculation or iteration. The iterative procedure is repeated until the energy reaches a minimum.

The main feature of the INO method is that one must carry out several (at least 3–4) CI calculations before the final result is obtained. However, the final energy is likely to be lower than would be possible with any single CI calculation which is less than complete.

c) **The complete active space self consistent field (CASSCF) method.** The most satisfactory MCSCF wavefunction includes all possible configurations in a limited set of orbitals known as the 'active space'. This concept was introduced by *Ruedenberg* and coworkers as the FORS (full optimized reaction space) method [8.59, 60]. They pointed out that, in general, no simple configuration selection scheme yields a balanced description of reactive systems. However, it is possible to select a set of 'active' orbitals and consider correlation only in this space. Thus, once the inactive and active orbitals are chosen, the wavefunction is completely specified.

Such a model leads to simplifications in the computational procedures used to obtain optimal orbitals and CI coefficients. The major technical difficulty inherent in the CASSCF model is the size of the complete CI expansion, N_{CAS}, which is given by the Weyl formula

$$
N_{\mathrm{CAS}} = \frac{2S+1}{n+1} \left(\begin{array}{c} n+1 \\ \frac{N}{2} - S \end{array} \right) \left(\begin{array}{c} n+1 \\ \frac{N}{2} + S + 1 \end{array} \right) ,
\tag{8.127}
$$

where n is the number of active orbitals and N is the number of active electrons, and the total spin is S. Obviously N_{CAS} increases strongly as a function of the size of the active orbital space. The size of the CI expansion used in a CASSCF calculation is almost always much larger than those normally used in the earlier applications of the MCSCF model. It was only when the graphical unitary group approach [8.61–65] for full CI calculations was invented in the years 1975–78 that an efficient computational procedure for CASSCF calculations could be developed.

8.4.5 Spin-Coupled Valence Bond Theory

In the valence bond (VB) theory the wavefunctions for the electrons in a molecule are constructed directly from the wavefunctions of the constituent atoms, which are non-orthogonal to each other. This non-orthogonality lies at the heart of the valence bond approach.

Non-orthogonal orbitals are almost always localized. As a result we obtain a simple picture within which the tendency of electrons to avoid one another is preserved while, at the same time, still allowing for the necessary constructive interference between them. Technically, however, the calculation of matrix elements of the Hamiltonian, and of other operators between wavefunctions constructed from non-orthogonal orbitals, is difficult.

While the simplest VB wavefunction yields the correct qualitative picture of a particular molecular state, its quantitative accuracy is rather limited. In order to improve matters, further structures must be included, formed from wavefunctions of the constituent atoms in excited states. Such expansion of the molecular wavefunction begins to resemble the MO-CI expansion, particularly in its slow convergence properties. As a result, almost all of the original clarity and insight of the valence bond approach tends to become somewhat obscured.

In their seminal paper, *Coulson* and *Fischer* [8.66] showed that the wavefunction for H_2, which consists of a linear combination of the Heitler-London covalent plus ionic structures, is equivalent to a single covalent structure formed from orbitals which are deformed atomic orbitals. In other words, the disconcerting intervention of so many ionic structures in the general case is due to the fact that the contributing atomic or fragment states deform on formation of the molecule. Thus, it would be appropriate to incorporate into VB theory the idea of expanding the orbitals in basis sets with complete optimization of the expansion coefficients in order to allow the constituent orbitals to undergo a general deformation from the isolated atom form.

Consider a wavefunction for an n-electron system in which each electron is described by a distinct spatial orbital $\phi_\mu (\mu = 1, 2, \ldots, n)$. These orbitals are, in general, all non-orthogonal. The spins of the individual electrons are coupled together so as to form a definite overall resultant spin S. The form of the wavefunction, referred to as a 'spin-coupled' wavefunction, is

$$\Psi_{S,M} = (n!)^{-1/2} \sum_{k=1}^{f_S^n} C_{Sk} \mathcal{A}(\phi_1, \phi_2, \ldots, \phi_n \Theta_{S,M;k}^n) . \tag{8.128}$$

In this equation, $\Theta_{S,M;k}^n$ are the n-electron spin functions, eigenfunctions of $\hat{S}^2$ and of $\hat{S}_z$, with eigenvalues $S(S+1)$ and M, respectively and $\mathcal{A}$ is the antisymmetrizing operator. The index k denotes the mode of coupling between the individual electron spins, the total number of allowed couplings being given by

$$f_S^n = \frac{(2S+1)n!}{(\frac{1}{2}n + S + 1)!(\frac{1}{2}n - S)!} . \tag{8.129}$$

One way of visualising the different possible spin couplings is by means of the branching diagram which provides a graphical representation of the construction of $\Theta_{S,M;k}^n$ by successively coupling the electron spins according to the rules for coupling angular momenta [8.67]. The set of spin functions constructed in this way is normalized and orthogonal, and is known as the Yamanouchi-Kotani basis.

The occupied orbitals are expressed as a linear combination of basis functions, χ_p

$$\phi_\mu(\mathbf{r}) = \sum_{p=1}^m C_{\mu p} \chi_p(\mathbf{r}) . \tag{8.130}$$

The functions χ_p are almost always a set of atomic functions, usually of Gaussian or Slater form, centred on all the nuclei of the molecule. The expansion coefficients $C_{\mu p}$ are fully optimized simultaneously with the spin coupling coefficients C_{Sk} using a very efficient second-order procedure referred to as the 'stabilized Newton-Raphson' method [8.68].

At convergence for each occupied ϕ_μ a set of excited state, or virtual, orbitals $\phi_\mu^{(i)}$, $i = 1, 2, \ldots, m$, is determined. Each set is orthonormal, but is not orthogonal, to the virtual orbitals derived from the other occupied orbitals. If $\mu = \nu$,

$$\langle \Phi_\mu^{(i)} | \Phi_\nu^{(j)} \rangle \delta_{ij} \tag{8.131}$$

and, otherwise,

$$\langle \Phi_\mu^{(i)} | \Phi_\nu^{(j)} \rangle = \Delta_{ij}^{\mu\nu} . \tag{8.132}$$

Excited spin-coupled configurations are obtained by replacing one, two, or more occupied orbitals by these virtual orbitals. The total wavefunctions are in the form of a linear combination of spin-coupled functions and the excited configurations

$$\Psi_{S;M} \equiv A_0 [\phi_1 \phi_2 \dots \phi_n] + \sum_{i_1 i_2 i_3 \dots i_n} A_{i_1 i_2 i_3 \dots i_n} [\phi_1^{(i_1)} \phi_2^{(i_2)} \dots \phi_n^{(i_n)}] . \tag{8.133}$$

The final energies of the various states and the corresponding eigenfunctions are determined by forming the matrix of the Hamiltonian over the spin-coupled structures and diagonalizing.

The spin-coupled valence bond method has recently been developed to a very high level of sophistication; details of the method and its applications to various molecular systems can be found in a succinct review paper by *Gerratt* et al. [8.69].

8.5 A Glimpse into the Real World of Multiply Charged Molecules: Ambiguities and Controversies

The experimental and theoretical difficulties associated with studies of multiply charged molecular ions have already been briefly referred to at the beginning of this chapter. Manifestations of these difficulties can be found in the numerous ambiguities and controversies that continue to dog investigations of not only relatively complex polyatomic species but also simple diatomic and triatomic systems. By way of illustration we focus attention on a few selected examples which should convince the reader of the continuing necessity for developing new experimental and theoretical techniques in order to gain a deeper and more reliable insight into the physical properties of this class of molecular species.

8.5.1 Diatomic Ions

a) **Metastable He_2^{2+} ions.** Results of Coulomb-explosion experiments (Sect. 8.2.8) on singly and doubly charged helium dimer ions, carried out by *Balkacem* et al. [8.15], have provided experimental evidence for the ultrashort bond length (0.75 Å) of He_2^{2+} and have enabled the placing of a lower limit of 5 μs on the lifetime of this doubly charged molecule. However, despite the structural simplicity of this two-electron system, quantum-chemical predictions of the properties of He_2^{2+} are not in total accord with the experimental observations of Balkacem et al. We present below some results of molecular orbital calculations of the potential energy (PE) curves of low-lying electronic states of He_2^{2+} which indicate that experimental data on the

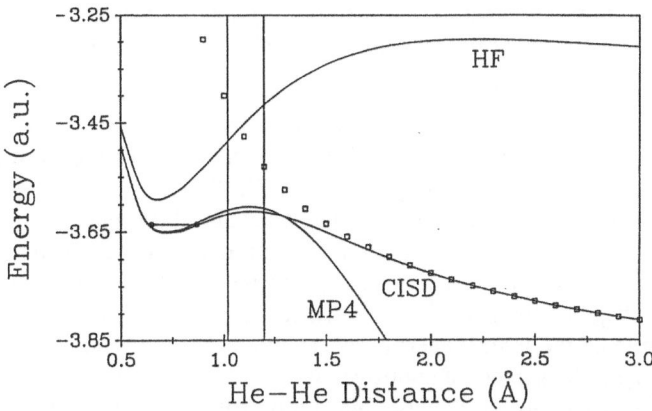

Fig. 8.5. Potential energy curves of He_2^{2+}. $X^1 \Sigma_g^+$ states are shown by solid lines and the lowest $^3 \Sigma_g^+$ state by the squares. Vertical lines indicate the Franck-Condon region from the $\nu = 0$ level of $X^2 \Sigma_g^+$ He_2^+

metastability of this di-cation, as reported by *Balkacem* et al. [8.15] and, in an earlier charge stripping experiment conducted by *Guilhaus* et al. [8.70], are not compatible with information deduced from contemporary quantum-chemical computations.

All-electron, ab initio configuration interaction calculations of PE curves of singlet and triplet states of He_2^{2+} have been carried out using the Hartree-Fock (HF) technique [8.71]; the basis functions used had double zeta representation of 1s, 2s and 2p orbitals, with appropriate splittings of the wavefunctions given by *Poirier* et al. [8.72]. Electron correlation energies were initially evaluated using Möller-Plesset perturbation theory up to fourth order (UMP4). Full configuration interaction (CI) calculations were also carried out, taking into account all single and double excitations (CISD).

Computed PE curves for the ground $^1 \Sigma_g^+$ and the lowest $^3 \Sigma_g^+$ state are shown in Fig. 8.5. In the case of the ground state curves, all three calculations (HF without CI, MP4 and CISD) yield an equilibrium He-He distance of 0.75 Å, in excellent accord with experimental data. However, the HF calculation is clearly unreliable as it yields a grossly incorrect $He^+ + He^+$ dissociation limit and an unrealistically large (11.24 eV) well depth. The MP4 and the CISD curves coincide at low values of internuclear separation and yield well depths of 1.04 eV and 0.91 eV, respectively. However, at internuclear separations larger than 1 Å, the crossing of the HF curve by the triplet state suggests that the first excited state of $^1 \Sigma$ symmetry lies so close to the ground state that a perturbative MP4 type of treatment ceases to be effective. The CISD curve, which represents a *complete* MO calculation for the ground state, is in qualitative agreement with the earlier computations. As is evident from Fig. 8.5, no He_2^{2+} vibrational level having μs lifetimes will be accessible in vertical transitions from the He_2^+ ground state. The first excited singlet state of the di-cation also possesses a shallow minimum, but it lies at an He-He distance of 3.6 Å, also outside the region vertically accessible from the He_2^+ ground state. As the He_2^+-He_2^{2+} transition in a charge stripping collision follows the Franck-Condon principle, it is clear that accord between experiment and contemporary MO calculations is still not total.

b) Quantal description of metastable CO^{2+}. Ambiguities surrounding experimental and theoretical studies of doubly charged molecular ions, even in the case of relatively simple species such as CH^{2+}, have already been referred to in Chap. 6. In this section we focus attention on another simple species, CO^{2+}.

Metastable CO^{2+} ions were the first multiply charged molecular species to be experimentally observed [8.2] in 1932. Since then, a large number of experimentalists and theoreticians have studied this ion, and perhaps the only point of agreement to date between all investigators is that CO^{2+} is metastable for at least a few μs. The quantal identification of the lowest energy metastable electronic state of this molecule, however, remains the subject of controversy.

The earliest theoretical attempt at constructing the potential energy curve of CO^{2+} was by *Hurley* [8.73], who devised a semi-empirical method of utilizing the quantum-mechanical virial theorem to derive the potential energy curve of the doubly charged ion from the self-consistent-field type of curves of the isoelectronic BN radical; $^3\Pi$ symmetry was assigned to the ground state curve so derived. However, subsequent studies [8.74] of photon emission from excited states of CO^+ product ions produced in electron capture collisions between CO^{2+} and H_2 yielded results which could not be reconciled with Hurley's potential function. Subsequent to Hurley's calculations, a number of modern ab initio quantum-chemical computations have also indicated that $^3\Pi$ is the lowest energy metastable state of CO^{2+} (see [8.75, 76] for a brief overview and compilation of pertinent references). However, there exists an alternative body of theoretical information [8.77–80] which suggests that the lowest energy metastable CO^{2+} state possesses $^1\Sigma^+$ symmetry. Despite this lack of clarity, the results of almost all collision experiments involving CO^{2+} ions (electron capture studies, collision-induced dissociation, excitation experiments) have continued to be interpreted on the basis that $^3\Pi$ is the lowest energy metastable state. Furthermore, some of these experiments have also yielded conflicting information on whether or not CO^{2+} also possesses a metastable, electronically excited state [8.81].

Recently we have carried out a detailed quantum-chemical study of this molecular ion using a number of different methodologies for calculation of correlation energy available on the GAUSSIAN 88 series of programs. The basis set used in these calculations was of 6-311G** type, containing 4s, 3p and 1d functions for both C and O, with a total of 36 atomic orbitals. The results of total energy for the internuclear distance 1.25 Å using various CI methods are shown in Table 8.1.

Using perturbative CI procedures it becomes clear that the correlation contribution to the total energy is high (particularly for the $^1\Sigma$ state) and, moreover, that this contribution oscillates as higher orders of perturbation are included. This indicates that there are many low-lying $^1\Sigma$ states close to the ground state and perturbation calculations carried out at many orders higher than at the MP4 level would be necessary in order to obtain a reliable energy value for the $^1\Sigma$ state. The relative positions of $^1\Sigma^+$ and $^3\Pi$ also interchanged as higher order perturbations were included

Table 8.1. Total energies of $^1\Sigma^+$ and $^3\Pi$ states of CO^{2+} at an internuclear distance of 1.125 Å

	Energy (H)		Time taken (h:min:s)	
	$^1\Sigma^+$	$^3\Pi$	$^1\Sigma^+$	$^3\Pi$
SCF	−111.181457	−111.342398		
MP2	−111.577851	−111.566170		
MP3	−111.491575	−111.572448		
MP4	−111.653904	−111.597167	0:28:48	1:06:32
CID	−111.529182	−111.578718	0:40:59	1:42:55
CISD	−111.559154	−111.593346	0:59:48	2:22:19
CCD	−111.529424	−111.576254		
ST4CCD	−111.590750	−111.597510	1:42:59	3:28:36
QCISD	−111.569138	−111.591996		
QCISD(T)	−111.588481	−111.602670	2:55:17	4:02:03
QCISD(w)	−111.383967	−111.417764	0:58:52	2:10:55

(see Table 8.1). It is, consequently, clear that recourse has to be made to variational methods of incorporating CI effects in order to generate potential energy functions of low-lying states of CO^{2+}.

As indicated in Table 8.1, various variational methods (CID, CISD, CCD, ST4CCD, QCISD, QCISD(T)) also show significant fluctuation in the magnitudes of the correlation contributions. Since the QCISD(T) method is the most sophisticated of the ones used, it may not be unreasonable to assume that the result obtained using this method yields the most reliable of the energy values shown in Table 8.1. It is of interest to note that all the variational methods consistently indicate that $^3\Pi$ is lower in energy than $^1\Sigma$ at the internuclear distance of 1.25 Å. In order to correlate this with the available experimental information it becomes necessary to carry out energy calculations over a range of internuclear distances, in other words, to calculate potential energy curves. Results of such a calculation, carried out at the QCISD(T) level, for the $^3\Pi$, $^1\Sigma$, $^3\Sigma^-$ and $^3\Sigma^+$ states of CO^{2+} are shown in Fig. 8.6.

In assessing the quantal identity of the lowest CO^{2+} electronic state which is accessed in charge stripping, electron impact or double electron capture experiments, cognisance has to be taken of the Franck-Condon principle; the lowest energy doubly charged state which will be accessed will be that which has the maximum overlap in the Franck-Condon region with the ground electronic state of CO^+ (in the case of charge stripping experiments), or neutral CO (in other experiments). In both CO and CO^+, the vertical transition to CO^{2+} occurs with maximum probability at an internuclear separation of 1.1 Å. It is clear from the curves shown in Fig. 8.6 that at an internuclear distance of 1.1 Å the $^3\Pi$ state cannot yield metastable CO^{2+} ions due to rapid predissociation into $C^+ + O^+$ induced by the curve crossing between the $^3\Pi$ and $^3\Sigma^-$ states. The $^1\Sigma^+$ state, on the other hand, is expected to present a substantial barrier towards dissociation. Even the higher vibrational levels in this state are likely to be fairly long-lived as a crossing with the $^3\Sigma^-$ is forbidden on two

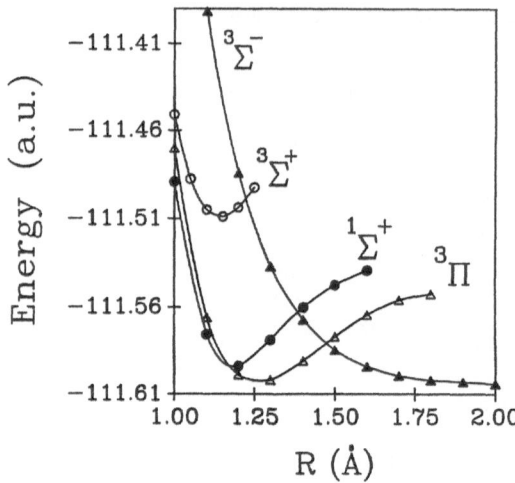

Fig. 8.6. Potential energy curves of low-lying states of CO^{2+} computed at the QCISD(T) level

grounds: firstly, dipole selection rules prohibit a crossing between singlet and triplet states and, secondly, even if spin-orbit coupling is considered to be substantial so that the above rule can be violated, the spin-orbit operator will be ineffective in the case of a two-electron transition such as $^1\Sigma^+$-$^3\Sigma^-$.

It is pertinent to draw attention to the time required to carry out such calculations. On a MicroVAX II computer the QCISD(T) method required 3 hours for calculating a single point on the $^1\Sigma^+$ curve, and 4 hours in the case of a point on the $^3\Pi$ curve. For larger molecules, or for calculations on diatomics carried out with larger basis sets, such a method would clearly be impractical. We have therefore carried out a series of CI calculations on CO^{2+} at the QCISD(T) level using a 'window' which includes only the valence orbitals of C and O. The results of these computations, which are indicated in the last column of Table 8.1, show a significant reduction of correlation energy compared to the calculations carried out without a window; the effect of carrying out these calculations at various internuclear distances has also been studied. Although the global features describing the three low-lying states of CO^{2+} are the same whether a window is used or not, there is a significant difference in the relative positions of the curves in the two cases, the heights of the potential barriers and the crossing points between the curves. This is contrary to the expectation that consideration of valence orbitals in CI calculations ought to be sufficient to predict accurate interaction curves for diatomic and polyatomic molecules. Indeed, this is a basic assumption in most MCSCF calculations using a discrete active space.

8.5.2 Triatomic Ions

Ambiguities surrounding the quantal identification of metastable states present only one manifestation of the difficulties encountered in the real world of multiply charged molecules. As an example we focus attention on another type of difficulty by considering electron capture reactions by CS_2^{2+} and CS_2^{3+} ions.

Charge transfer reactions have attracted considerable experimental as well as theoretical attention, mainly for single electron capture processes by multiply charged atomic ions. Experiments covering a wide energy range have provided data that indicate that the transfer of charge in such collisions occurs with high probability in moderately exoergic reactions. Such reactions can be readily interpreted in terms of avoided crossings of adiabatic potential energy curves describing the initial and final pseudomolecular systems over a range of internuclear separations (R_c) between 3 and 10 Å (see Chap. 3). This has given rise to the concept of a 'reaction window' within which selective electron capture takes place into only a limited number of excited electronic states of the product ions. Relative cross sections have been measured [8.83] for a large number of reactant ground state to product ground state electron capture reactions involving CS_2^{2+} projectile ions and a variety of atomic and molecular target gases, using the technique of ion translational energy spectrometry; the variation of cross section with the internuclear separation where the avoided crossing occurs (Fig. 8.7) indicates that the relative cross section has a maximum in the region of 6–7 Å and falls off at lower and higher values of R_c.

As in the case of multiply charged atomic ions, the functional form shown in Fig. 8.7 is qualitatively understood by considering the variation with internuclear distance of the interaction matrix element H_{12}, which couples the incoming and outgoing potential energy functions which define the trajectories followed by the reactants and products. At large values of internuclear distance, R_c, at which a curve crossing occurs, H_{12} decreases exponentially with R_c; consequently, the cross section for transitions between the incoming and outgoing potential energy curves will also decrease at larger values of internuclear separation. On the other hand, at shorter internuclear distances, that is, for low values of R_c, the cross section falls off as the magnitude of the classical cross section ($\pi R_c^2 / 2$) decreases. Experimental data [8.82, 83] indicates that the validity of the reaction window picture also extends to doubly

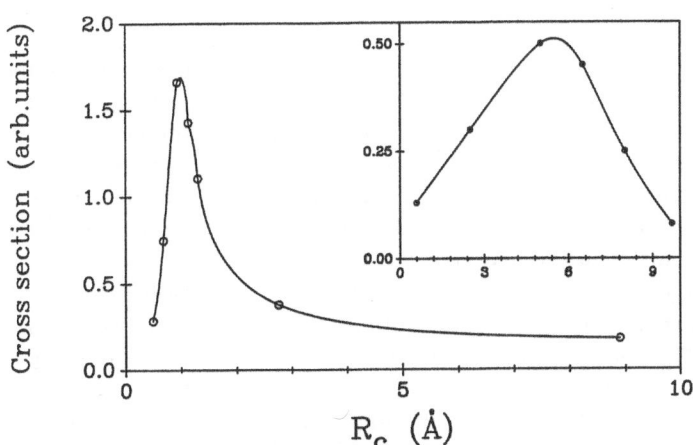

Fig. 8.7. An anomalous reaction window in the dependence of the relative cross for single electron capture by CS_2^{3+} with avoided crossing distance, R_c. Inset shows the usual type of window obtained in the case of electron capture by CS_2^{2+} ions

charged molecular ions. However, in the case of triply charged ions, CS_2^{3+}, the functional dependence of the ground state to ground state electron capture cross section with R_c behaves in an anomalous fashion (Fig. 8.7). The conventional reaction window picture appears to require modification in that a considerably narrower window is observed experimentally, with a maximum cross section occurring for those reactions that result from an avoided crossing at surprisingly short internuclear distances, in the region of 1–2 Å.

The anomalous reaction window obtained for triply charged CS_2 molecules remains unexplained. However, in discussing this problem, *Mathur* et al. [8.82] have drawn attention to analogous electron capture cross section measurements involving highly charged Kr ions (Kr^{n+}, $n = 10$–25) colliding with He. In these collisions [8.82], the measured cross sections do not have a maximum at any value of R_c but appear, on the contrary, to follow the classical $\pi R_c^2/2$ behaviour. It was postulated that, in the case of such highly charged atomic ions there exists a large density of electronic states in the product ion channel; electron capture may therefore occur at large internuclear distances and into large principal quantum number orbitals of the product ion around which a number of levels and sublevels are located. In such circumstances, a single avoided crossing picture may break down and several crossings may have to be taken into account. Even if the transition probability is small for a single crossing, the total electron capture cross section may become large because of multiple crossing effects. Consequently, the capture cross section would be expected to rise with R_c, although why it should tend to approach the maximum cross section, πR_c^2, is not immediately obvious.

Of course, electron capture reactions involving CS_2^{3+} projectiles simulate those involving Kr^{n+} ($n = 10$–25) because both systems possess a large density of states in the product ion channel. In both systems, the reaction window picture appears to require modification, although the precise nature of the modification remains unclear.

8.5.3 Polyatomic Ions

a) **Doubly charged methane.** Here we present examples of ambiguities concerning doubly charged polyatomic molecules which concern the formation energetics and structural properties of such species. As in all other cases, including the instances involving diatomic and triatomic species cited above, the interplay between theory and experiment continues to be strong even for somewhat larger molecular systems. We consider here doubly charged ions of a highly symmetrical molecule, CH_4, and those of larger linear molecules containing a number of C-H bonds.

The doubly charged molecular ion of methane, CH_4^{2+}, has been the subject of many investigations [8.84]. However, in spite of the relatively large body of experimental as well as theoretical data that has now been accumulated, there remains a glaring ambiguity concerning the experimentally determined value of a fundamental property of this doubly charged molecular ion, namely the energy required to form it, and the theoretically determined geometrical structure of the lowest-energy metastable state of the ion.

Conventional electron impact ionization experiments have failed to produce stable or metastable CH_4^{2+} ions. Experimental information on this species is from Auger spectroscopy experiments and from photoionization and ion impact studies. Whilst Auger spectroscopy experiments yield the highest values of double ionization energy, IE_2, (40.7 eV [8.23], 42 eV [8.85]), photoionization measurements [8.86–88] yield values which range from 33.9 eV to 36.5 eV. Ion impact experiments, on the other hand, yield widely divergent results. Double charge transfer experiments yield IE_2 values ranging from 35.6 eV for a singlet state [8.84] to 38.6 eV for a triplet state [8.89] whilst charge stripping experiments [8.84] produce the most divergent range (30.6 eV to 38.9 eV).

It is clear that different types of experiment access different electronic states of CH_4^{2+}. Recourse to theoretical information ought, in principle, to enable us to unscramble the apparently conflicting experimental data. Unfortunately, the theoretical information itself appears to be somewhat ambiguous. The ground electronic state of neutral methane has tetrahedral geometry and, in T_d point-group symmetry, has an electronic configuration and symmetry $1a_1^2 2a_1^2 1t_2^6,^1A_1$. Removing two electrons from the outermost $1t_2$ orbital gives rise to a multiplet of possible CH_4^{2+} electronic states: $^3T_1$, ^1E, $^1T_2$ and $^1A_1$. The energies of several of these states have been calculated using various theoretical techniques. To illustrate the type of problems encountered in reconciling theoretical information with experimental data (obtained using experimental and theoretical methods described in Sects. 8.2–8.4), we consider below the results of one recent theoretical exercise.

Ab initio molecular orbital calculations were performed using the GAUSSIAN 86 set of programs. Configuration interaction effects were included using Möller-Plesset correlation energy calculations at the second order perturbation level (MP2) and the fourth order perturbation level, including single, double, triple and quadruple excitations (MP4/SDTQ). All the initial geometry optimizations were performed using the MP2 procedure while the MP4/SDTQ procedure was finally used to calculate total energies (HF+CI) for the optimized geometries. GTO basis sets used were of 6-311G** type, consisting of 4s, 3p and 1d functions for carbon and 3s and 1p functions for hydrogen.

The geometry optimization process for neutral CH_4 correctly yielded minimum energy for T_d symmetry with the computed C-H equilibrium distance being 1.092 Å, in excellent accord with the experimental value of 1.092 Å. Using this geometry, MP4/SDTQ calculations were also performed for CH_4^+, CH_4^{2+} (S = 0) and CH_4^{2+} (S = 1) ions. Geometry optimization for CH_3^+ indicated D_{3h} symmetry with a C-H bond distance of 1.091 Å and a minimum energy which was found to be much lower than for either of the spin states of CH_4^{2+}. Furthermore, the computed dissociation pathway of CH_4^{2+} to CH_3^+ along C_{3v} symmetry was found to have no potential barrier, indicating that doubly charged methane ions produced in the double charge transfer experiments instantaneously dissociate to $CH_3^+ + H^+$.

With reference to the charge stripping experiments that have been carried out [8.84], we consider the possibility that the projectile CH_4^+ ions may have relaxed to a minimum energy configuration rather than maintaining the same T_d geometry as in neutral CH_4. The multivariable potential energy surface of CH_4 possesses a

Table 8.2. MP4SDTQ energies (in atomic units) for CH_4^+ and CH_4^{2+} ions for various symmetries. Vertical ionization energies (in eV) are given in brackets

Symmetry	Energy [a.u.]		
	CH_4^+	CH_4^{2+} (S=0)	CH_4^{2+} (S=1)
D_{4h}	−39.902085	−39.202988 (19.0)	−38.976808 (25.2)
D_{2d}	−39.936102	−39.136184 (21.8)	−39.034063 (24.5)
C_{3v}	−39.921580	−39.046981 (23.8)	−39.041956 (23.9)

number of energy minima with different point group symmetries. In a recent study, we have considered all the possible symmetries for which minima occur; geometry optimization for various symmetries gave the following results.

For D_{4h} symmetry (square planar arrangement) the C-H equilibrium distance was 1.123 Å. For D_{2d} symmetry, the corresponding C-H distance was 1.118 Å, while the angle between the symmetry axis and the C-H bond was 70.8°. In the case of C_{3v} symmetry, the equilibrium distance between the axial hydrogen, H_{ax}, and carbon is 1.321 Å while the distance between equitorial hydrogen, H_{eq}, and carbon is 1.095 Å; the angle H_{ax}-C-H_{eq} has an optimum value of 94.6°. The energies of the singly charged ion for these optimized geometries, shown in Table 8.2, agree well with earlier calculations [8.90] except for the minimum energy configuration with C_{2v} symmetry. Our attempt to locate local minima within this symmetry always yielded the equilibrium geometry with D_{2d} symmetry, in accord with the CASSCF calculations of *Siegbahn* [8.91], who also obtained minimum energy for CH_4^+ ions with D_{2d} symmetry.

Our calculations indicate that there are only two possible geometries for which the CH_4^{2+} ion is metastable; local minima are obtained for D_{4h} symmetry with a C-H distance of 1.177 Å with S = 0 and for D_{2d} symmetry with S = 1 and a C-H distance of 1.223 Å, with the angle between the symmetry axis and the C-H bond being 35.76°. The computed energies for the doubly charged ion are also shown in Table 8.2. Our findings indicate that the minimum energy geometry for CH_4^{2+} in D_{2d} symmetry is so significantly different from any of the three symmetry states of CH_4^+ that vertical ionization is unlikely to lead to a doubly charged state which will relax to a metastable (experimentally detectable) doubly charged state with D_{2d} symmetry. Hence, the most probable metastable states of CH_4^{2+} which have been observed in charge stripping experiments should be spin singlets with D_{4h} symmetry.

The potential energy contour has also been evaluated for dissociation of CH_4^{2+} (D_{4h}) to CH_3^+ (D_{3h}) along the C_{2v} symmetry pathway. Calculations involved optimization of the bond distances R_1 and R_2 (see Fig. 8.8) and bond angle θ for every value of R_3. Plots of R_1 against X, θ against X and E against X are shown in Fig. 8.9 where $X = (R_3 - R_1)/(R_3 + R_1)$. At $X = 0$, $R_1 = R_2 = R_3 = 1.177$ Å and

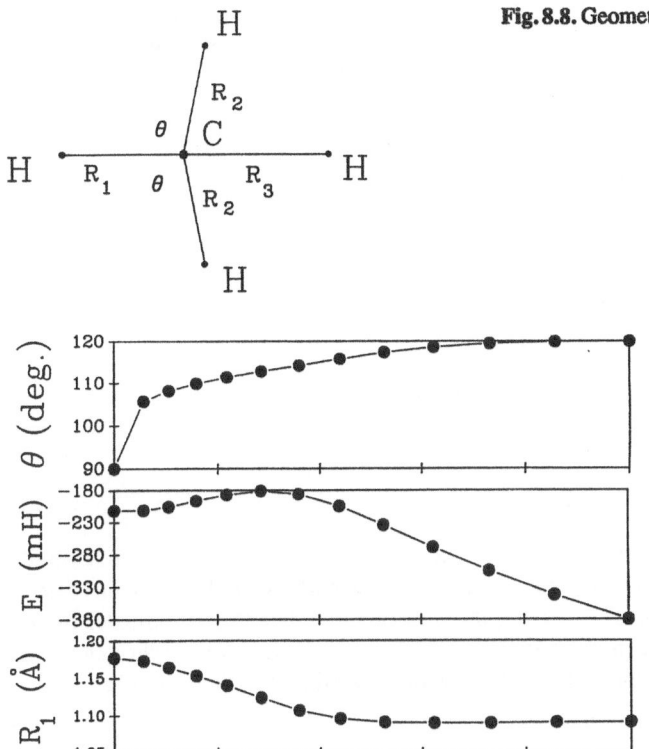

Fig. 8.8. Geometrical configuration of CH_4^{2+}

Fig. 8.9. Dissociation of CH_4^{2+} (D_{4h}) to CH_3^+ (D_{3h}) + H^+ along the C_{2v} symmetry pathway. Geometrical parameters θ and R are defined in Fig. 8.8.

$\theta = 90°$ while at $X = 1$, $R_1 = R_2 = 1.091$ Å, $R_3 = \infty$ and $\theta = 120°$. The resulting dissociation energy is 4.57 eV and the potential barrier along the dissociation pathway is 0.83 eV. Although results of this theoretical treatment are in good accord with the CASSCF calculations [8.91], the theoretical information obtained still does not permit totally unambiguous insight into the energetic and structural properties of metastable CH_4^{2+} ions.

b) Large, doubly charged hydrocarbon molecules. In an exhaustive and systematic charge stripping study *Rabrenović* et al. [8.92] have measured the ionization energies of a large number of singly charged hydrocarbon molecules and observed some interesting and, as yet, unexplained, relationships between double ionization energies and the number of carbon and hydrogen atoms in each molecular species. The dependence of single ionization energies of ions of the type $C_xH_y^+$ on the number, y, of hydrogen atoms in the ions with even and odd values of x is depicted in Fig. 8.10. In the case of odd carbon number ions, there appear distinct minima at even hydrogen numbers, suggesting that the energy required to remove an unpaired

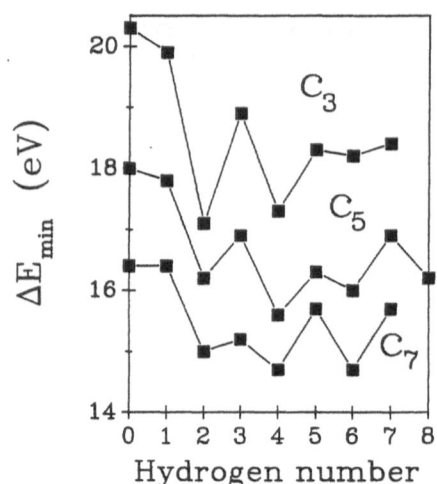

Fig. 8.10. Minimum energy required to ionize $C_x H_y^+$ as a function of y for odd carbon number ions

electron from ions is less than that required to remove a paired electron. The same situation would, of course, be expected for ions of even carbon number with increasing number of hydrogen atoms. However, the experimental data [8.92] indicates only a single sharp minimum which occurs for each set of ions when four hydrogen atoms are present. Rationalization of these observations continues to be elusive.

Acknowledgements. It is a pleasure to acknowledge the skilful and enthusiastic support we have enjoyed from our colleagues F. A. Rajgara, C. Badrinathan, U. T. Raheja, S. Mazumdar, V. Krishnamurthy and K. Nagesha.

References

8.1   A. Carrington, B. A. Thrush: Phil. Trans. R. Soc. London A324, 73 (1988)
8.2   E. Friedlander, H. Kallman, W. Lasereff, B. Rosen: Z. Phys. 76, 60 (1932)
8.3   D. Mathur, C. Badrinathan: J. Phys. B 20, 1517 (1987)
8.4   T. G. Heil, S. E. Butler, A. Dalgarno: Phys. Rev. A 27, 2365 (1983)
8.5   D. L. Cooper, M. J. Ford, J. Gerratt, M. Raimondi: Phys. Rev. A 34, 1752 (1986)
8.6   S. S. Prasad, D. R. Furman: J. Geophys. Res. 80, 1360 (1975)
8.7   D. Mathur, C. Badrinathan, F. A. Rajgara, U. T. Raheja: Chem. Phys. 103, 447 (1986)
8.8   S. Mazumdar, F. A. Rajgara, V. R. Marathe, C. Badrinathan, D. Mathur: J. Phys. B 21, 2815 (1988)
8.9   D. Mathur, C. Badrinathan: J. de Physique C1 50, 137 (1989)
8.10 R. G. Cooks, T. Ast, J. H. Beynon: Int. J. Mass Spectrom. Ion Processes 11, 490 (1973)
8.11 J. Appell: "Double Electron Transfer and Related Reactions", in *Collision Spectroscopy*, ed. by R. G. Cooks (Plenum, New York 1978) Chap. 4
8.12 M. Barber, D. J. Bell, M. Morris, L. W. Tetler, M. D. Woods, J. J. Monaghan, W. E. Morden: Org. Mass Spectrom. 24, 504 (1989)
8.13 B. E. Jones, L. E. Abbey, H. L. Chatham, A. W. Hanner, L. A. Teleshefsky, E. M. Burgess, T. F. Moran: Org. Mass Spectrom. 18, 282 (1982)

8.14   A. Galindo-Uribarri, H. W. Lee, H. Chang: J. Chem. Phys. **83**, 3685 (1985)

8.15   A. Balkacem, E. P. Kanter, R. E. Mitchell, Z. Vager, B. J. Zabransky: Phys. Rev. Lett. **63**, 2555 (1989)

8.16   D. Mathur, E. Krishnakumar, F. A. Rajgara, U. T. Raheja, C. Badrinathan: Int. J. Mass Spectrom. Ion Processes **99**, 237 (1990)

8.17   H. Tawara, T. Tonuma, K. Baba, M. Kase, T. Kambara, H. Kumagai, I. Kohno: Nucl. Instrum. Methods **B23**, 203 (1987)

8.18   H. Tawara, T. Tonuma, H. Shabita, M. Kase, T. Kambara, S. H. Be, H. Kumagai, I. Kohno: Phys. Rev. A **33**, 1385 (1986)

8.19   A. K. Edwards, R. M. Wood, M. F. Steur, Phys. Rev. A. **15**, 48 (1977)

8.20   A. K. Edwards, R. M. Wood: J. Chem. Phys. **76**, 2938 (1982)

8.21   R. L. Ezell, A. K. Edwards, R. M. Wood: J. Chem. Phys. **81**, 1341 (1984)

8.22   M. L. Langford, D. Mathur, F. M. Harris: Rapid Comm. Mass Spectrom. **2**, 167 (1988)

8.23   R. Spohr, T. Bergmark, N. Magnusson, L. O. Werme, C. Nordling, K. Siegbahn: Phys. Scr. **67**, 31 (1970)

8.24   T. A. Carlson: *Photoelectron and Auger Spectroscopy* (Plenum, New York 1975)

8.25   P. Lablanquie, I. Nenner, P. Millie, P. Morin, J. H. D. Eland, M. Hubin-Franskin, J. Delwiche: J. Chem. Phys. **82**, 2951 (1985)

8.26   G. Dujardin, S. Leach, O. Dutuit, P.-M. Guyon, M. Richard-Viard: Chem. Phys. **88**, 339 (1984)

8.27   P. J. Richardson, J. H. D. Eland, P. G. Fournier, D. L. Cooper: J. Chem. Phys. **84**, 3189 (1986)

8.28   J. H. D. Eland, F. S. Wort, R. N. Royds: J. Electron Spectrosc. Relat. Phenom. **41**, 297 (1986)

8.29   P. K. Carroll: Can. J. Phys. **36**, 1585 (1958)

8.30   P. C. Cosby, R. Moller, H. Helm: Phys. Rev. A **28**, 766 (1983)

8.31   D. Cossard, F. Launay, J. M. Robbe, G. Gandara: J. Molec. Spectrosc. **113**, 142 (1985)

8.32   M. J. Besnard, L. Hellner, Y. Malinovich, G. Dujardin: J. Chem. Phys. **85**, 1316 (1986)

8.33   K. Tohji, D. M. Hanson, B. X. Yang: J. Chem. Phys. **85**, 7492 (1986)

8.34   F. L. Mohler: "Survey of Multiply Charged Ions"; NBS Technical Note 243, National Bureau of Standards, Washington (1964)

8.35   W. Koch, N. Heinrich, H. Schwarz, F. Maquin, D. Stahl: Int. J. Mass Spectrom. Ion Proc. **67**, 305 (1985)

8.36   J. H. D. Eland, F. S. Wort, P. Lablanquie, I. Nenner: Z. Phys. D **4**, 31 (1986)

8.37   B. Brehm, G. de Frenes: Int. J. Mass Spectrom. Ion Phys. **26**, 251 (1978)

8.38   F. Sasaki, K. Ohno: J. Math. Phys. **4**, 1140 (1963)

8.39   P. O. Löwdin: Rev. Mod. Phys. **32**, 328 (1960)

8.40   P. O. Löwdin: Rev. Mod. Phys. **36**, 966 (1964)

8.41   W. A. Goddard III: Phys. Rev. **182**, 48 (1969)

8.42   S. F. Boys: Proc. R. Soc. London A**200**, 542 (1950)

8.43   S. Rothenberg, H. F. Schaefer III: J. Chem. Phys. **54**, 2765 (1971)

8.44   J. Vladimiroff: J. Phys. Chem. **77**, 1983 (1973)

8.45   J. W. Viers, F. E. Harris, H. F. Schaefer III: Phys. Rev. A **1**, 24 (1970)

8.46   M. H. Johnson: Phys. Rev. **39**, 197 (1932)

8.47   R. K. Nesbet: J. Chem. Phys. **43**, 311 (1965)

8.48   I. Shavitt: J. Comput. Phys. **6**, 124 (1970)

8.49   S. R. Langhoff, E. R. Davidson: Int. J. Quantum Chem. **8**, 61 (1974)

8.50   J. A. Pople, R. Seeger, R. Krishnan: Int. J. Quantum Chem. **11**, 149 (1977)

8.51   J. Cizek: J. Chem. Phys. **45**, 4256 (1966)

8.52   J. Paldus, J. Cizek, I. Shavitt: Phys. Rev. A **5**, 50 (1972)

8.53   K. Raghavachari: J. Chem. Phys. **82**, 4607 (1985)

8.54   J. A. Pople, M. Head-Gordon, K. Raghavachari: J. Chem. Phys. **87**, 5968 (1987)

8.55   C. Möller, M. S. Plesset: Phys. Rev. **46**, 618 (1934)

8.56   A. C. Wahl, G. Das: "The multiconfiguration self-consistent field method" in *Methods of Electronic Structure Theory*, ed. by H. F. Schaeffer III (Plenum, New York 1977) pp. 51–78

8.57   P. O. Löwdin: Phys. Rev **97**, 1474 (1955)

8.58   C. F. Bender, E. R. Davidson: J. Phys. Chem. **70**, 2675 (1966)

8.59   K. Ruedenberg, M. V. Schmidt, M. M. Gilbert, S. T. Elbert: Chem. Phys. **71**, 41 (1982)

8.60   K. Ruedenberg, M. V. Schmidt, M. M. Gilbert: Chem. Phys. **71**, 51 (1982)

8.61   I. Shavitt: Int. J. Quantum Chem. **11**, 133 (1977)

8.62   I. Shavitt: Int. J. Quantum Chem. **12**, 5 (1978)

8.63   B. O. Roos, P. R. Taylor, P. E. M. Siegbahn: Chem. Phys. **48**, 157 (1980)

8.64   P. E. M. Siegbahn, J. Almlöf, A. Heiberg, B. O. Roos: J. Chem. Phys. **74**, 2384 (1981)

8.65   B. O. Roos: Int. J. Quantum Chem. **14**, 175 (1980)

8.66   C. A. Coulson, I. Fischer: Phil. Mag. **40**, 386 (1949)

8.67   M. Kotani, A. Amemiya, E. Ishigura, T. Kimura: *Tables of Molecular Integrals*, 2nd ed. (Maruzen, Tokyo 1963)

8.68   J. Gerratt, M. Raimondi: Proc. Roy. Soc. Lond. **A371**, 525 (1980)

8.69   J. Gerratt, D. L. Cooper, M. Raimondi: "The spin-coupled valence bond theory of molecular electronic structure", in *Studies in Physical and Theoretical Chemistry*, Vol. 64, ed. by D. J. Klein, N. Trinajstic (Elsevier, Amsterdam 1990) pp. 287–349

8.70   M. Guilhaus, A. G. Brenton, J. H. Beynon, M. Rabrenovic, P. von Rague Schleyer, J. Phys. B **17**, L605 (1984)

8.71   D. Mathur, V. Krishnamurthy: (to be published)

8.72   R. Poirier, R. Kari, I. G. Csizmadia, *Handbook of Gaussian Basis Sets* (Elsevier, Amsterdam, 1985)

8.73   A. C. Hurley, J. Chem. Phys. **54**, 3656 (1971)

8.74   G. H. Bearman, F. Ranjbar, H. H. Harris, J. J. Leventhal: Chem. Phys. Lett. **42**, 335 (1976)

8.75   M. Larsson, B. J. Olsson, P. Sigray: Chem. Phys. **139**, 457 (1989)

8.76   V. R. Marathe, D. Mathur: Chem. Phys. Lett. **163**, 189 (1989)

8.77   N. Correia, A. Flores-Riveros, H. Agren, K. Helenelund, L. Asplund, U. Gelius: J. Chem. Phys. **83**, 2035 (1985)

8.78   G. E. Laramore: Phys. Rev. A **29**, 23 (1984)

8.79   C.-M. Liegener: Chem. Phys. Lett. **106**, 201 (1984)

8.80   D. Mathur, V. R. Marathe, S. Mazumdar: J. Phys. B **22**, L385 (1989)

8.81   D. Mathur, F. A. Rajgara: Phys. Rev. A **41**, 4824 (1990)

8.82   D. Mathur, R. G. Kingston, F. M. Harris, J. H. Beynon: J. Phys. B **19**, L575 (1986)

8.83   D. Mathur, F. M. Harris: Mass Spectrom. Reviews **8**, 269 (1989)

8.84   W. J. Griffiths, D. Mathur, F. M. Harris: J. Phys. B **20**, L493 (1987)

8.85   R. Rye, T. E. Madey, J. E. Houston, P. H. Holloway: J. Chem. Phys. **69**, 1504 (1978)

8.86   P. G. Fournier, J. Fournier, F. Salama, P. J. Richardson, J. H. D. Eland: J. Chem. Phys. **83**, 241 (1985)

8.87   G. Dujardin, D. Winkoun, S. Leach: Phys. Rev. **A3**, 3027 (1985)

8.88   P. A. Hatherly, M. Stankiewicz, L. J. Franski, K. Codling, M. A. McDonald: Chem. Phys. Lett. **159**, 355 (1989)

8.89   J. Appell, J. A. Horsley: J. Chem. Phys. **60**, 3445 (1974)

8.90   W. J. Meyer: Chem. Phys. **58**, 1017 (1973)

8.91   P. E. M. Siegbahn: Chem. Phys. **66**, 443 (1982)

8.92   M. Rabrenović, C. J. Proctor, T. Ast, C. G. Herbert, A. G. Brenton, J. H. Beynon: J. Phys. Chem. **87**, 3305 (1983)

9. Dissociative Recombination in Ion-Electron Collisions: New Directions

J.B.A. Mitchell

With 6 Figures

The dissociative recombination of molecular ions, which can be represented as

$$e + AB^+ \rightarrow AB^{**} \rightarrow A^* + B , \tag{9.1}$$

where A and B can be atoms or molecules, is a difficult process to study experimentally for it is only significant at low collision energies. Two main approaches to the problem have been used: plasma decay methods and intersecting beam experiments. In the former, a plasma containing the ion to be studied is created and the decay of the electron concentration is monitored as a function of time. If the particular ion under study dominates the plasma and competing electron removal mechanisms such as diffusion to the walls of the apparatus are minimised, then the rate coefficient for the recombination can be determined from the rate of the electron density decay. In the latter approach, a mass selected ion beam is made to intersect an electron beam and the resulting recombination products are detected directly. In this way, the recombination cross section can be determined. A variety of individual variants of these approches have been developed, including the microwave afterglow technique [9.1], the flowing afterglow Langmuir probe method (FALP) [9.2], the merged [9.3], inclined [9.4], and crossed [9.5] beam techniques. Hybrid techniques such as the quadrupole ion trap [9.6] and the electron beam trap techniques [9.7] have also been applied to this subject.

A critical review of all these methods has been given by *Mitchell* and *McGowan* [9.8] and so this chapter will not address the details of these techniques specifically. The intention is, rather, to address the subject of new techniques which are currently under development as well as some that are still at the conceptual stage.

9.1 New Developments in the Merged Beam Technique

The merged beam method involves creating an electron beam of known energy, causing it to interact with a mass analyzed ion beam, also of known energy, in a collinear

Springer Series in Chemical Physics, Vol. 54
Deepak Mathur (ed.): Physics of Ion Impact Phenomena
© Springer-Verlag Berlin Heidelberg 1991

MEIBE 1 apparatus

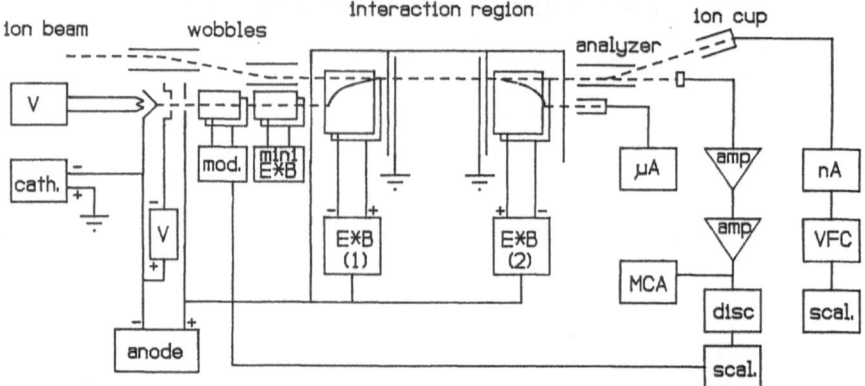

Fig. 9.1. Schematic outline of the merged beam apparatus at the University of Western Ontario

configuration, and monitoring the resulting collision products using a suitable detector. This is illustrated Fig. 9.1. The main advantages of the technique are:

1. Collision processes can be studied at very low interaction energies;
2. The technique is capable of operating at very high energy resolution;
3. The ion under study is generally unambiguously identified by its mass-to-charge ratio;
4. The internal energy of the ion can be determined from the measurement of suitable excitation thresholds;
5. The products have high laboratory-frame energies since they are derived from fast ions. This allows them to be detected with high efficiency using conventional nuclear detectors.

Up to now, the majority of experiments performed using this method have been concerned with measurements of the total electron-ion capture cross section. The exceptions are two measurements of polyatomic ion recombination branching ratios [9.10, 11]. Of great interest however is the identity of the excitation state of the collision products. For example, the recombination of H_2^+ proceeds as follows

$$e + H_2^+(\nu) \rightleftharpoons H_2^{**}(^1\Sigma_g^+) \rightarrow H(1s) + H(nl) \tag{9.2}$$

The $^1\Sigma_g^+$ state does not proceed directly to neutral hydrogen atom pairs. Its actual dissociation limit, at large internuclear separation, is $H^+ + H^-$ (Fig. 9.2). Experimental measurement of the cross section for the recombination of H_2^+ to yield ion pairs, however, indicates that this accounts for less than 1 % of the total recombination cross section [9.12]. The production of H atom pairs therefore arises from a curve crossing between the $^1\Sigma_g^+$ state and a suitable Rydberg state of H_2. The branching ratios for the possible decay channels are not known. Crossed beam studies [9.5, 13] in which the identity of the excitation states of the products were determined spectroscopically showed that, for the case of hydrogen molecular ions with a full

276

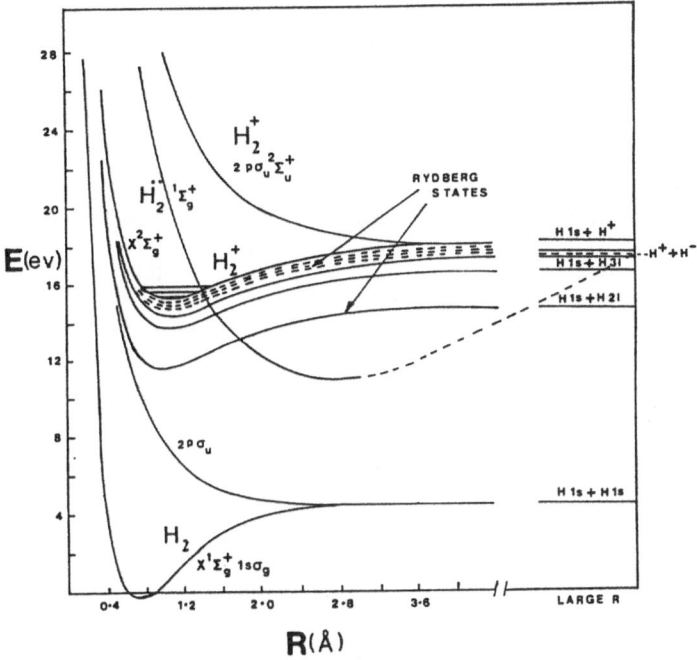

Fig. 9.2. Potential energy curves for H$^+$ and H$_2$ (from [9.9])

spectrum of vibrational states ($\nu = 0$–18) populated, interacting with electrons with more than 0.5 eV collision energy, the partial cross sections for the formation of H(2s) and H($n = 4$) each accounted for about 10 % of the total recombination cross sections. No information concerning the distribution of the remaining 80 % of the recombination products was obtained.

Two methods are currently under development which should allow the excitation states of the recombination products to be identified using the merged beam technique. The first employs a time and position sensitive detector to measure the kinetic energy and, hence also, the potential energy of the products in low principal quantum numbers while the latter employs a field ionization detector to selectively detect products in states with high principal quantum numbers. These techniques are described separately below.

The products of dissociative recombination are released with kinetic energies determined by the initial energy of the electron-ion system (electronic, vibrational and kinetic) and the potential energy of the products. Amounts of kinetic energy released for a variety of initial and final states and for an incident electron energy of 1 eV, are listed in Table 9.1. In the merged beam technique, the parent ion is moving at high velocities in the laboratory frame of reference and the products also have this velocity, with increments determined by the kinetic energy release. In the case of H$_2^+$ recombination, conservation of momentum requires that the hydrogen atoms move apart with velocities equal in magnitude but opposite in direction. If they move in a

277

Table 9.1. Final channel kinetic energy release for e + $H_2^+ (v)$ recombination at $E_{cm} = 1.0$ eV

	Kinetic energy released [eV] Final channel						
	$v = 0$	$v = 1$	$v = 2$	$v = 3$	$v = 4$	$v = 5$	$v = 6$
H(2l) + H(1s)	1.75	2.02	2.26	2.51	2.72	2.96	3.13
H(3l) + H(1s)	–	0.13	0.37	0.625	0.835	1.08	1.25
H(4l) + H(1s)	–	–	–	–	0.17	0.41	0.58
H(5l) + H(1s)	–	–	–	–	–	0.11	0.28
	More channels open up as E_{cm} is increased						

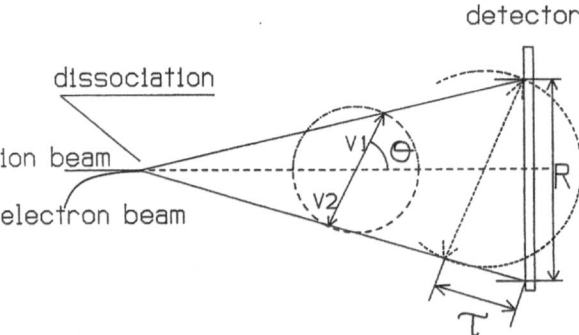

Fig. 9.3. Schematic diagram illustrating the dissociation of an H_2 molecule. The fragments fly apart in opposite directions with equal kinetic energy, arriving at the detector with spatial and temporal separations R and τ, respectively

direction perpendicular to the beam axis, then they will arrive simultaneously at the detector but spatially separated from each other by an amount R, tracing out circular locus (see Fig. 9.3). If, on the other hand, the dissociation occurs along the beam axis, then the particles will arrive at the same point in space but with mutual time delay $\Delta\tau$. It can be shown that the kinetic energy, E_d, of the recombination products can be measured using the relation

$$E_d = \frac{E_0}{4L^2}[v_0\Delta\tau^2 + R^2], \tag{9.3}$$

where E_0 and v_0 are the primary beam energy and axial velocity, respectively, and L is the distance between the collision centre and the detector. Using coincidence techniques to select particle pairs with the given values of $\Delta\tau$ and the multianode detector illustrated in Fig. 9.4, developed by *Los* and coworkers at the FOM Institute in Amsterdam [9.14], R and, therefore E_d, can be determined.

Application of this technique to the study of dissociative recombination using a merged beam apparatus is currently in progress at the University of Western Ontario. The fact that the energy of the ion beam is high (100–400 keV) in this experiment

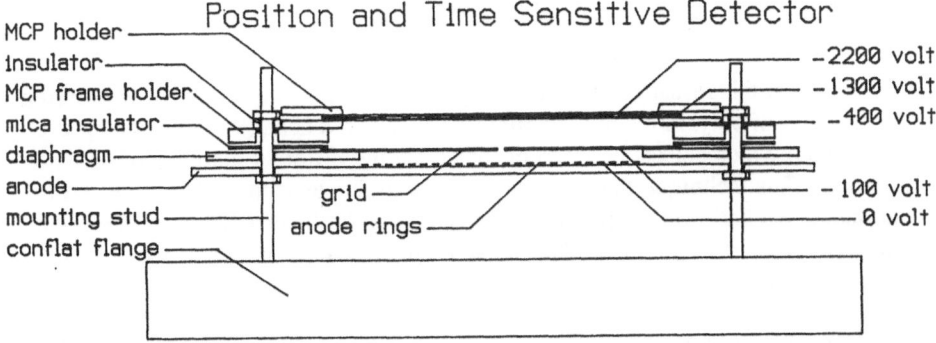

Fig. 9.4. Multianode time and position sensitive detector for the measurement of recombination product kinetic energies. (Based on the design of [9.14])

causes a problem in that the required time resolution is of the order of 1 ns or better. Also, in order to achieve a sufficient separation of the products in space, a long flight path (ca. 4 m) is required. Preliminary tests with this detector indicate that it has a time resolution of approximately 0.5 ns and a spatial resolution of about 100 μm. This is sufficient to provide an energy resolution of better than 0.1 eV of final product kinetic energy.

9.2 Detection of Highly Excited States

When the molecular ions under study are highly vibrationally excited, or when the electron energy is high, then there is the possibility of producing product atoms in highly excited states. This will complicate the analysis of the data from the time and position sensitive detector and, generally, the kinetic energy released in such cases will be too small to allow contributions from competing channels to be distinguished. There is, however, an alternative means of detecting highly excited states, and that is via field ionization. It can be shown [9.15] that a highly excited atom will be ionized when it enters an electric field E (expressed in volts m^{-1}) provided

$$E > \frac{3.2 \times 10^{10}}{n^4} . \tag{9.4}$$

By arranging for the neutral products to pass through a region of electric field strength sufficient to ionize a given principal quantum number state, n, such atoms can then be identified by electrostatically deflecting the ions thus produced into a particle detector. This technique is known as field ionization detection and has been successfully used by a number of authors [9.15]. It is possible to identify a particular quantum state by measuring the ion flux produced by two field strengths sufficient to ionize n and $n + 1$ respectively. A field ionization detector is currently being installed in the MEIBE-II apparatus at the University of Wesern Ontario [9.16].

9.3 Measurements of Branching Ratios

One aspect of this research which is sorely in need of attention is that of identifying the products of polyatomic ion recombination. *Mitchell* et al. [9.10] have employed a technique which selects between channels having different numbers of collision products. This technique does not, however, provide any information concerning the state of excitation of the products; nor can it discriminate between channels with the same numbers of products. Thus it cannot provide information on the branching ratio for reactions such as:

$$H_3O^+ + e \rightarrow H_2O + H , \tag{9.5}$$

$$H_3O^+ + e \rightarrow OH + H_2 . \tag{9.6}$$

The density of reactants in an afterglow experiment is much greater than in merged beam experiments and so it becomes possible to perform a spectroscopic examination of the recombination products. Ultraviolet resonance absorption detection techniques have been recently employed by *Rowe* et al. [9.18] to determine the atomic products of the recombination of H_2O^+.

In a recent paper, *Adams* et al. [9.19] have described the application of a laser induced fluorescence (LIF) technique to measure the density of OH radicals produced following the recombination of HCO_2^+ in a FALP apparatus. This experiment involves a clever calibration procedure in which the density of H atoms in a molecular hydrogen afterglow is first determined using vacuum ultraviolet absorption spectroscopy. N_2O is then added to the flow and the reaction

$$H + NO_2 \rightarrow OH + NO \tag{9.7}$$

converts all the H atoms into OH radicals. Since the concentration of H atoms is known, the number density of OH radicals in the discharge can be determined. (An intermediate step involves the addition of N_2 gas to the hydrogen discharge which is initially dominated by H_3^+ ($\nu = 0$) ions which have a small recombination rate coefficient. The H_3^+ ions react with the N_2 to form N_2H which rapidly removes the electrons in the apparatus, thus creating a neutral flow). The OH radicals are then detected using the LIF technique where the ultraviolet radiation from a frequency-doubled dye laser excites the (1,0) band of the OH ($A^2\Sigma \leftarrow X^2\Pi$) transition and the subsequent fluorescent decay radiation is detected. Since the concentration of OH radicals is known, this procedure allows the LIF equipment (dye laser and filtered photomultiplier) to be calibrated. The experiment is then performed by adding CO_2 gas to a hydrogen afterglow. HCO_2^+ ions are formed via the proton transfer reaction

$$H_3^+ + CO_2 \rightarrow HCO_2^+ + H_2 , \tag{9.8}$$

and any excited ions thus formed are quenched via the proton transfer reaction

$$(HCO_2^+)^* + CO_2 \rightarrow (CO_2)^* + HCO_2^+ . \tag{9.9}$$

OH radicals produced from the recombination reactions

$$HCO_2^+ + e \rightarrow H + CO_2 , \qquad (9.10)$$

$$HCO_2^+ + e \rightarrow OH + CO , \qquad (9.11)$$

$$HCO_2^+ + e \rightarrow O + H + CO , \qquad (9.12)$$

are then detected using the calibrated LIF equipment, and hence the branching ratio for OH production can be determined. *Adams* et al. [9.19] have shown that $34 \pm 6\%$ of the HCO_2^+ ions recombine to form OH $X^2\Pi$ radicals, half of which are in the $\nu = 0$ vibrational state.

The percentages of OH radicals resulting from the recombination of other ions, including H_3O^+, N_2OH^+ and O_2H^+ have also been determined, and by using reaction (9.7) to convert H atom products to OH radicals which can be detected using the calibrated LIF equipment, the rates for the production of H atoms following the recombination of ions such as N_2H^+, HCO^+ and CH_5^+ have also been determined.

Clearly this technique will prove to be very valuable in elucidating the recombination chemistry of complex molecules.

9.4 Recombination Studies at Storage Rings

The technology of ion storage rings was developed by the high energy physics community in order to maximise the utility of beams of short lived 'expensive' particles such as antiprotons. In order to ensure the longest possible storage time for the particles, the energy spread of the beam must be reduced using 'cooling' techniques. One such method involves merging a beam of electrons with the ion beam over part of its orbit so that energy is exchanged in collisions between the electrons and the ions. The velocities of the electrons and ions are matched but since the electrons are so much lighter, their energy, as well as their absolute energy spread, are very much smaller. The electrons and ions thermalize over a period of time, T_c, resulting in an ion beam with a very small energy spread. This arrangement is, of course, very similar to that of the MEIBE experiment described earlier and so an electron cooler is an excellent device for studying electron-ion collisions at moderate energies with good energy resolution. A number of ion storage rings are currently at or near operational status (LEAR at CERN, the INDIANA COOLER RING, ASTRID at Aarhus, TSR at Heidelberg, ESR at GSI Darmstadt, CRYRING at Stockholm) or under development (HISTRAP at Oak Ridge) and atomic physics experiments are an important component of these projects [9.20–23].

Anderson et al. [9.22] have recently reported some very remarkable results from the measurement of the dielectronic recombination of helium-like oxygen ions using a pilot scale electron cooler at the ASTRID facility. The apparatus used is shown schematically in Fig. 9.5, and the measured dielectronic recombination rate coefficient for O^{6+} is shown as a function of electron energy in Fig. 9.6. The energy

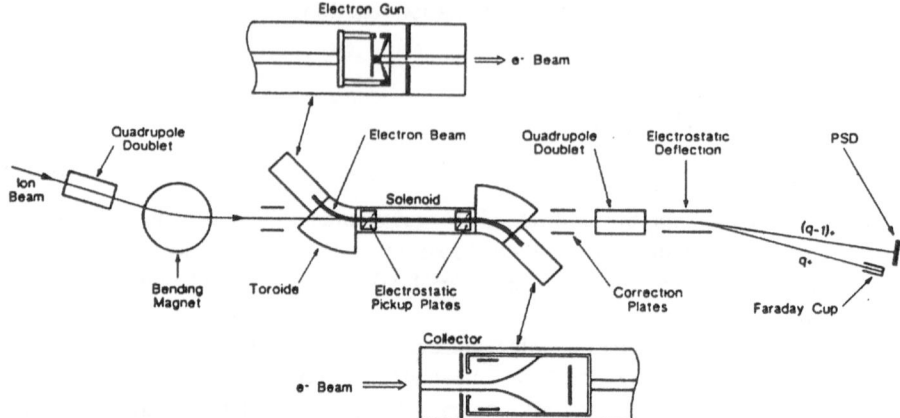

Fig. 9.5. The electron cooler, merged beam apparatus of *Andersen* et al. [9.22] used for the measurement of dielectronic recombination cross sections

resolution of this experiment was determined to be ca. 0.135 eV, with an electron current of 5–10 mA. This is essentially the spread imposed by the transverse energy of the electrons leaving the hot (1000 K) cathode. Generally, electron coolers in storage rings operate at high electron currents where space charge effects would increase the energy spread; however, presumably, once the ions have been cooled, the current could be reduced for a time to allow the recombination to be studied.

While it is very interesting to study electron-ion collisions in such a device, it must also be remembered that the charge changing collisions can lead to ion loss from the ring. A detailed discussion of this effect for the case of atomic ions is given by *Beyer* [9.21]. The time constant, T_{loss}, for ion loss due to collision effects is given by

$$T_{\text{loss}} = (\alpha n_e \eta)^{-1} , \qquad (9.13)$$

where α is the collision rate coefficient, n_e is the electron number density and η is the fraction of the ring intercepted by the electron beam. Typical values are $n_e \sim 10^{-8}$ cm^3 and $\eta = 0.02$.

For atomic ions with low charge states, the recombination rates will tend to be small ($< 10^{-8}$ cm^{3}s^{-1}) except at very low electron energies. This would yield loss rate time constants in excess of 50 seconds. For the case of molecular ions however, one can expect to have recombination rates at least an order of magnitude larger than this, thus increasing the loss rate proportionately. This problem can be overcome, in principle, by reducing the electron cooler beam current, possibly using alternate cooling methods (laser, stochastic, etc.) to reduce the ion beam energy spread. In fact such a reduction would be required in any event in order to reduce the energy spread caused by the space charge depression across the beam so that the low centre-of-mass energy region, where dissociative recombination dominates, could be accessed. The experimental details will have to be worked out for each individual case. The ma-

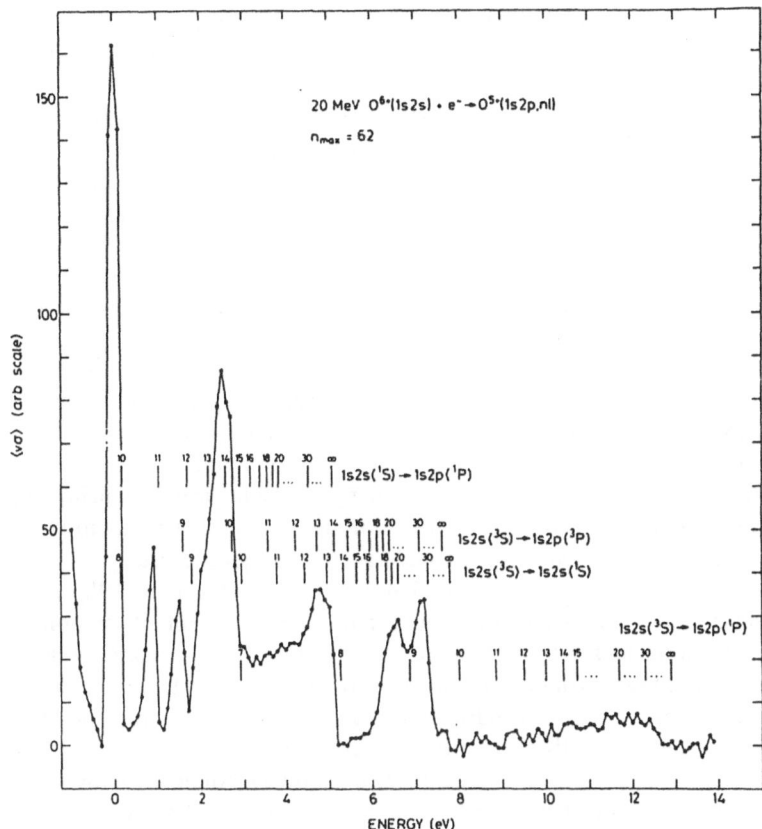

Fig. 9.6. The rate coefficient $\langle \sigma v \rangle$ as a function of relative electron energy for dielectronic recombination of O^{6+}. To obtain a lower limit for the rate coefficient, multiply the ordinate by 10^{-11} cm^{3}s^{-1} (from [9.22])

jor advantage, however, in studying molecular ion recombination at a storage ring is that the long storage lifetime of the ions will allow vibrationally excited states of heteropolar molecular ions to decay leading, therefore, to well defined initial conditions. No doubt, such experiments will be attempted in due course.

9.5 Heavy Ions

Very little is known concerning the recombination of complex ions. *Biondi* [9.24] and *Adams* and *Smith* [9.25] have examined some heavy diatomics, cluster ions and complex hydrocarbon species formed in afterglows, but such measurements are limited to species which can be produced as dominant ions in the plasma. As currently implemented, the MEIBE apparatus cannot handle ions with molecular weights greater than 32 due to an upper limit of 400 keV laboratory energy for the ion beam produced from the Van de Graaff accelerator used as the injector for this machine. One can of course envisage moving the apparatus to a higher energy accelerator but such a move is inconvenient for a variety of reasons, including limited access to beam time in such a facility. As mentioned earlier, attempts to examine the excitation state of the products in the MEIBE apparatus are complicated by the high velocity of the products in the laboratory reference frame and these would, of course, be aggravated with a higher energy injector.

An alternative approach is being examined. This involves creating a target of highly excited atoms through which the ion beam is passed. When the electron in such an atom is in a high principal quantum number state, it appears to the projectile ion as if it were free and so the collisions can be treated as an electron-ion event rather than as an ion-atom collision [9.26]. The relative velocity of the electron-ion collision, however, is determined by the centre-of-mass energy of the ion-atom collision. In other words, low electron-ion relative collision velocities, and therefore, low centre-of-mass electron-ion collision energies, can be produced using this technique. This phenomenon has been exploited by a number of workers to study electron attachment and electron-molecule inelastic scattering processes.

The approach used is similar to that implemented by *Kondow* et al. [9.27]. An electron impact Rydberg atom generator is used to produce an essentially stationary target of Rydberg atoms and this would be bombarded by a fast ion beam. Following collisions, the ion beam would be deflected into a Faraday cup and the neutrals formed in the interaction region would be detected with a surface barrier detector using a time and position sensitive detector as described above.

The success of such an appoach will be determined by the efficiency with which the high Rydberg state atoms can be produced. A generator has been constructed and is currently under test using a CO_2 cluster beam as diagnostic probe. The aim is to study the rate of formation of negative ion clusters formed as a result of electron capture from the Rydberg atoms. If this proves successful, then the generator will be incorporated into a recombination experiment as described above.

One of the problems of this method is that it is not a clean collision event, that is, there will be many target atoms in ground or low excited states and these can act as charge exchange targets with the incoming ion beam. The effect of this can, however, be minimized by using a high incident beam energy to limit the charge exchange cross section and by modulating the Rydberg state population by applying a suitable field ionization voltage to the generator and counting in and out of phase with this voltage. In other words, the charge exchange process acts as the background while the Rydberg capture is the signal. The advantage of the technique is that, contrary to the merged beam approach, the heavier the ion, the lower the relative collision velocity and, therefore, the lower the attainable centre-of-mass collision energy.

9.6 Epilogue

John Hasted has contributed greatly to the field of recombination research, having played an important role in the development of both FALP [9.28] and the electron beam trap [9.7] methods. John's work is characterized by the simplicity of the apparatus which he used. In his own words he liked the 'string and sealing wax' approach. He perhaps, therefore, would not approve of some of the techniques, particularly those employing multi-million dollar storage rings. For this apologies are offered in advance. Of course John's crowning achievement, which will go down in the annals of Recombination History, is his skillful and witty rendition of the now-famous ballad, '*Ions Fare Ye Well*', the words of which are reproduced below with the permission of their author.

Ions Fare Ye Well

Song text by J. B. Hasted.
Tune – Stenka Razin and Clementine.

High above the acid raindrops
Where the sun must sometimes shine
There the ions in their glory
Ionize and recombine.

O my darlings, O my darlings,
*   O my darlings ions mine*
Thou art lost and gone forever,
*   if but once you recombine.*

Alpha rays from radium bromide
In the dark so brightly shine
Turn to helium, then Sir William
Crookes the spectrum, every line.

In a cavity, in a discharge
Back in nineteen forty-nine
Fred Biondi waved his wand, he
Made the ions recombine.

Quoth the ions in the flowing afterglow
As he drifted down the line
Hey, buster, there's a cluster
Now we all can recombine.

Tokamaks and Stellarators
Gulping taxes, yours and mine
But all that they are heating
Is tungsten twenty-nine.

From the Synchrotrons of others
From the linacs past their prime
We improve our physics
It don't cost you but a dime.

In a dusty laboratory
Mid the sealing wax and twine
Dwelt a poet; he don't know it
But he's only got one rhyme.

Fare ye well, ye sticks of sealing wax
Fare ye well, ye ball of twine
Let the air into the system
For the last and final time.

Reproduced from *Electronic and Atomic Collisions* eds. H. B. Gilbody, W. R. Newell, F. H. Read, A. C. H. Smith (North-Holland, Amsterdam 1988)

References

9.1   F. J. Mehr, M. A. Biondi: Phys. Rev. 181, 264 (1969)
9.2   E. Alge, N. G. Adams, D. Smith: J. Phys. B. 16, 1433 (1983)
9.3   D. Auerbach, R. Cacak, S. R. Caudano, C. J. Keyser, T. D. Gally, J. B. A. Mitchell, J. Wm. McGowan, S. F. J. Wilk: J. Phys. B. 18, 3797, (1977)

9.4   B. Peart, K. T. Dolder: J. Phys. B. **7**, 236 (1974)
9.5   R. A. Phaneuf, D. H. Crandall, G. H. Dunn: Phys. Rev. **11**, 528 (1975)
9.6   R. A. Heppner, F. L. Walls, W. T. Armstrong, G. H. Dunn: Phys. Rev. **13**, 1000 (1976)
9.7   D. Mathur, S. U. Khan, J. B. Hasted, J. Phys. B. **11**, 3615 (1978)
9.8   J. B. A. Mitchell, J. Wm. McGowan: 'Experimental Studies of Electron-Ion Collisions' in *Physics of Ion-Ion and Electron-Ion Collisions* ed. by F. Brouillard, J. Wm. McGowan, (Plenum, New York, 1983) p. 279
9.9   H. Hus, F. Yousif, C. Noren, A. Sen, J. B. A. Mitchell: Phys. Rev. Lett. **60**, 1006 (1988)
9.10 J. B. A. Mitchell, J. L. Forand, C. T. Ng, D. P. Levac, R. E. Mitchell, P. M. Mul, W. Claeys, A. Sen, J. Wm. McGowan: Phys. Rev. Lett. **51**, 885 (1983)
9.11 J. B. A. Mitchell, F. B. Yousif: 'Molecular Ion Recombination: Branching Ratio Measurements', in *Microwave and Particle Beam Sources and Directed Energy Concepts*, ed. by H. E. Brandt, SPIE Proceedings Vol. 1061, 1989, p. 61
9.12 B. Peart, K. T. Dolder: J. Phys. B **8**, 1570 (1975)
9.13 M. Vogler, G. H. Dunn: Phys. Rev. **11**, 1983 (1975)
9.14 D. P. de Bruijn, J. Los: Rev. Sci. Instrum. **53**, 1020 (1982)
9.15 F. Brouillard: 'Rydberg States', in *Atomic and Molecular Processes and Controlled Thermonuclear Fusion*, ed. by C. J. Joachain, D. E. Post, (Plenum Press, New York, 1983) p. 313
9.16 T. J. Morgan, J. B. A. Mitchell: 'Laser Stimulated Radiative Recombination: a Field Ionization Detection Approach', in *Dissociative Recombination: Theory, Experiment and Applications*, ed. by J. B. A. Mitchell and S. L. Guberman, (World Scientific, Singapore, 1989) p. 175
9.17 J. L. Forand, J. B. A. Mitchell, J. Wm. McGowan: J. Phys. E. **18**, 623 (1985)
9.18 B. R. Rowe, F. Valee, J. L. Queffelec, J. C. Gomet, M. Morlais: J. Chem Phys. **88**, 845 (1988)
9.19 N. G. Adams, C. R. Herd, D. Smith: J. Chem. Phys. **91**, 963 (1989)
9.20 P. Hvelplund: J. de Physique C1 **50**, 459 (1989)
9.21 H. F. Beyer: J. de Physique C1 **50**, 471 (1989)
9.22 L. H. Andersen, P. Hvelplund, H. Knudsen, P. Kvistgaard: Phys. Rev. Lett. **62**, 2656 (1989)
9.23 J. A. Tanis, E. M. Berndtein, S. Chantrenne, M. W. Clark, S. Datz, P. F. Dittner, C. F. Foster, W. G. Graham, W. W. Jacobs, J. B. A. Mitchell, J. R. Mowat, D. Schneider, M. P. Stockli: Nuc. Instr. Meth. **B43**, 290 (1989)
9.24 M. A. Biondi: Comments. At. Mol. Phys. **5**, 85 (1975)
9.25 N. G. Adams, D. Smith: Chem. Phys. Lett. **144**, 11 (1988)
9.26 L. J. Wang, M. King, T. J. Morgan: J. Phys. B. **19**, L623 (1986)
9.27 T. Kondow: 'Negative Ion Formation by Collision of Rydberg Atoms with Clusters', in *Electronic and Atomic Collisions* ed. by D. C. Lorents, W. E. Meyerhof, J. R. Peterson, (North-Holland, Amsterdam, 1986) p. 517
9.28 M. R. Mahadevi, J. B. Hasted, M. M. Nakshbandi: J. Phys. B. **4**, 1726 (1971)

Subject Index

Resonance strength 26
Resonant electron capture 22
Resonant-excitation-auto-double-ionization
  (READI) 28, 54
Resonant-excitation-double-autoionization
  (REDA) 28, 59, 64, 82
Resonant-excitation-quadruple-autoionization
  (REQA) 31
Resonant-excitation-triple-autoionization
  (RETA) 31, 83
Roothaans equations 248, 250

S ions 66, 105, 217
Saha equation 13
Scaling rule 16
Scattering wave potential 71
Semi-empirical formulae 21
Sequential ionization 35
SF_6 ions 211
SH ions 192
Shake-off 30, 72
Si ions 59, 66
Sodium-like ions 59
Spin conservation 145, 175, 204, 233
Sr ions 64

State-selective electron capture 99, 105, 112,
  116
Statistical ratio 128, 175
Stochastic heating 7
Storage rings 281
Sudden approximation 72
Superelastic scattering 166, 180, 235

Term dependence 69
Threshold ionization 22
Ti ions 64
Transition
  metals 63
  probability matrix 142
Trielectronic-capture-auto-double-
  ionization 52

U ions 38, 47

Weighted transition probability 144

Xe
  ions 7, 38, 76, 110, 123, 174
  isonuclear sequence 76

Zr ions 64